The
Colossal Book of
Short Puzzles
& Problems

Also by Martin Gardner

Also by Martin Gardner

The Magic Numbers of Dr. Matrix
Knotted Doughnuts and Other Mathematical
    Entertainments
The Wreck of the Titanic Foretold?
Riddles of the Sphinx
The Annotated Innocence of Father Brown
The No-Sided Professor (short stories)
Time Travel and Other Mathematical
    Bewilderments
The New Age: Notes of a Fringe Watcher
Gardner's Whys and Wherefores
Penrose Tiles to Trapdoor Ciphers
How Not to Test a Psychic
The New Ambidextrous Universe
More Annotated Alice
The Annotated Night Before Christmas
Best Remembered Poems (ed.)
Fractal Music, Hypercards, and More
The Healing Revelations of Mary Baker Eddy
Martin Gardner Presents
My Best Mathematical and Logic Puzzles
Classic Brainteasers
Famous Poems of Bygone Days (ed.)
Urantia: The Great Cult Mystery
The Universe Inside a Handkerchief
On the Wild Side
Weird Water and Fuzzy Logic
The Night Is Large
Science Magic
Mental Magic
The Annotated Thursday
Visitors from Oz (novel)
Gardner's Workout
The Annotated Alice: The Definitive Edition
Did Adam and Eve Have Navels?
The Colossal Book of Mathematics
Mind-Boggling Word Puzzles
Are Universes Thicker than Blackberries?
Smart Science Tricks

Martin Gardner

The
COLOSSAL
Book of
SHORT PUZZLES
and
PROBLEMS

**Edited by
Dana Richards**

Combinatorics

Probability

Algebra · Geometry

Topology · Chess

Logic · Cryptarithms

Wordplay · Physics

and Other Topics
of Recreational
Mathematics

W. W. Norton & Company

New York · London

Copyright © 2006 by Martin Gardner
Copyright © 2006 by Martin Gardner and Dana Richards

Manufacturing by The Maple-Vail Book Manufacturing Group
Book design by Charlotte Staub
Production manager: Anna Oler

Library of Congress Cataloging-in-Publication Data

Gardner, Martin, 1914–
    The colossal book of short puzzles and problems / Martin Gardner;
    edited by Dana Richards.
       p. cm.
    Includes index.
    ISBN 0-393-06114-0 (hardcover)
    1. Scientific recreations. 2. Mathematical recreations.
I. Richards, Dana, 1955– . II. Title.
Q164.G248 2006
793.8—dc22   2005024080

W. W. Norton & Company, Inc.
500 Fifth Avenue, New York, N.Y. 10110
www.wwnorton.com

W. W. Norton & Company Ltd.
Castle House, 75/76 Wells Street, London W1T 3QT

1 2 3 4 5 6 7 8 9 0

For my granddaughter Katie Gardner

Contents

Preface

During the more than 25 years that I wrote the "Mathematical Games" column in *Scientific American*, I made it a practice of devoting one or two columns each year to short elementary problems, preferably puzzles that had surprising, aha!, solutions. Answers were withheld until the following month. With few exceptions none of these puzzles was original. Some were picked up from books and periodicals. Others came from readers of the column.

Most of the puzzles were of unknown origin, circulating among mathematicians and puzzle buffs the way anonymous jokes, riddles, and limericks circulate. Occasionally a reader would inform me that he was the inventor of a puzzle I had published without credit.

In 2004 friend Dana Richards, a computer scientist at George Mason University, went through all my short-problem columns as they appear in book collections, copied the puzzles, and classified them by the kind of mathematics involved. The book you now hold is the result of those efforts. I hope that readers unfamiliar with my columns will find the puzzles fresh and intriguing, and that readers who followed my columns will enjoy meeting the puzzles again. Here and there I have added new material to the answers. At the book's close I provide twelve of the best brainteasers to have come my way since I stopped writing the column in 1986.

Martin Gardner
Norman, Oklahoma
November 2004

Introduction

Martin Gardner has been called the "Dean of American Puzzlers." His "Mathematical Games" column in *Scientific American* inspired an entire generation to appreciate mathematics. In his monthly essays he showed that math was not only entertaining but also surprising, magical, ubiquitous, and beautiful. During his 25-year tenure his essays touched every corner of math, starting out with simple time-tested topics and broadening to areas never mentioned in the popular literature.

Surprisingly, Martin Gardner is not a mathematician. His strength as a writer is his ability to communicate to a broad audience. Since he had to learn each topic himself he knew what path to follow to draw the public into the mysteries of mathematics. I say "mystery" because Gardner knew that the allure of math was not so much in *solving* problems but in *discovering* how to solve problems. The pleasure is in knowing that you can solve it, not in knowing the answer. Before coming to *Scientific American* he had been contributing to the magic literature for over 25 years, and the ability to explain hidden principles was second nature to him.

Over time Gardner came to rely less on popular math sources and drew more from the mainstream math journals (though rarely from the most technical ones). More importantly, he came to be the focus of everyone interested in recreational mathematics. (His math books have been translated into at least 17 languages, and his readership is worldwide.) As a result, his columns were fueled by contributions of his loyal readers, more and more as the years went by.

Gardner selected the best 50 essays for his *The Colossal Book of Mathematics* (Norton, 2001). The present volume is a companion to that collection. The column was usually an essay but, occasionally, he would substitute a collection of puzzles. These appeared irregularly, approximately once a year, and contained about eight problems. His readers would eagerly wait all year for these problems, because Gardner would bide his time until he had a fine collection. Almost always the problems were new, usually contributed by his correspondents. If a problem in this volume seems familiar it is probably because Gardner made it famous.

This volume is divided into four parts, each covering a broad topic. Originally the puzzles were mixed up, each column bridging many topics. I have reordered all the puzzles so that, say, all the geometric problems are together. I have added a short introduction to each chapter. This will help the reader with a special focus to find a specific puzzle without a detailed index. Puzzle books are rarely organized this way. (Henry Ernest Dudeney, the greatest British puzzler, did organize his books in this fashion.) Of course, you are free to browse the book at will.

Sometimes a puzzle is just a puzzle, but usually these problems are doors that open into mathematical corridors, and in the answer section Gardner takes you down the path of discovery. Not all of the problems are equally difficult. Occasionally, a column would be devoted to "ridiculous questions" or "catch questions" which are not hard; in fact the hard part is realizing how easy they are! In each chapter the puzzles are *progressively more difficult*, starting with catch questions and ending with posers that have led to research topics.

This book marks the first time these problems have been collected together. Since they represent the core of the column for many of Gardner's fans, it is a pleasure to showcase them. A new generation will learn that mathematics is entertaining, magical, and beautiful.

Dana Richards
Fairfax, Virginia
November 2004

Problems worthy of attack
prove their worth by hitting back.

— Piet Hein

part one

Combinatorial & Numerical Problems

chapter 1

Combinatorics

Combinatorics is the core of "discrete mathematics," which is concerned with collections of objects that satisfy specified criteria. A famous example is the topic of magic squares, arrangements of integers in square arrays so that the sums of rows and columns are equal. Discrete mathematics is central to most of recreational mathematics. After all, only the simplest concepts of continuous math (i.e., calculus) can be grasped by a large audience, while discrete math problems can usually be understood by everyone (even if the solutions to those problems cannot). In this chapter you will find a diverse set of problems. In later chapters we will see further combinatorial problems that belong to more focused topic areas.

Problems

PROBLEM 1.1—Ambiguous Dates

In this country a date such as July 4, 1971, is often written 7/4/71, but in other countries the month is given second and the same date is written 4/7/71. If you do not know which system is being used, how many dates in a year are ambiguous in this two-slash notation?

PROBLEM 1.2—Ten-Digit Numbers

How many different 10-digit numbers, such as 7,829,034,651, can be written by using all 10 digits? Numbers starting with zero are excluded.

PROBLEM 1.3—Magic Squares and Cards

Nine heart cards from an ordinary deck are arranged to form a magic square so that each row, column, and main diagonal has the largest possible constant sum, 27. (Jacks count 11, queens 12, kings 13.) An example is seen in Figure 1.1. Drop the requirement that each value must be different. Allowing duplicate values, what is the largest constant sum for an order-3 magic square that can be formed with nine cards taken from a deck?

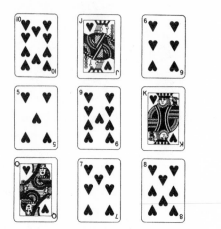

Figure 1.1

PROBLEM 1.4—Two-Cube Calendar

In Grand Central Terminal in New York I saw in a store window an unusual desk calendar [see Figure 1.2]. The day was indicated simply by arranging the two cubes so that their front faces gave the date. The face of each cube bore a single digit, 0 through 9, and one could arrange the cubes so that their front faces indicated any date from 01, 02, 03, . . . , to 31. What are the four digits that cannot be seen on the left cube and the three that cannot be seen on the right cube? It is a bit trickier than one might expect.

Figure 1.2

PROBLEM 1.5—Mutilated Chessboard

The props for this problem are a chessboard and 32 dominoes. Each domino is of such size that it exactly covers two adjacent squares on the board. The 32 dominoes therefore can cover all 64 of the chessboard squares. But now suppose we cut off two squares at diagonally opposite corners of the board and discard one of the dominoes, as shown in Figure 1.3. Is it possible to place the 31 dominoes on the board so that all the remaining squares are covered? If so, show how it can be done. If not, prove it impossible.

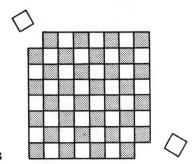

Figure 1.3

PROBLEM 1.6—Termite and 27 Cubes

Imagine a large cube formed by gluing together 27 smaller wooden cubes of uniform size [see Figure 1.4]. A termite starts at the center of the face of any one of the outside cubes and bores a path that takes him once through every cube. His movement is always parallel to a side of the large cube, never diagonal.

Is it possible for the termite to bore through each of the 26 outside cubes once and only once, then finish his trip by entering the central cube for the first time? If possible, show how it can be done; if impossible, prove it.

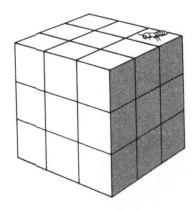

Figure 1.4

Combinatorics   5

It is assumed that the termite, once it has bored into a small cube, follows a path entirely within the large cube. Otherwise it could crawl out on the surface of the large cube and move along the surface to a new spot of entry. If this were permitted, there would, of course, be no problem.

PROBLEM 1.7—The Blank Column

A secretary, eager to try out a new typewriter, thought of a sentence shorter than one typed line, set the controls for the two margins, and, then, starting at the left and near the top of a sheet of paper, proceeded to type the sentence repeatedly. She typed the sentence exactly the same way each time, with a period at the end followed by the usual two spaces. She did not, however, hyphenate any words at the end of a line: When she saw that the next word (including whatever punctuation marks may have followed it) would not fit in the remaining space on a line, she shifted to the next line. Each line, therefore, started flush at the left with a word of her sentence. She finished the page after typing 50 single-spaced lines.

Without experimenting on a typewriter, answer this question: Is there sure to be at least one perfectly straight column of blank spaces on the sheet, between the margins, running all the way from top to bottom?

PROBLEM 1.8—Digit-Placing Problem

This perplexing digital problem, inventor unknown, was passed on to me by L. Vosburgh Lyons of New York City. The digits from 1 to 8 are to be placed in the eight circles shown in Figure 1.5, with this pro-

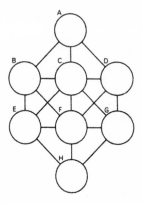

Figure 1.5

viso: no two digits directly adjacent to each other in serial order may go in circles that are directly connected by a line. For example, if 5 is placed in the top circle, neither 4 nor 6 may be placed in any of the three circles that form a horizontal row beneath it, because each of these circles is joined directly to the top circle by a straight line.

There is only one solution (not counting a rotation or mirror reflection as being different), but if you try to find it without a logical procedure, the task will be difficult.

PROBLEM 1.9—The Folded Sheet

Mathematicians have not yet succeeded in finding a formula for the number of different ways a road map can be folded, given n creases in the paper. Some notion of the complexity of this question can be gained from the following puzzle invented by the British puzzle expert Henry Ernest Dudeney.

Divide a rectangular sheet of paper into eight squares and number them on one side only, as shown in the top left of Figure 1.6. There are 40 different ways that this "map" can be folded along the ruled lines to form a square packet which has the "1" square face-up on top and all other squares beneath. The problem is to fold this sheet so that the squares are in serial order from 1 to 8, with the 1 face-up on top.

If you succeed in doing this, try the much more difficult task of doing the same thing with the sheet numbered in the manner pictured at the bottom of the illustration.

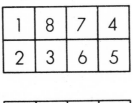

Figure 1.6

PROBLEM 1.10—Pohl's Proposition

Frederik Pohl, a top writer of science fiction, thought of this stunt, which appeared in a magic magazine called *Epilogue*. Computer programmers are likely to solve it more quickly than others.

Ask someone to draw a horizontal row of small circles on a sheet of paper to indicate a row of coins. Your back is turned while he does this. He then places the tip of his right thumb on the first circle so that his thumb and hand completely cover the row of circles. You turn around and bet you can immediately put on the sheet a number that will indicate the total number of combinations of heads and tails that are possible if each coin is flipped. For example, two coins can fall in four different ways, three coins in eight different ways, and so on.

You have no way of knowing how many coins he drew and yet you win the bet easily. How?

Problem 1.11—Coins of the Realm

In this country at least eight coins are required to make the sum of 99 cents: a half-dollar, a quarter, two dimes, and four pennies. Imagine yourself the leader of a small, newly independent nation. You have the task of setting up a system of coinage based on the cent as the smallest unit. Your objective is to issue the smallest number of different coins that will enable any value from 1 to 100 cents (inclusive) to be made with no more than two coins.

For example, the objective is easily met with 18 coins of the following values: 1, 2, 3, 4, 5, 6, 7, 8, 9, 10, 20, 30, 40, 50, 60, 70, 80, 90. Can you do better? Every value must be obtainable either by one coin or as the sum of two coins. The two coins need not, of course, have different values.

Problem 1.12—Digital Problem

In the 10 cells of Figure 1.7 inscribe a 10-digit number such that the digit in the first cell indicates the total number of zeros in the entire number, the digit in the cell marked "1" indicates the total number of 1's in the number, and so on to the last cell, whose digit indicates the total number of 9's in the number. (Zero is a digit, of course.) The answer is unique.

Figure 1.7

COMBINATORIAL & NUMERICAL PROBLEMS

PROBLEM 1.13—Langford's Problem

Many years ago C. Dudley Langford, a Scottish mathematician, was watching his little boy play with colored blocks. There were two blocks of each color, and the child had piled six of them in a column in such a way that one block was between the red pair, two blocks were between the blue pair, and three were between the yellow pair. Substitute digits 1, 2, 3 for the colors and the sequence can be represented as 312132.

This is the unique answer (not counting its reversal as being different) to the problem of arranging the six digits so that there is one digit between the 1's and there are two digits between the 2's and three digits between the 3's.

Langford tried the same task with four pairs of differently colored blocks and found that it too had a unique solution. Can you discover it? A convenient way to work on this easy problem is with eight playing cards: two aces, two deuces, two threes, and two fours. The object then is to place them in a row so that one card separates the aces, two cards separate the deuces, and so on.

There are no solutions to "Langford's problem," as it is now called, with five or six pairs of cards. There are 26 distinct solutions with seven pairs. No one knows how to determine the number of distinct solutions for a given number of pairs except by exhaustive trial-and-error methods, but perhaps you can discover a simple method of determining whether there is a solution.

PROBLEM 1.14—Colored Bowling Pins

A wealthy man had two bowling lanes in his basement. In one lane 10 dark-colored pins were used; in the other, 10 light-colored pins. The man had a mathematical turn of mind, and the following problem occurred to him one evening as he was practicing his delivery:

Is it possible to mix pins of both colors, then select 10 pins that can be placed in the usual triangular formation in such a way that no three pins of the same color will mark the vertices of an equilateral triangle?

If it is possible, show how to do it. Otherwise prove that it cannot be done. A set of checkers will provide convenient pieces for working on the problem.

PROBLEM 1.15—Ten Soldiers

Ten soldiers, no two of them the same height, stand in a line. There are 10!, or 3,628,800, different ways the men can arrange themselves, but in every arrangement at least four soldiers will form a series of ascending or descending heights. If all but those four leave the line, the four will stand like a row of panpipes.

You can convince yourself of this by experimenting with 10 playing cards bearing values from ace to 10. The values represent the height order of the soldiers. No matter how you arrange the 10 cards in a row, it will always be possible to pick out at least four cards (there may, of course, be more) in ascending or descending order. Suppose, for instance, you arrange the cards in the following order: 5, 7, 9, 2, 1, 4, 10, 3, 8, 6. The set 5, 7, 9, 10 is in ascending order. Can you eliminate such a set by moving, say, the 10 to between the 7 and the 9? No, because you then create the set 10, 9, 8, 6, which is in descending order.

Let p (for panpipe) be the number of the largest set of soldiers that will always be found in order in a row of n soldiers of n different heights, no matter how they arrange themselves. The problem—and it is not easy—is to prove that if n equals 10, p equals 4. In doing so you are likely to discover the general rule by which the p number is easily computed for every n.

PROBLEM 1.16—Handshakes and Networks

Prove that at a recent convention of biophysicists the number of scientists in attendance who shook hands an odd number of times is even. The same problem can be expressed graphically as follows. Put as many dots (biophysicists) as you wish on a sheet of paper. Draw as many lines (handshakes) as you wish from any dot to any other dot. A dot can "shake hands" as often as you please, or not at all. Prove that the number of dots with an odd number of lines joining them is even.

PROBLEM 1.17—Blue-Empty Graph

Six Hollywood stars form a social group that has very special characteristics. Every two stars in the group either mutually love each other or mutually hate each other. There is no set of three individuals who mutually love one another. Prove that there is at least one set of three individuals who mutually hate each other. The problem leads into a fascinating field of graph theory, "blue-empty chromatic graphs," the nature of which will be explained when the answer is given.

PROBLEM 1.18—Five Couples

My wife and I recently attended a party at which there were four other married couples. Various handshakes took place. No one shook hands with himself (or herself) or with his (or her) spouse, and no one shook hands with the same person more than once.

After all the handshakes were over, I asked each person, including my wife, how many hands he (or she) had shaken. To my surprise each gave a different answer. How many hands did my wife shake?

PROBLEM 1.19—Bowling-Ball Pennies

Kobon Fujimura, the leading puzzle authority of Japan, devised this tricky little puzzle, which appears in one of his books. Arrange 10 pennies in the familiar bowling-pin formation [see Figure 1.8]. What is the smallest number of coins you must remove so that no equilateral triangle, of any size, will have its three corners marked by the centers of three pennies that remain? Not counting rotations and reflections as different, there is only one pattern for the removal of the minimum number of pennies.

Note that the pattern contains two equilateral triangles that are tipped so that their bases are not horizontal.

Figure 1.8

PROBLEM 1.20—Killing Squares and Rectangles

This harmless-looking problem in combinatorial geometry, which I found on page 49 of *Sam Loyd and His Puzzles* (Barse and Co., 1928), has more to it than first meets the eye. Forty toothpicks are arranged as shown in Figure 1.9 to form the skeleton of an order-4 checker-

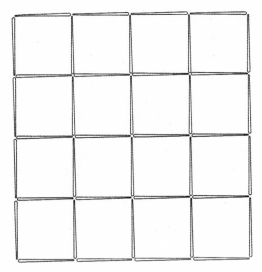

Figure 1.9

board. The problem is to remove the smallest number of toothpicks that will break the perimeter of every square. "Every square" means not just the 16 small ones but also the nine order-2 squares, the four order-3 squares, and the one large order-4 square that is the outside border—30 squares in all.

(On any square checkerboard with n^2 cells the total number of different rectangles is

$$\frac{(n^2 + n)^2}{4},$$

of which

$$\frac{n(n + 1)(2n + 1)}{6}$$

are squares. "It is curious and interesting," wrote Henry Ernest Dudeney, the noted British puzzle expert, "that the total number of rectangles is always the square of the triangular number whose side is n.")

The answer given in the old book was correct, and the reader should have little difficulty finding it. But can he go a step further and state a simple proof that the answer is indeed minimum?

This far from exhausts the puzzle's depth. The obvious next step is to investigate square boards of other sizes. The order-1 case is trivial. It is easy to show that three toothpicks must be removed from the order-

2 board to destroy all squares, and six from the order-3. The order-4 situation is difficult enough to be interesting; beyond that the difficulty seems to increase rapidly.

The combinatorial mathematician is not likely to be content until he has a formula that gives the minimum number of toothpicks as a function of the board's order and also a method for producing at least one solution for any given order. The problem can then be extended to rectangular boards and to the removal of a minimum number of unit lines to kill all rectangles, including the squares. I know of no work that has been done on any of these questions.

Try your skill on squares with sides from four through eight. A minimal solution for order-8, the standard checkerboard (it has 204 different squares), is not easy to find.

PROBLEM 1.21—Dodecahedron-Quintomino Puzzle

John Horton Conway defines a "quintomino" as a regular pentagon whose edges (or triangular segments) are colored with five different colors, one color to an edge. Not counting rotations and reflections as being different, there are 12 distinct quintominoes. Letting 1, 2, 3, 4, 5 represent the five colors, the 12 quintominoes can be symbolized as follows:

A. 12345	E. 12534	J. 13425
B. 12354	F. 12543	K. 13524
C. 12435	G. 13245	L. 14235
D. 12453	H. 13254	M. 14325

The numbers indicate the cyclic order of colors going either clockwise or counterclockwise around the pentagon [see left illustration, in Figure 1.10]. In 1958 Conway asked himself if it was possible to color the edges of a regular dodecahedron (middle illustration) in such a way that each of the 12 quintominoes would appear on one of the solid's 12 pentagonal faces. He found that it was indeed possible. Can you find a way to do it?

Those who like to make mechanical puzzles can construct a cardboard model of a dodecahedron with small magnets glued to the inside of each face. The quintominoes can be cut from metal and colored on both sides (identical colors opposite each other) so that any piece can be "reflected" by turning it over. The magnets, of course, serve to hold the quintominoes on the faces of the solid while one

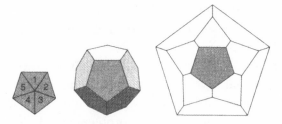

Figure 1.10

works on the puzzle. The problem is to place the 12 pieces in such a way that the colors match across every edge.

Without such a model, the Schlegel diagram of a dodecahedron (right illustration) can be used. This is simply the distorted skeleton of the solid, with its back face stretched to become the figure's outside border. The edges are to be labeled (or colored) so that each pentagon (including the one delineated by the pentagonal perimeter) is a different quintomino.

PROBLEM 1.22—Frieze Patterns

A frieze is a pattern that endlessly repeats itself along an infinite strip. Such patterns can exhibit different kinds of basic symmetry, but here we shall be concerned only with what is called "glide symmetry." A glide consists of a slide (more technically a "translation") combined with mirror reflection and a half-turn. For example, repeatedly gliding the letter R to the right along a strip generates the following frieze:

RЯRЯRЯRЯRЯRЯRЯRЯ. . .

H.S.M. Coxeter, a geometer at the University of Toronto, investigated in depth a remarkable class of frieze patterns that can be constructed very simply by using nonnegative integers, if the lack of symmetry in the shapes of the numerals is ignored [see Figure 1.11]. Think of the numerals as representing spots of colors, all 1's the same color, all 2's another color, and so on. In this instance any rectangular portion of the frieze that is nine columns wide, such as the shaded one shown here, can be regarded as the unit pattern. By gliding it left or right—that is, sliding and simultaneously reflecting and inverting—the infinite frieze pattern is generated.

To produce this type of frieze pattern, begin with infinite borders of 0's and 1's at top and bottom, and a path of numbers from top to bottom such as the zigzag path of eight 1's shown on the left between

```
... 0   0   0   0   0   0   0   0   0 ...

...   1   1   1   1   1   1   1   1   ...

...   1   2   5   2   1   3   2   2 ...

...     1   9   9   1   2   5   3   ...

...   1   4  16   4   1   3   7   7 ...

...   1   3   7   7   3   1   4  16   ...

...1  2   5   3   5   2   1   9   9 ...

...   1   3   2   2   3   1   2   5   ...

...   1   1   1   1   1   1   1   1 ...

...   0   0   0   0   0   0   0   0   ...
```

Figure 1.11

the borders of 0's. The numbers in such a path (which may be straight, or crooked as it is here), as well as the length of the path, can be varied to produce different patterns. A simple formation rule, common to all such patterns, is now applied to obtain all the other integers. The surprising glide symmetry that results is a nontrivial consequence of this rule.

Our puzzle, suggested by Coxeter, is to guess the simple rule. Hint: It can be written as an equation with three terms involving nothing more than multiplication and addition, and no exponents. When Coxeter first showed the pattern given here to the mathematician Paul Erdös, Erdös guessed the rule in 20 seconds. A discussion of the properties of such friezes, their fascinating historical background, and their applications to determinants, continued fractions, and geometry can be found in Coxeter's "Frieze Patterns" in *Acta Arithmetica* (vol. 18, 1971, pages 297–310). On friezes in general and their seven basic kinds of symmetry see Coxeter's modern classic, *Introduction to Geometry* (Wiley, 1961, pages 47–49).

PROBLEM 1.23—Rotating Roundtable

In 1969, after 10 weeks of haggling, the Vietnam peace negotiators in Paris finally decided on the shape of the conference table: a circle seating 24 people, equally spaced. Assume that place cards on such a table bear 24 different names and that on one occasion there is such confusion that the 24 negotiators take seats at random. They discover that no one is seated correctly. Regardless of how they are seated, is it always possible to rotate the table until at least two people are simultaneously opposite their place cards?

A much more difficult problem arises if just one person finds his correct seat. Will it then always be possible to rotate the table to bring at least two people simultaneously opposite their cards?

PROBLEM 1.24—Pool-Ball Triangles

Colonel George Sicherman of Buffalo tells me that while he was watching a game of pool, the following problem occurred to him: Is it possible to form a "difference triangle" in arranging the 15 balls in the usual triangular configuration before the start of a game? In a difference triangle the numbers 1 through 15 are arranged in such a way that each number below a pair of numbers is the positive difference between that pair.

It is evident from Figure 1.12 that the problem is trivial with the balls numbered 1, 2, and 3 and that two solutions can be obtained with them. The illustration also shows the four solutions for six balls and the four for 10 balls. To Sicherman's surprise the 15 pool balls have only one basic solution. (It can, of course, be reflected.) Can you find it?

Searching for the solution is considerably simplified by first exploring triangular patterns of even and odd to see which patterns have exactly eight odd and seven even spots. It does not take long to

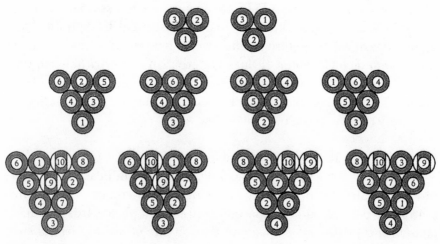

Figure 1.12

COMBINATORIAL & NUMERICAL PROBLEMS

discover that there are only five arrangements for the triangle's top row: EEOEO, OEEEO, OOOEE, OOOOE, and OOEEO. The 15 ball obviously must be in the top row and the 14 ball must be in the same row or below 15 and 1. Other dodges shorten an exhaustive analysis.

The problem is related to one posed by Hugo Steinhaus in his book *One Hundred Problems in Elementary Mathematics* (Dover, 1979, a translation from an earlier Polish edition). Given a triangular array with an even number of spots, is it always possible to form an even-odd difference pattern in which the number of even spots equals the number of odd ones? The problem remained unsolved for more than a decade until Heiko Harborth, in the *Journal of Combinatorial Theory: Series A* (vol. 12, 1972, pages 253–59), proved that the answer is yes.

As far as I know, no work has been done on what we might call the general pool-ball problem. Given any triangular number of balls, numbered consecutively from 1, is it always possible to form a difference triangle? If not, is there a largest triangle for which a solution is possible? If there is, what is it? We now know that odd-even patterns for such solutions exist for all triangles with an even number of balls. Do they also exist for all triangles with an odd number of balls?

Let me add the following joke problem for those of you who succeed in solving the 15-ball problem. Suppose the balls bear the 15 consecutive even numbers from 2 through 30. Is it possible to arrange the set in a difference triangle?

Answers

ANSWER 1.1—(July 1971)

Each month has 11 ambiguous dates (a date such as 8/8/71 is not ambiguous), making 132 in all. (Contributed by David L. Silverman.)

ANSWER 1.2—(July 1971)

Ten digits can be permuted in 10! = 3,628,800 different ways. A 10-digit number cannot start with zero, so that we must subtract 3,628,800/10 = 362,880 to obtain the answer: 3,265,920.

ANSWER 1.3—(July 1971)

The highest constant is 36, as shown in shown in Figure 1.13. C. C. Cousins, Charles W. Bostick, and others noticed that four of the court cards in the illustration for the answer to this problem are incorrectly

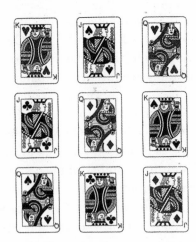

Figure 1.13

drawn. The Jack of Diamonds and the Jack of Clubs should be one-eyed, and the King of Spades and King of Hearts should face the other way. Some readers thought the Queen of Diamonds should face the other way, but Bostick took the trouble to examine 30 different decks made in the United States and found that in 18 of them the Queen of Diamonds faced right, and in 12 cases she faced left, so this card can not be considered wrong.

ANSWER 1.4—(April 1969)

Each cube must bear a 0, 1, and 2. This leaves only six faces for the remaining seven digits, but fortunately the same face can be used for 6 and 9, depending on how the cube is turned. The picture shows 3, 4, 5 on the right cube, and therefore its hidden faces must be 0, 1, and 2. On the left cube one can see 1 and 2, and so its hidden faces must be 0, 6 or 9, 7, and 8.

John S. Singleton wrote from England to say that he had patented the two-cube calendar in 1957/8 (British patent number 831572), but allowed the patent to lapse in 1965. (For a variation of this problem, in which three cubes provide abbreviations for each month, see my *Scientific American* column for December 1977.)

ANSWER 1.5—(February 1957)

It is impossible to cover the mutilated chessboard (with two opposite corner squares cut off) with 31 dominoes, and the proof is easy. The two diagonally opposite corners are of the same color. Therefore their removal leaves a board with two more squares of one color than

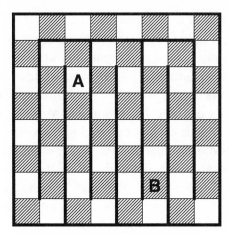

Figure 1.14

of the other. Each domino covers two squares of opposite color, since only opposite colors are adjacent. After you have covered 60 squares with 30 dominoes, you are left with two uncovered squares of the same color. These two cannot be adjacent, therefore they cannot be covered by the last domino.

An interesting question now arises. Suppose that any two squares of *opposite* color are removed from a chessboard, say squares A and B in Figure 1.14. Can the remaining 62 squares always be completely covered by 31 dominoes? The answer is yes, and Ralph Gomory hit upon a simple, elegant proof.

Removing any two squares of opposite color divides the closed path (one square wide) into two segments. Each segment must contain an *even* number of squares that alternate colors. Obviously each segment can be covered with n dominoes, where n is half the number of squares in the segment. Think of the dominoes as little boxcars that can be placed end to end along each of the two segments. Hence the task is always solvable.

ANSWER 1.6—(October 1960)

It is not possible for the termite to pass once through the 26 outside cubes and end its journey in the center one. This is easily demonstrated by imagining that the cubes alternate in color like the cells of a three-dimensional checkerboard, or the sodium and chlorine atoms in the cubical crystal lattice of ordinary salt. The large cube will then consist of 13 cubes of one color and 14 of the other color. The termite's path is always through cubes that alternate in color along the way; therefore if the path is to include all 27 cubes, it must begin and end with a cube belonging to the set of 14. The central cube, however, belongs to the 13 set; hence the desired path is impossible.

The problem can be generalized as follows: A cube of even order (an even number of cells on the side) has the same number of cells of one color as it has cells of the other color. There is no central cube, but complete paths may start on any cell and end on any cell of opposite color. A cube of odd order has one more cell of one color than the other, so a complete path must begin and end on the color that is used for the larger set. In odd-order cubes of orders 3, 7, 11, 15, 19, . . . , the central cell belongs to the smaller set, so it cannot be the end of any complete path. In odd-order cubes of 1, 5, 9, 13, 17, . . . , the central cell belongs to the larger set, so it can be the end of any path that starts on a cell of the same color. No closed path, going through every unit cube, is possible on any odd-order cube because of the extra cube of one color.

Many two-dimensional puzzles can be solved quickly by similar "parity checks." For example, it is not possible for a rook to start at one corner of a chessboard and follow a path that carries it once through every square and ends on the square at the diagonally opposite corner.

ANSWER 1.7—(November 1970)

There is certain to be at least one column of blank spaces on that typed page. Assume that the sentence is n spaces long, including the first space following the final period. This chain of n spaces will begin each typed line, although the chain may be cyclically permuted, beginning with different words in different lines. Consequently the first n spaces of every line will be followed by a blank space. (T. Robert Scott originated this problem, which was sent to me by his friend W. Lloyd Milligan of Columbia, South Carolina.)

ANSWER 1.8—(February 1962)

If the numbers from 1 to 8 are placed in the circles as shown in the Figure 1.15, no number will be connected by a line to a number immediately above or below it in serial order. The solution (including its upside-down and mirror-image forms) is unique.

L. Vosburgh Lyons solved it as follows. In the series 1, 2, 3, 4, 5, 6,

7, 8 each digit has two neighboring numbers except 1 and 8. In the diagram, circle C is connected to every circle except H. Therefore if C contains any number in the set 2, 3, 4, 5, 6, 7, only circle H will remain to accommodate both neighbors of whatever number goes in C. This is impossible, so C must contain 1 or 8. The same argument applies to circle F. Because of the pattern's symmetry, it does not matter whether 1 goes in C or F, so let us place it in C. Circle H is the only circle available for 2. Similarly, with 8 in circle F, only circle A is available for 7. The remaining four numbers are now easily placed.

Thomas H. O'Beirne and Herb Koplowitz each solved the problem by drawing a new diagram in which the old network of lines is replaced by lines connecting all circles that were not connected before. The original problem now takes the form of placing the digits in the circles so that a connected path can be traced from 1 to 8, taking the digits in order. It is easy to see, by inspection of the new diagram, that only four ways of placing the digits are possible, and they correspond to the rotations and reflections of the unique solution.

Fred Gruenberger of the Rand Corporation, Santa Monica, wrote to say that he had encountered this problem about a year earlier "through a friend at the Walt Disney Studios, where it had already consumed a fair amount of Mr. Disney's staff time." Gruenberger had used it as the basis of a West Coast television show, "How a Digital Computer Works," to illustrate the difference between the way a human mathematician approaches such a problem and the brute-force approach of a digital computer that finds the solution by running through all possible permutations of the digits, in this case 40,320 different arrangements.

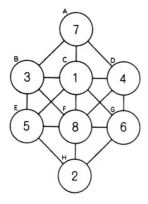

Figure 1.15

This problem was discussed and generalized in "The No-Touch Puzzle and Some Generalizations," by D. K. Cahoon, in *Mathematics Magazine* (vol. 45, November 1972, pages 261–65).

ANSWER 1.9—(May 1959)

The first sheet is folded as follows. Hold it face down so that when you look down on it the numbered squares are in this position:

$$\frac{2365}{1874}$$

Fold the right half on the left so that 5 goes on 2, 6 on 3, 4 on 1, and 7 on 8. Fold the bottom half up so that 4 goes on 5 and 7 on 6. Now tuck 4 and 5 between 6 and 3, and fold 1 and 2 under the packet.

The second sheet is first folded in half the long way, the numbers outside, and held so that 4536 is uppermost. Fold 4 on 5. The right end of the strip (squares 6 and 7) is pushed between 1 and 4, then bent around the folded edge of 4 so that 6 and 7 go between 8 and 5, and 3 and 2 go between 1 and 4.

ANSWER 1.10—(February 1970)

To win the bet, draw a 1 to the left of the tip of the thumb that is covering the row of circles. When the thumb is removed, the paper will show a binary number consisting of 1 followed by a row of 0's. If we assume the 0's represent n coins, this binary number will be equivalent to the decimal number 2^n, the number of ways n coins can fall heads or tails.

ANSWER 1.11—(June 1964)

With as few as 16 different coins one can express any value from 1 cent to 100 cents as the sum of no more than 2 coins. The coins are: 1, 3, 4, 9, 11, 16, 20, 25, 30, 34, 39, 41, 46, 47, 49, 50. This solution is given, without proof that it is minimal, in Problem 19 of Roland Sprague's *Recreation in Mathematics*, translated from the German by T. H. O'Beirne (London: Blackie and Son, 1963). Sprague's solution has a range of only 100. A 16-integer solution with the higher range of 104 was provided by Peter Wegner of the University of London: 1, 3, 4, 5, 8, 14, 20, 26, 32, 38, 44, 47, 48, 49, 51, 52.

ANSWER 1.12—(February 1970)

The only answer is 6,210,001,000. I do not have the space for a detailed proof, but a good one by Edward P. DeLorenzo is in Allan J. Gottlieb's puzzle column in the Massachusetts Institute of Technology's *Technical Review* for February 1968. The same column for June 1968 has a proof by Kenneth W. Dritz that for fewer than 10 cells the only answers in bases 1 through 9 are 1,210; 2,020; 21,200; 3,211,000; 42,101,000, and 521,001,000.

See the *Journal of Recreational Mathematics* (vol. 11, 1978–79, pages 76–77) for Frank Rubin's general solution. He shows that for all bases above 6 there is one solution, of the form $R21(0 \ldots 0)1,000$, where R is four less than the base, and the number of zeroes inside the parentheses is seven less than the base.

Two papers on the digital problem and its generalizations appeared in *The Mathematical Gazette*: "Self-Descriptive Strings," by Michael McKay and Michael Waterman (vol. 66, March 1982, pages 1–4), and "Self-Descriptive Lists—a Short Investigation," by Tony Gardner (vol. 68, March 1984, pages 5–10).

ANSWER 1.13—(November 1967)

The unique solution to Langford's problem with four pairs of cards is 41312432. It can be reversed, of course, but this is not considered a different solution.

If n is the number of pairs, the problem has a solution only if n is a multiple of 4 or one less than such a multiple. C. Dudley Langford posed his problem in *The Mathematical Gazette* (vol. 42, October 1958, page 228). For subsequent discussion see C. J. Priday, "On Langford's Problem (I)," and Roy O. Davies, "On Langford's Problem (II)," both in *The Mathematical Gazette* (vol. 43, December 1959, pages 250–55).

The 26 solutions for $n = 7$ are given in *The Mathematical Gazette* (vol. 55, February 1971, page 73). Numerous computer programs have confirmed this list and found 150 solutions for $n = 8$. E. J. Groth and John Miller independently ran programs which agreed on 17,792 sequences for $n = 11$ and 108,144 for $n = 12$.

R. S. Nickerson, in "A Variant of Langford's Problem," *American Mathematical Monthly* (vol. 74, May 1967, pages 591–95), altered the rules so that the second card of a pair, each with value k, is the kth card after the first card; put another way, each pair of value k is separated by k-1 cards. Nickerson proved that the problem was solvable if and only if the number of pairs is equal to 0 or 1 (modulo 4). John Miller ran a program which found three solutions for $n = 4$ (they are 11423243, 11342324, and 41134232); five solutions for $n = 5$; 252 solutions for $n = 8$; and 1,328 for $n = 9$.

Frank S. Gillespie and W. R. Utz, in "A Generalized Langford Problem," *Fibonacci Quarterly* (vol. 4, April 1966, pages 184–86), extended the problem to triplets, quartets, and higher sets of cards. They were unable to find solutions for any sets higher than pairs. Eugene Levine, writing in the same journal ("On the Generalized Langford Problem," vol. 6, November 1968, pages 135–38), showed that a necessary condition for a solution in the case of triplets is that n (the number of triplets) be equal to -1, 0, or 1 (modulo 9). Because he found solutions for $n = 9$, 10, 17, 18, and 19, he conjectured that the condition is also sufficient when n exceeds 8. The nonexistence of a solution for $n = 8$ was later confirmed by a computer search.

You may enjoy finding a solution to the triplet problem for $n = 9$. Take from a deck all the cards of three suits which have digit values (ace through nine). Can you arrange these 27 cards in a row so that for each triplet of value k cards there are k cards between the first and second card, and k cards between the second and third? It is an extremely difficult combinatorial puzzle. (Three solutions are given below.)

D. P. Roselle and T. C. Thomasson, Jr., "On Generalized Langford Sequences," *Journal of Combinatorial Theory* (vol. 11, September 1971, pages 196–99), report on some new nonexistence theorems and give one solution each for triplets when $n = 9$, 10, and 17. At this point, no Langford sequence had yet been found for sets of integers higher than three, nor had anyone proved that such sequences do or do not exist.

A breakthrough on Langford's problem was later obtained by two Japanese mathematicians at the Miyagi Technical College. Takanois Hayasaka and Sadao Saito used a special-purpose calculator to search for quartets that would form Langford sequences. They found three, each of length $4 \times 24 = 96$ numbers. They proved that the minimum value of n (the number of quartets) is 24. The three sequences are given in their note on "Langford Sequences: A Progress Report," in *Mathematical Gazette* (vol. 63, December 1979, pages 261–62). That same year they reported the results of a computer search for pentuplets through $n = 24$, and sextuplets through $n = 21$. No solutions were found.

In 1980 at Lewis and Clark College, John Miller completed a computer search on the original Langford problem, with doublets, for solutions when $n = 15$. He reported that he found 39,809,640 chains, excluding reversals. Later, on Nickerson's variant, he found 227,968 solutions for $n = 12$ and 1,520,280 solutions for $n = 13$.

Eugene Levine found only one solution for triplets when $n = 9$, but Miller made an exhaustive computer search that found three solutions:

191618257269258476354938743

191218246279458634753968357

181915267285296475384639743

Another exhaustive search by Miller turned up five solutions for triplets when $n = 10$:

110161793586310753968457210429824

110121429724810564793586310753968

410171418934710356839752610285296

811013196384731064958746251029725

131101349638457106495827625102987

Finally, Jaromir Abram wrote a paper on the general problem: "Exponential Lower Bounds for the Numbers of Skolem and Extremal Langford Sequences," in *Ars Combinatoria* (vol. 22, 1986, pages 187–98).

ANSWER 1.14—(February 1962)

It is not possible to mix bowling pins of two different colors and set up a triangular formation of 10 pins in such a way that no three pins of the same color mark the corners of an equilateral triangle. There are many ways to prove this; the following is typical.

Assume that the two colors are red and black and that the 5 pin [see Figure 1.16] is red. Pins 4, 9, 3 form an equilateral triangle, so at least one of these pins must be red. It does not matter which we make red, because of the figure's symmetry, so let us make it the 3 pin. Pins 2 and 6 must therefore be black. Pins 2, 6, 8 form a triangle, forcing us to make 8 a red pin. This in turn makes the 4 and 9 pins black. Pin 10 cannot be black, for this would form a black triangle with 6 and 9, nor can it be red, because this would form a red triangle with 3 and 8.

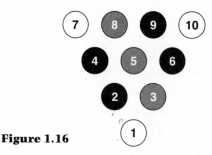

Figure 1.16

Therefore pin 5, with which we started, cannot be red. Of course, the same argument will show that it cannot be black.

ANSWER 1.15—(February 1967)

If n soldiers of differing height stand in a row, at least p soldiers will be in either ascending or descending order. The number p is the square root of the smallest perfect square that is not less than n.

To prove this, label each soldier with a pair of letters, a and d. Let a be the maximum number of men on the soldier's left, including himself, who are in ascending height order. (By "left" I mean to your right as you face the row of soldiers.) Let d be the maximum number on his left, including himself, in descending order. It is easy to show (this is left to you) that no two soldiers can have the same pair of numbers. Their a numbers or their d numbers may be the same, but not both.

Assume that 10 soldiers are so arranged that their largest subset in ascending or descending order has p members, the lowest possible. No soldier can have an a or a d number greater than p. Since no two soldiers have identical pairs of a and d numbers, p must be large enough to provide at least 10 different pairs of a and d numbers.

Can p equal 3? No, because this provides only $3^2 = 9$ pairs of numbers:

$$a \quad 1\ 1\ 1\ 2\ 2\ 2\ 3\ 3\ 3$$
$$d \quad 1\ 2\ 3\ 1\ 2\ 3\ 1\ 2\ 3$$

Any number p will provide p^2 pairs of a and d numbers. Since 3^2 is 9, we do not have enough pairs to associate with the 10 soldiers. But 4^2 is 16, more than enough. We conclude that no matter how 10 soldiers arrange themselves, at least four must be in order. Four remains the p number for sets of soldiers up to and including 16. But 17 soldiers have a p number of 5 because we have to go to the next-highest p number to find enough a and d number pairs.

If 100 soldiers of different heights stand in a row, it is not possible for less than 10 to be in panpipe order. But add one more soldier and the p number jumps to 11.

A good discussion of the problem appears in Section 7 of "Combinatorial Analysis," by Gian-Carlo Rota, in *The Mathematical Sciences*, a collection of essays edited by the National Research Council's Committee on Support of Research in the Mathematical Sciences (MIT Press, 1969). For generalizations and extensions of the problem see two articles: "Monotonic Subsequences," by J. B. Kruskal, Jr., in *Proceedings of the American Mathematical Society* (vol. 4, 1953, pages 264–74) and "Longest Increasing and Decreasing Subsequences," by Craige Schensted, in *Canadian Journal of Mathematics* (vol. 13, 1961, pages 179–91).

ANSWER 1.16—(August 1958)

Because two people are involved in every handshake, the total score for everyone at the convention will be evenly divisible by two and therefore even. The total score for the men who shook hands an even number of times is, of course, also even. If we subtract this even score from the even total score of the convention, we get an even total score for those men who shook hands an odd number of times. Only an even number of odd numbers will total an even number, so we conclude that an even number of men shook hands an odd number of times.

There are other ways to prove the theorem; one of the best was sent to me by Gerald K. Schoenfeld, a medical officer in the U.S. Navy. At the start of the convention, before any handshakes have occurred, the number of persons who have shaken hands an odd number of times will be zero. The first handshake produces two "odd persons." From now on, handshakes are of three types: between two even persons, two odd persons, or one odd and one even person. Each even-even shake increases the number of odd persons by two. Each odd-odd shake decreases the number of odd persons by two. Each odd-even shake changes an odd person to even and an even person to odd, leaving the number of odd persons unchanged. There is no way, therefore, that the even number of odd persons can shift its parity; it must remain at all times an even number.

Both proofs apply to a graph of dots on which lines are drawn to connect pairs of dots. The lines form a network on which the number of dots that mark the meeting of an odd number of lines is even.

ANSWER 1.17—(October 1962)

Every two people in a set of six people either mutually love or mutually hate each other, and there is no set of three who mutually love one another. The problem is to prove that there is a set of three who mutually hate one another.

The problem is easily solved by a graph technique. Six dots represent the six individuals [see Figure 1.17]. All possible pairs are connected by a broken line that stands for either mutual love or mutual hate. Let blue lines symbolize love and red lines symbolize hate.

Consider dot A. Of the five lines radiating from it, at least three must be of the same color. The argument is the same regardless of which color or which three lines we pick, so let us assume three lines are red [shown solid black in the illustration]. If the lines forming triangle BCE are all blue, then we have a set of three people who mutually love one another. We are told no such set exists; therefore at least one side of this triangle must be red. No matter which side we pick for red, we are sure to form an all-red triangle (i.e., three people who mutually hate one another). The same result is obtained if we choose to make the first three lines blue instead of red. In that case the sides of triangle BCE must all be red; otherwise a blue side would form an all-blue triangle. In brief, there must be at least one triangle that is either all-blue or all-red. The problem rules out an all-blue triangle, so there must be an all-red one.

Actually, a stronger conclusion is obtainable. If there is no all-blue triangle, it can be shown (by more complicated reasoning) that there are at least two all-red triangles. In graph theory, a two-color graph of this sort, with no blue triangles, is called a blue-empty chromatic

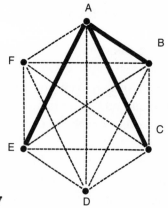

Figure 1.17

graph. If the number of points is six, as in this problem, the minimum number of red triangles is two.

When the number of points in a blue-empty graph is less than six, it is easy to draw such graphs with no red triangles. When the number of points is seven, there must be at least four red triangles. For an eight-point blue-empty graph the minimum number of red triangles is eight; for a nine-point graph it is thirteen. Anyone wishing to go deeper into the topic can consult the following papers:

- R. E. Greenwood and A. M. Gleason, "Combinatorial Relations and Chromatic Graphs," *Canadian Journal of Mathematics* (vol. 7, 1955, pages 1–7).
- Leopold Sauve, "On Chromatic Graphs," *American Mathematical Monthly* (vol. 68, February 1961, pages 107–11).
- Gary Lorden, "Blue-empty Chromatic Graphs," *American Mathematical Monthly* (vol. 69, February 1962, pages 114–20).
- J. W. Moon and Leo Moser, "On Chromatic Bipartite Graphs," *Mathematics Magazine* (vol. 35, September 1962, pages 225–27).
- J. W. Moon, "Disjoint Triangles in Chromatic Graphs," *Mathematics Magazine* (vol. 39, November 1966, pages 259–61).
- Martin Gardner, "Ramsey Theory," *The Colossal Book of Mathematics*, Norton, 2001, pages 437–54.

ANSWER 1.18—(May 1973)

Among the five married couples no one shook more than eight hands. Therefore if nine people each shake a different number of hands, the numbers must be 0, 1, 2, 3, 4, 5, 6, 7, and 8. The person who shook eight hands has to be married to whoever shook no hands (otherwise he could have shaken only seven hands). Similarly, the person who shook seven hands must be married to the person who shook only one hand (the hand of the person who shook hands only with the person who shook eight hands). The person who shook six must be married to the person who shook two, and the person who shook five must be married to the person who shook three. The only person left, who shook hands with four, is my wife.

The above reasoning, which makes use of the familiar "pigeonhole principle," can be clarified by diagramming the problem [see Figure 1.18]. Every graph that lacks loops and multiple edges must contain at least two points that have the same number of lines attached to them. In this case the graph has only two such points, those representing me and my wife. (From Lars Bertil Owe of Lund, Sweden.)

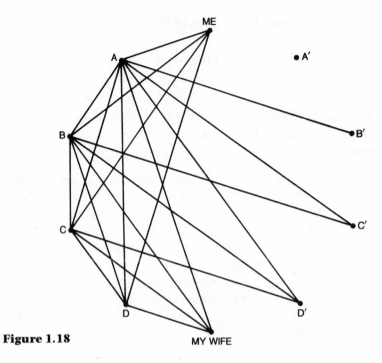

Figure 1.18

ANSWER 1.19—(February 1970)

The four pennies shown shaded [see Figure 1.19] are the fewest that must be removed from the 10 so that no three remaining coins mark the corners of an equilateral triangle. Barring rotations, the pattern is unique; it is, of course, identical with its reflection.

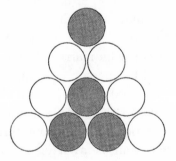

Figure 1.19

ANSWER 1.20—(November 1965)

The smallest number of unit line segments that can be removed from a four-by-four checkerboard to render it "square-free" is nine. One way to do this is shown on the interior four-by-four square of the first diagram in Figure 1.20.

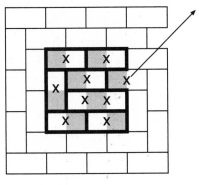
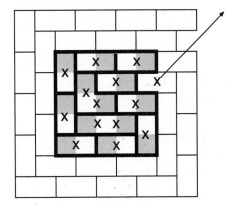

Figure 1.20

To prove this minimum, note that the eight shaded cells have no side in common; to break the perimeters of all eight, at least eight unit lines must be removed. The same argument applies to the eight white cells. We can, however, "kill" all 16 cells with the same eight lines if we pick lines shared by adjacent cells so that each erased line kills a white and a shaded cell simultaneously. But if we do this, not one of the removed lines will be on the board's outside border, which forms the largest square. Therefore at least nine cells must be removed to kill the 16 small cells plus the outer border. As the solution shows, the same nine will eliminate all 30 squares on the board.

The same argument proves that every even-order square must have a solution at least equal to $\frac{1}{2}n^2 + 1$, being the square's order. Can this be achieved on all even-order squares? A proof by induction is implicit in the procedure shown in the illustration. We merely plug a domino in the open cell on the border of the four-by-four, then run a chain of dominoes around the border as shown. This provides a minimum solution of 19 for the order-6 board. The same procedure is applied once more to give the minimum solution of 33 for the eight-by-eight board. It is obvious that this procedure can be repeated endlessly, with each new border of dominoes raising the open cell one step, as shown by the arrow.

On the order-5 board the situation is complicated by the fact that there is one more shaded cell than there are white cells. At least 12 lines would have to be removed to kill 12 shaded and 12 white cells simultaneously. This would, of course, form 12 dominoes. If the remaining shaded cell were on the outside border, both this cell and the border could be killed by taking one more line, which suggests that odd-order squares might have a minimum solution of $\frac{1}{2}(n^2 + 1)$. To achieve this, however, the dominoes would have to be arranged so as not to form an unbroken square higher than order-1. It can be shown that this is never possible, so that the minimum is raised to to $\frac{1}{2}(n^2 + 1) + 1$. The second drawing in Figure 1.20 shows a procedure that achieves this minimum for all odd-order squares.

D. J. Allen, George Brewster, John Dickson, John W. Harris, and Andrew Ungar were the first readers to draw a single diagram that could be extended to display all solutions rather than separate diagrams as I had found, for the odd and even cases.

David Bienenfeld, John W. Harris, Matthew Hodgart, and William Knowlton, attacking the companion problem of creating rectangle-free patterns, discovered that the L-tromino plays in this problem the same role the domino plays in the square-free problem. For squares of orders 2 through 12, the minimum number of lines that must be removed to kill all rectangles are, respectively: 3, 7, 11, 18, 25, 34, 43, 55, 67, 82, 97. Perhaps at some future time I can comment on the formulas and algorithms that were developed. Figure 1.21 shows a pattern for order-8.

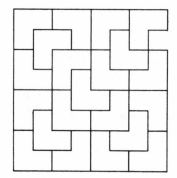

Figure 1.21

ANSWER 1.21—(November 1970)

The three essentially distinct solutions of the dodecahedron-quintomino puzzle are shown on Schlegel diagrams in Figure 1.22. They were first published by John Horton Conway, the inventor of the problem, in the British mathematical journal *Eureka* (October 1959, page 22). Each solution has a mirror reflection, of course, and colors can be interchanged without altering the basic pattern. The letters correspond to those assigned to the 12 quintominoes. The letter outside each diagram denotes the quintomino on the solid's back face, repre-

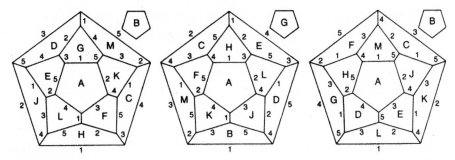

Figure 1.22

sented by the diagram's perimeter. Conway found empirically that whenever the edges of 11 faces were correctly labeled, the twelfth face was automatically labeled to correspond with the remaining quintomino. He did not prove that this must always be true. Because the regular dodecahedron is the "dual" of the regular icosahedron, the problem is equivalent to coloring the edges of the regular icosahedron so that at its 12 vertexes the color permutations corresporrd to the permutations of colors on the 12 quintominoes. In 1972 a white plastic version of Conway's puzzle was on sale in the United States under the name "Enigma." Patterns of black dots were used instead of colors.

ANSWER 1.22—(April 1972)

The formation rule for H.S.M. Coxeter's frieze patterns is that every four adjacent numbers

$$
\begin{matrix}
 & b & \\
a & & d \\
 & c &
\end{matrix}
$$

satisfy the equation $ad = bc + 1$.

ANSWER 1.23—(April 1969)

If a circular table seats an even number of people, equally spaced, with place cards marking their spots, no matter how they seat themselves it is always possible to rotate the table until two or more are seated correctly. There are two initial situations to be considered:

a. *No person sits correctly.* The easy proof is based on what mathematicians call the "pigeonhole principle": If n objects are placed in $n-1$ pigeonholes, at least one hole must contain two objects. If the table seats 24 people, and if every person is incorrectly seated, it

Combinatorics

clearly is possible to bring each person opposite his card by a suitable rotation of the table. There are 24 people but only 23 remaining positions for the table. Therefore at least two people must be simultaneously opposite their cards at one of the new positions. This proof applies regardless of whether the number of seats is odd or even.

b. *One person sits correctly.* Our task is to prove that the table can be rotated so that at least two people will be correctly seated. Proofs of this can be succinctly stated, but they are technical and require a knowledge of special notation. Here is a longer proof that is easier to understand. It is based on more than a dozen similar proofs that were supplied by readers.

Our strategy will be the *reductio ad absurdum* argument. We first assume it is *not* possible to rotate the table so that two persons are correctly seated, then show that this assumption leads to a contradiction.

If our assumption holds, at no position of the table will all persons be incorrectly seated because then the situation is the same as the one treated above, and which we disposed of by the pigeonhole principle. The table has 24 positions, and there are 24 people, therefore at each position of the table exactly one person must be correctly seated.

Assume that only Anderson is at his correct place. For each person there is a "displacement" number indicating how many places he is clockwise from his correct seat. Anderson's displacement is 0. One person will be displaced by 1 chair, another by 2 chairs, another by 3, and so on, to one person who is displaced by 23 chairs. Clearly no two people can have the same displacement number. If they did it would be possible to rotate the table to bring them both simultaneously to correct positions—a possibility ruled out by our assumption.

Consider Smith, who is improperly seated. We count chairs counterclockwise around the table until we get to Smith's place card. The count is equal, of course, to Smith's displacement number. Now consider Jones who is sitting where Smith is supposed to be. We continue counting counterclockwise until we come to Jones's place card. Again, the count equals Jones's displacement number. Opposite Jones's place card is Robinson. We count counterclockwise to Robinson's place card, and so on. Eventually, our count will return to Smith. If Smith and Jones had been in each other's seats, we would have returned to Smith after a two-person cycle that would have carried us just once around the table. If Smith, Jones, and Robinson are occupying one another's seats, we return to Smith after a cycle of three counts. The cycle may involve any number of people from 2 through 23 (it cannot include Anderson because he is correctly seated), but eventually the counting will return to where it started after circling the table an integral number of times. Thus the sum of all the displacement numbers in the cycle must equal 0 modulo 24— that is, the sum must be an exact multiple of 24.

If the cycle starting with Smith does not catch all 23 incorrectly seated persons, pick out another person not in his seat and go through the same procedure. As before, the counting must eventually return to where it started, after an integral number of times around the circle. Therefore the sum of the displacements in this cycle also is a multiple of 24. After one or more of such cycles, we will have counted every person's displacement. Since each cycle count is a multiple of 24, the sum of all the cycle counts must be a multiple of 24. In other words, we have shown that the sum of all the displacement numbers is a multiple of 24.

Now for the contradiction. The displacements are 0, 1, 2, 3, . . . , 23. This sequence has a sum of 276, which is not a multiple of 24. The contradiction forces us to abandon our initial assumption and conclude that at least two people must have the same displacement number.

The proof generalizes to any table with an even number of chairs. The sum of $0 + 1 + 2 + 3 + \cdots + n$ is $n(n + 1)/2$, which is a multiple of n only when n is odd. Thus the proof fails for a table with an odd number of chairs.

George Rybicki solved the general problem this way. We start by assuming the contrary of what we wish to prove. Let n be the even number of persons, and let their names be replaced by the integers 0 to $n - 1$ "in such a way that the place cards are numbered in sequence around the table. If a delegate d originally sits down to a place card p, then the table must be rotated r steps before he is correctly seated, where $r = p - d$, unless this is negative, in which case $r = p - d + n$. The collection of values of d (and of p) for all delegates is clearly the integers 0 to $n - 1$, each taken once, but so also is the collection of values of r, or else two delegates would be correctly seated at the same time. Summing the above equations, one for each delegate, gives $S - S + nk$, where k is an integer and $S = n(n - 1)/2$, the sum of the inte-

gers from 0 to $n - 1$. It follows that $n = 2k + 1$, and odd number."
This contradicts the original assumption.

"I actually solved this problem some years ago," Rybicki writes, "for a different but completely equivalent problem, a generalization of the nonattacking 'eight queens' problem for a cylindrical chessboard where diagonal attack is restricted to diagonals slanting in one direction only. I proved that this was insoluble for any board of even order. The above is the translation of my proof into the language of the table problem. Incidentally, the proof is somewhat easier if one is allowed to use congruences modulo n."

Donald E. Knuth also called attention to the equivalence of the round table and the queens problem, and cited an early solution by George Polya. Several readers pointed out that when the number of persons is odd, a simple arrangement that prevents more than one person from being seated correctly, regardless of how the table is rotated, is to seat them counterclockwise to their place-card order.

ANSWER 1.24—(April 1977)

Except for reflection, the only solution of the problem with the 15 pool balls is shown in Figure 1.23. Col. George Sicherman of the State University of New York at Buffalo, who invented this problem, has found by computer that there are no solutions for similar triangles of orders 6, 7, and 8. Sicherman also found a simple parity proof of impossibility for all orders $2^n - 2$, when n is greater than 2.

This is how the proof goes for order 6, the smallest case. Call the triangle's top row a, b, c, d, e, f. Since addition is the same as subtraction (modulo 2) we can express the other numbers (modulo 2) by adding. The second row is $a + b$, $b + e$, $e + d$, $d + e$, $e + f$. The next row begins $a + 2b + e$, $b + 2e + d,\ldots$. Continue this way to the bottom number, which is $a + 5b + 10e + 10d + 5e + f$. The triangle con-

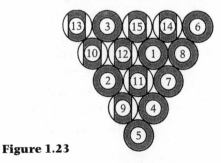

Figure 1.23

tains 6 a's, 20 b's, 34 e's, 34 d's, 20 e's and 6 f's. All the numbers are even; therefore the triangle's parity is even. The triangle contains 11 odd and 10 even numbers, however, giving it an odd parity, so that we have a contradiction. The above row of figures (6, 20, 34, 34, 20, 6) is the same as the seventh row of Pascal's triangle (1, 7, 21, 35, 35, 21, 7, 1) when the numbers are diminished by 1. Sicherman's general proof hinges on the well-known theorem that only rows numbered $2^n - 1$ of Pascal's triangle consist entirely of odd numbers.

Charles Trigg, in the paper "Absolute Difference Triangles," *Journal of Recreational Mathematics* (vol. 9, 1976–77, pages 271–75), proved the uniqueness of the order-5 triangle and discussed some of its unusual characteristics, such as the five consecutive digits along its left side. Among other things, Trigg showed that any absolute-difference triangle of consecutive integers must begin with 1 and that each of the first n consecutive n integers (where n is the triangle's order) must lie in a different horizontal row. It follows that the number at the triangle's bottom corner cannot exceed n. On the basis of computer results and certain arguments he conjectured that no absolute-difference triangles are higher than order 5.

The generalization of the pool-ball problem to triangles of order n, bearing consecutive numbers starting with 1, has now been solved. Herbert Taylor found an ingenious way to prove that no TAD (triangle of absolute differences) could be made with triangular arrays of order 9 or higher. Computer programs eliminated TADs of orders 6, 7, and 8; therefore the unique solution for the 15 pool balls is the largest TAD of this type. The only published proof known to me is given by G. J. Chang, M. C. Hu, K. W. Lih, and T. C. Shieh, in "Exact Difference Triangles," *Bulletin of the Institute of Mathematics*, Academia Sinica, Taipei, Taiwan (vol. 5, June 1977, pages 191–97).

Harry Nelson, who in 1977 edited the *Journal of Recreational Mathematics*, suggested a more general problem. Is there an absolute-difference triangle of order 5 that is formed with integers from the set 1 through 17? His computer program found 15 solutions. If the numbers are restricted to the interval 1 through 16, there are two solutions in addition to the 1-through-15 solution. With 8 missing, the top row is 5, 14, 16, 3, 15. With 9 missing, the top row is 8, 15, 3, 16, 14.

Solomon W. Golomb proposed three candidates for further investigation:

1. If all numbers in a TAD of order greater than 5 are distinct but not consecutive, how big is the largest number forced to be? (Example: An order-6 TAD is possible with the largest number as low as 22.)
2. Using all numbers from 1 to k, but allowing repeats, how big can k be in a TAD of order n? (Example: An order-6 TAD is possible with k as high as 20.)
3. For what orders is it possible to form a TAD modulo m, where m is the number of elements in the triangle and the numbers are consecutive from 1 to m? Each difference is expressed modulo m. Such triangles can be rotated so that every element below the top row is the sum (modulo m) of the two numbers above it. Here, in rotated form, are the four order-4 solutions:

```
    1   6   9   4           2   7   8   3
      7   5   3               9   5   1
        2   8                   4   6
          0                       0

    6   1   4   9           7   2   3   8
      7   5   3               9   5   1
        2   8                   4   6
          0                       0
```

A backtrack program by Golomb and Taylor found no solution for order 5. Colonel Sicherman reports a computer proof of impossibility for order 6. Higher orders remain open.

chapter 2

Probability

Probability is the one area of mathematics in which experts have often made mistakes. The difficulty is not with the formulas but in figuring out what exactly is being asked! If you find the wording ambiguous, then a second (or third) reading may be needed; the problem below about Smith's two children is arguably too ambiguous.

You need grasp only a few ideas about probabilities to solve the problems here. First is the notion of uniform probability, *if each of* n *events is equally likely to occur, then each occurs with probability 1/n. Second is the notion of* expected value; *if each event is associated with a value, then the expected value is the sum of each value multiplied by the probability of the corresponding event occurring.*

Often, a probability question is reduced to figuring out how often an event can occur. Therefore, probability often reduces to combinatorics. Several of the tasks in this chapter involve probabilities only indirectly, but they seem best suited to this chapter.

Problems

PROBLEM 2.1—Bridge Hands
Which situation is more likely after four bridge hands have been dealt: you and your partner hold all the clubs or you and your partner have no clubs?

PROBLEM 2.2—Stuffing Envelopes
A secretary types four letters to four people and addresses the four envelopes. If she inserts the letters at random, each in a different envelope, what is the probability that exactly three letters will go into the right envelopes?

PROBLEM 2.3—A View of the Pentagon

A man arrives at a random spot several miles from the Pentagon. He looks at the building through binoculars. What is the probability that he will see three of its sides? (From F. T. Leahy, Jr.)

PROBLEM 2.4—How Many Hemispheres?

Three points are selected at random on a sphere's surface. What is the probability that all three lie on the same hemisphere? It is assumed that the great circle, bordering a hemisphere, is part of the hemisphere.

PROBLEM 2.5—Cut Cards

A standard deck of 52 cards is shuffled and cut and the cut is completed. The color of the top card is noted. This card is replaced on top, the deck is cut again, and the cut is completed. Once more the color of the top card is noted. What is the probability that the two cards noted are the same color?

PROBLEM 2.6—Two Children

Mr. Smith has two children. At least one of them is a boy. What is the probability that both children are boys?

Mr. Jones has two children. The older child is a girl. What is the probability that both children are girls?

PROBLEM 2.7—Bronx vs. Brooklyn

A young man lives in Manhattan near a subway express station. He has two girlfriends, one in Brooklyn, one in the Bronx. To visit the girl in Brooklyn he takes a train on the downtown side of the platform; to visit the girl in the Bronx he takes a train on the uptown side of the same platform. Since he likes both girls equally well, he simply takes the first train that comes along. In this way he lets chance determine whether he rides to the Bronx or to Brooklyn. The young man reaches the subway platform at a random moment each Saturday afternoon. Brooklyn and Bronx trains arrive at the station equally often—every 10 minutes. Yet for some obscure reason he finds himself spending most of his time with the girl in Brooklyn: in fact on the average he goes there 9 times out of 10. Can you think of a good reason why the odds so heavily favor Brooklyn?

PROBLEM 2.8—Blue-Eyed Sisters

If you happen to meet two of the Jones sisters (this assumes that the two are random selections from the set of all the Jones sisters), it is an exactly even-money bet that both girls will be blue-eyed. What is your best guess as to the total number of blue-eyed Jones sisters?

PROBLEM 2.9—Lady or Tiger?

Frank Stockton's famous short story "The Lady or the Tiger?" tells of a semibarbaric king who enjoyed administering a curious kind of justice. The king sat on a high throne at one side of his public arena. On the opposite side were twin doors. The prisoner on trial could open either door, guided only by "impartial and incorruptible chance." Behind one door was a hungry tiger; behind the other, a desirable young lady. If the tiger sprang through the door, the man's fate was considered a just punishment for his crime. If the lady stepped forth, the man's innocence was rewarded by a marriage ceremony performed on the spot.

The king, having discovered his daughter's romance with a certain courtier, has placed the unfortunate young man on trial. The princess knows which door conceals the tiger. She also knows that behind the other door is the fairest lady of the court, whom she has observed making eyes at her lover. The courtier knows the princess knows. She makes a "slight, quick movement" of her hand to the right. He opens the door on the right. The tale closes with the tantalizing question: "Which came out of the opened door—the lady or the tiger?"

After extensive research on this incident, I am able to make the first full report on what happened next. The two doors were side by side and hinged to open toward each other. After opening the door on the right the courtier quickly pulled open the other door and barricaded himself inside the triangle formed by the doors and the wall. The tiger emerged through one door, entered the other and ate the lady.

The king was a bit nonplused, but, being a good sport, he allowed the courtier a second trial. Not wishing to give the wily young man another 50-50 chance, he had the arena reconstructed so that instead of one pair of doors there were now three pairs. Behind one pair he placed two hungry tigers. Behind the second pair he placed a tiger and a lady. Behind the third pair he placed two ladies who were identical twins and who were dressed exactly alike.

The cruel scheme was as follows. The courtier must first choose a pair of doors. Then he must select one of the two and a key would be tossed to him for opening it. If the tiger emerged, that was that. If the lady, the door would immediately be slammed shut. The lady and her unknown partner (either her twin sister or a tiger) would then be secretly rearranged in the same two rooms, one to a room, according to a flip of a special gold coin with a lady on one side and a tiger on the other. The courtier would be given a second choice between the same two doors, without knowing whether the arrangement was different or the same as before. If he chose a tiger, that was that again; if a lady, the door would be slammed shut, the coin-flipping procedure repeated to determine who went in which room, and the courtier given a third and final choice of one of the same two doors. If successful in his last choice, he would marry the lady and his ordeal would be over.

The day of the trial arrived and all went according to plan. Twice the courtier selected a lady. He tried his best to determine if the second lady was the same as the first but was unable to decide. Beads of perspiration glistened on his forehead. The face of the princess—she was ignorant this time of who went where—was as pale as white marble.

Exactly what probability did the courtier have of finding a lady on his third guess?

PROBLEM 2.10—Two Games in a Row

A certain mathematician, his wife, and their teenage son all play a fair game of chess. One day when the son asked his father for 10 dollars for a Saturday-night date, his father puffed his pipe a moment and replied:

"Let's do it this way. Today is Wednesday. You will play a game of chess tonight, another tomorrow and a third on Friday. Your mother and I will alternate as opponents. If you win two games in a row, you get the money."

"Whom do I play first, you or Mom?"

"You may have your choice," said the mathematician, his eyes twinkling.

The son knew that his father played a stronger game than his mother. To maximize his chance of winning two games in succession, should he play father-mother-father or mother-father-mother?

Leo Moser, a mathematician at the University of Alberta, is responsible for this amusing question in elementary probability theory. Of course you must prove your answer, not just guess.

PROBLEM 2.11—Poker Puzzle

As every poker player knows, a straight flush [Figure 2.1, left] beats four of a kind [Figure 2.1, right].

How many different straight flushes are there? In each suit a straight flush can start with an ace, a deuce, or any other card up to a 10 (the ace may rank either high or low), making 10 possibilities in all. Since there are four suits, there are four times 10, or 40, different hands that are straight flushes.

How many different four-of-a-kind hands are there? There are only 13. If there are 13 four-of-a-kind hands and 40 straight flushes, why does a straight flush beat four of a kind?

Figure 2.1

PROBLEM 2.12—Penny Bet

Bill, a student in mathematics, and his friend John, an English major, usually spun a coin on the bar to see who would pay for each round of beer. One evening Bill said: "Since I've won the last three spins, let me give you a break on the next one. You spin *two* pennies and I'll spin one. If you have more heads than I have, you win. If you don't, I win."

"Gee, thanks," said John.

On previous rounds, when one coin was spun, John's probability of winning was, of course, 1/2. What are his chances under the new arrangement?

PROBLEM 2.13—Bills and Two Hats

"No," said the mathematician to his 14-year-old son, "I do *not* feel inclined to increase your allowance this week by 10 dollars. But if you'll take a risk, I'll make you a sporting proposition."

The boy groaned. "What is it this time, Dad?"

"I happen to have," said his father, "10 crisp new 10-dollar bills and 10 crisp new one-dollar bills. You may divide them any way you please into two sets. We'll put one set into hat A, the other set into hat B. Then I'll blindfold you. I'll mix the contents of each hat and put one hat on the right and one on the left side of the mantel. You pick either hat at random, then reach into that hat and take out one bill. If it's a 10, you may keep it."

"And if it isn't?"

"You'll mow the lawn for a month, with no complaints."

The boy agreed. How should he divide the 20 bills between the two hats to maximize the probability of his drawing a 10-dollar bill, and what will that probability be?

PROBLEM 2.14—Families in Fertilia

> People who have three daughters try once more
> And then it's fifty-fifty they'll have four.
> Those with a son or sons will let things be.
> Hence all these surplus women. Q.E.D.

This "Note for the Scientist," by Justin Richardson, is from a Penguin collection called *Yet More Comic & Curious Verse*, selected by J. M. Cohen. Is the expressed thought sound?

No, although it is a commonly encountered type of statistical fallacy. George Gamow and Marvin Stern, in their book *Puzzle-Math* (Viking, 1958), tell of a sultan who tried to increase the number of women available for harems in his country by passing a law forbidding every mother to have another child after she gave birth to her first son; as long as her children were girls she would be permitted to continue childbearing. "Under this new law," the sultan explained, "you will see women having families such as four girls and one boy; ten girls and one boy; perhaps a solitary boy, and so on. This should obviously increase the ratio of women to men."

As Gamow and Stern make clear, it does nothing of the sort. Consider all the mothers who have had only one child. Half of their children will be boys, half girls. Mothers of the girls will then have a second child. Again there will be an even distribution of boys and girls. Half of those mothers will go on to have a third child and again there will be as many boys as there are girls. Regardless of the number of rounds and the size of the families, the sex ratio obviously will always be one to one.

Which brings us to a statistical problem posed by Richard G. Gould of Washington. Assume that the sultan's law is in effect and that parents in Fertilia are sufficiently potent and long-lived so that every family continues to have children until there is a son, and then stops. At each birth the probability of a boy is one-half. In the long run, what is the average size of a family in Fertilia?

PROBLEM 2.15—Three Coins

While your back is turned a friend places a penny, nickel, and dime on the table. He arranges them in any pattern of heads and tails provided that the three coins are not all heads or all tails.

Your object is to give instructions, without seeing the coins, that will cause all three to be the same (all heads or all tails). For example, you may ask your friend to reverse the dime. He must then tell you whether you have succeeded in getting all the coins alike. If you have not, you again name a coin for him to turn. This procedure continues until he tells you that the three coins are the same.

Your probability of success on the first move is 1/3. If you adopt the best strategy, what is your probability of success in two moves or fewer? What is the smallest number of moves that guarantees success on or before the final move?

You should find those questions easy to answer, but now we complicate the game a bit. The situation is the same as before, only this time your intent is to make all the coins show heads. Any initial pattern except all heads is permitted. As before, you are told after each move whether you have succeeded. Assuming that you use the best strategy, what is the smallest number of moves that guarantees success? What is your probability of success in two moves or fewer, in three moves or fewer, and so on up to the final move at which the probability reaches 1 (certainty)?

PROBLEM 2.16—Two Urn Problems

Probability theorists are fond of illustrating theorems with problems about identical objects of different colors that are taken from urns, boxes, bags, and so on. Even the simplest of such problems can be confusing. Consider, for instance, the fifth of Lewis Carroll's *Pillow Problems:* "A bag contains one counter, known to be either white or black. A white counter is put in, the bag shaken, and a counter drawn out, which proves to be white. What is now the chance of drawing a white counter?"

"At first sight," Carroll begins his answer, "it would appear that, as the state of the bag, *after* the operation, is necessarily identical with its state *before* it, the chance is just what it was, viz. 1/2. This, however, is an error."

Carroll goes on to show that the probability of a white counter remaining in the bag actually is 2/3. His proof is a bit long-winded. Howard Ellis, a Chicago reader, has done it differently. Let B and W_1 stand for the black or white counter that may be in the bag at the start and W_2 for the added white counter. After removing a white counter there are three equally likely states:

IN BAG	OUTSIDE BAG
W_1	W_2
W_2	W_1
B	W_2

In two of these states a white counter remains in the bag, and so the chance of drawing a white counter the second time is 2/3.

A recent problem of this kind, with an even more surprising answer, begins with a bag containing an unknown number of black counters and an unknown number of white counters. (There must be at least one of each color.) The counters are taken out according to the following procedure. A counter is chosen at random and discarded. A second counter is taken at random. If it is the same color as before, it too is discarded. A third is chosen. If it matches the two previously taken counters, it is discarded. This continues, with the counters discarded as long as they match the color of the first counter.

Whenever a counter is taken that has a *different* color from the previous one, it is replaced in the bag, the bag is shaken, and the entire process begins again.

To make this crystal clear, here is a sample of how the first 10 drawings might go:

1. First counter is black. Discard.
2. Next counter is black. Discard.
3. Next is white. Replace and begin again.
4. First is black. Discard.
5. Next is white. Replace and begin again.

COMBINATORIAL & NUMERICAL PROBLEMS

6. First counter is white. Discard.
7. Next is white. Discard.
8. Next is black. Replace and begin again.
9. First counter is black. Discard.
10. Next is white. Replace and begin again.

It turns out, amazingly, that regardless of the ratio of white to black counters at the start, there is a fixed probability that the last counter left in the bag is black. What is that probability?

PROBLEM 2.17—First Black Ace

A deck of 52 playing cards is shuffled and placed facedown on the table. Then, one at a time, the cards are dealt face up from the top. If you were asked to bet in advance on the distance from the top of the first black ace to be dealt, what position (first, second, third, . . .) would you pick so that if the game were repeated many times, you would maximize your chance in the long run of guessing correctly?

PROBLEM 2.18—Coin Game

Three coins are on the table: a quarter, a half-dollar, and a silver dollar. Smith owns one coin and Jones owns the other two. All three coins are tossed simultaneously.

It is agreed that any coin falling tails counts zero for its owner. Any coin falling heads counts its value in cents. The tosser who gets the larger score wins all three coins. If all three come up tails, no one wins and the toss is repeated.

What coin should Smith own so that the game is fair, that is, so that the expected monetary win for each player is zero?

David L. Silverman, author of the excellent book of game puzzles called *Your Move* (McGraw-Hill, 1971), is responsible for this problem. It has an amazing answer. Even more astonishing is a generalization, formally proved by Benjamin L. Schwartz, of which this problem is a special case.

PROBLEM 2.19—Irrational Probabilities

It is very easy to use a penny as a randomizer for deciding between two alternatives with probabilities expressed by rational fractions. Suppose you wish to decide between A and B, with a probability of 3/7 for A and 4/7 for B. The number of equally likely ways a penny can fall when flipped n times is 2^n, so three flips of the coin give eight possible triplets: *HHH, HHT, HTH*, and so on. Eliminate one triplet; then pick any three of the remaining seven and designate them triplets that decide for *A*. The other four triplets decide for *B*. Flip the penny three times. If the result is the eliminated triplet, ignore it and flip three more times. Eventually you will flip one of the seven triplets. The chance that this will be a triplet in the set of three is clearly 3/7, with 4/7 as the probability that it will be in the set of four.

The procedure is easily extended to a decision between *n* alternatives, each with a rational probability. Suppose there are three alternatives with the probabilities *A* = 1/3, *B* = 1/2, and *C* = 1/6. Use the above procedure to decide between 1/3 and 2/3 (the sum of 1/2 and 1/6). If the decision is for *A*, you are finished. Otherwise you must continue by deciding between *B* and *C*. To do so, divide 1/2 (*B*'s fraction) by 2/3 to obtain 3/4, and divide 1/6 (*C*'s fraction) by 2/3 to obtain 1/4. The penny is used as before to decide between *B* = 3/4 and *C* = 1/4. The procedure obviously generalizes to *n* alternatives, provided that the probabilities are rational fractions.

Moreover, the coin need not be a fair one. Suppose it is biased and falls heads with a probability of $1/\pi$. The probability of heads followed by tails remains equal to the probability of tails followed by heads, and so you simply flip doublets, ignoring *HH* and *TT*. Let *HT* count for heads and *TH* count for tails. With this new definition of heads and tails, each equally likely, the biased coin clearly can be used for deciding between *n* alternatives, each with rational probabilities.

Suppose, now, you wish to decide between *n* alternatives, each with an *irrational* probability. For example: *A* is the fractional part of the square root of 2. *B* is the fractional part of π, and *C* = 1 − (*A* + *B*). If you can decide between two irrational probabilities using a fair coin, you can do it with a biased coin by redefining heads and tails as explained; and if you can decide between two irrational alternatives, you can decide between any number of irrational alternatives by the method given for *n* rational alternatives.

But how can a coin be used to decide between two irrational probabilities? Let us focus the problem with a precise example. *A* = .1415926535 . . . , the fractional part of π. *B* = .8584073464 . . . , which is 1 − *A*. You wish to decide between *A* and *B* by flipping a fair coin. A delightful procedure for doing this, which applies to all irrational fractions, was devised by Persi Diaconis. It will be disclosed in the answer section. (Hint: The method makes use of binary notation.)

Answers

ANSWER 2.1—(August 1968)

The probabilities are the same. If you and your partner have no clubs, all the clubs will have been dealt to the other two players.

ANSWER 2.2—(August 1968)

Zero. If three letters match the envelopes, so will the fourth.

ANSWER 2.3—(July 1971)

One proof that the probability is 1/2: Suppose the man has a *Doppelgänger* directly opposite him on the other side of the Pentagon's center and the same distance away. If either man sees three sides, his double must see only two. Since there is an equal probability that either man is at either spot, the probability is 1/2 that he will see three sides.

The probability of 1/2 that a distant viewer will see three sides of the Pentagon is correct only as a limit as the viewer's distance from the Pentagon building approaches infinity. My solution ignores the fact that there are five infinitely long strips, each crossing the building, inside of which both the viewer and his *Doppelgänger* can see only two sides (and if very close to the Pentagon, only one side). This was pointed out by readers too numerous to list.

The probability is zero, commented P. H. Lyons, "if the smog in Washington is anything like what it is here in Toronto."

ANSWER 2.4—(August 1968)

The probability is certainty. Any three points on a sphere are on a hemisphere.

ANSWER 2.5—(August 1969)

Whatever the color of the first card cut, this card cannot be the top card of the second cut. The second cut selects a card randomly from 51 cards of which 25 are the same color as the first card, and therefore the probability of the two cards' matching in color is 25/51, a bit less than 1/2.

ANSWER 2.6—(May 1959)

If Smith has two children, at least one of which is a boy, we have three equally probable cases:

> Boy, boy
> Boy, girl
> Girl, boy

In only one case are both children boys, so the probability that both are boys is 1/3.

Jones's situation is different. We are told that his older child is a girl. This limits us to only two equally probable cases:

> Girl, girl
> Girl, boy

Therefore the probability that both children are girls is 1/2.

[This is how I answered the problem in my column. After reading protests from many readers, and giving the matter considerable further thought, I realized that the problem was ambiguously stated and could not be answered without additional data. For a later discussion of the problem, see Chapter 2 of *The Colossal Book of Mathematics*.]

ANSWER 2.7—(February 1957)

The answer to this puzzle is a simple matter of train schedules. While the Brooklyn and Bronx trains arrive equally often—at 10-minute intervals—it happens that their schedules are such that the Bronx train always comes to this platform one minute after the Brooklyn train. Thus the Bronx train will be the first to arrive only if the young man happens to come to the subway platform during this one-minute interval. If he enters the station at any other time—i.e., during a nine-minute interval—the Brooklyn train will come first. Since the young man's arrival is random, the odds are nine to one for Brooklyn.

ANSWER 2.8—(October 1960)

There are probably three blue-eyed Jones sisters and four sisters altogether. If there are n girls, of which b are blue-eyed, the probability that two chosen at random are blue-eyed is:

$$\frac{b(b-1)}{n(n-1)}$$

We are told that this probability is 1/2, so the problem is one of finding integral values for b and n that will give the above expression a value of 1/2. The smallest such values are $n = 4$, $b = 3$. The next highest values are $n = 21$, $b = 15$, but it is extremely unlikely that there would be as many as 21 sisters, so four sisters, three of them blue-eyed, is the best guess.

ANSWER 2.9—(February 1962)

The problem of the lady or the tiger is merely a dressed-up version of a famous ball-and-urn problem analyzed by the great French mathematician Pierre Simon de Laplace (see James R. Newman's *The World of Mathematics* [New York: Simon and Schuster, 1956], vol. 2, page 1332). The answer is that the young man on his third choice of a door has a probability of 9/10 that he will choose the lady. The pair of doors concealing two tigers is eliminated by his first choice of a lady, which leaves 10 equally probable possibilities for the entire series of three choices.

If the doors conceal two ladies:

Lady 1 — Lady 1 — Lady 1
Lady 1 — Lady 1 — Lady 2
Lady 1 — Lady 2 — Lady 1
Lady 1 — Lady 2 — Lady 2
Lady 2 — Lady 1 — Lady 1
Lady 2 — Lady 1 — Lady 2
Lady 2 — Lady 2 — Lady 1
Lady 2 — Lady 2 — Lady 2

If the doors conceal a lady and a tiger:

Lady 3 — Lady 3 — Lady 3
Lady 3 — Lady 3 — Tiger

Of the 10 possibilities in the problem's "sample space," only one ends with a fatal final choice. The probability of the man's survival is therefore 9/10.

ANSWER 2.10—(October 1962)

A plays a stronger chess game than *B*. If your object is to win two games in a row, which is better: to play against *A*, then *B*, then *A*; or to play *B*, then *A*, then *B*?

Let P_1 be the probability of your defeating A and P_2 the probability of your defeating B. The probability of your *not* winning against A will then be $1 - P_1$ and the probability of your not winning against B will be $1 - P_2$.

If you play your opponents in the order ABA, there are three different ways you can win two games in a row:

1. You can win all three games. The probability of this occurring is
$P_1 \times P_2 \times P_1 = P_1^2 P_2$.
2. You can win the first two games only. The probability of this is
$P_1 \times P_2 \times (1 - P_1) = P_1 P_2 - P_1^2 P_2$.
3. You can win the last two games only. The probability is
$(1 - P_1) \times P_2 \times P_1 = P_1 P_2 - P_1^2 P_2$.

The three probabilities are now added to obtain $P_1 P_2 (2 - P_1)$. This is the probability that you will win twice in a row if you play in the order ABA.

If the order is BAB, a similar calculation will show that the probability of winning all three games is $P_1 P_2^2$, of winning the first two games is $P_1 P_2 - P_1 P_2^2$, and of winning the last two games is $P_1 P_2 - P_1 P_2^2$. The sum of the three probabilities is $P_1 P_2 (2 - P_2)$. This is the probability of winning two games in a row if you play in the order BAB.

We know that P_2, which is the probability of your winning against B, is greater than P_1, the probability of your winning against A, so it is apparent that $P_1 P_2 (2 - P_1)$ must be greater than $P_1 P_2 (2 - P_2)$. In other words, you stand a better chance of winning twice in succession if you play ABA: first the stronger player, then the weaker, then the stronger.

Fred Galvin, Donald MacIver, Akiva Skidell, Ernest W. Stix, Jr., and George P. Yost were the first of many readers who reached the same conclusion by the following informal reasoning. To win two games in a row it is essential that the son win the second game, therefore it is to his advantage to play the second game against the weaker player. In addition, he must win at least once against the stronger player, therefore it is to his advantage to play the stronger player twice. Ergo, ABA. Galvin pointed out that if the problem can be solved without knowing the probabilities involved, the answer will be obtainable from any special case. Consider the extreme case when the son is certain to beat his mother. He then is sure to win two games in a row if he can beat his father once, so obviously he increases his chances by playing his father twice.

ANSWER 2.11—(December 1979)

The poker puzzle is answered when we consider the fact that a hand with four identical values always has a fifth card. For each four of a kind there are 48 different fifth cards. Consequently there are 48 × 13, or 624, different poker hands containing four of a kind, compared with 40 hands that are straight flushes. It is therefore much less likely that you will be dealt a straight flush, and for this reason a straight flush beats four of a kind. The problem was contributed by M. H. Greenblatt to *Journal of Recreational Mathematics* (vol. 5, no. 1, January 1972, page 39).

ANSWER 2.12—(November 1963)

Bill spins one penny, John two. John wins if he has more heads than Bill. A tabulation of the eight equally probable ways three coins can fall shows that John wins in four cases and loses in four, so his chances of winning are 1/2, which is what they would be if a single coin were spun. His probability of winning remains the same whenever he has one more coin than Bill. Thus if he has 51 coins and Bill has 50, each man still has an equal chance of winning.

This problem appeared in the Canadian magic magazine *Ibidem* (December 1961, page 24); it was contributed by "Ravelli," pen name of chemist Ronald Wohl.

ANSWER 2.13—(June 1964)

The boy maximizes his chance of drawing a 10-dollar bill by putting a single 10-dollar bill in one hat, the other 19 bills (9 ten-dollar bills and 10 one-dollar bills) in the other hat. His chance of picking the hat with the 10-dollar bill is 1 in 2, and the probability of picking a 10-dollar bill from that hat is 1 (certain). If he picks the other hat, there is still a probability of 9/19 that he will draw a 10-dollar bill from it.

This simple stochastic process is shown in Figure 2.2. The probability that he will draw a 10-dollar bill from hat A is 1/2 × 1, or 1/2. The probability that he will draw a 10-dollar bill from hat B is 1/2 ×

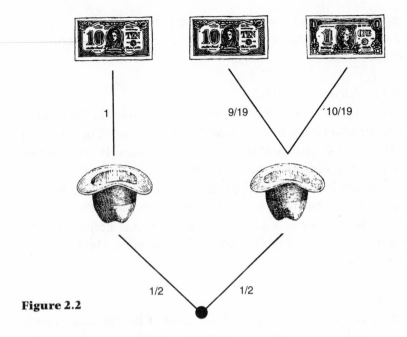

Figure 2.2

9/19, or 9/38. The sum of the two probabilities, 14/19 (or almost 3/4), is his overall probability of getting a 10-dollar bill.

ANSWER 2.14—(November 1967)

The first round of births in Fertilia produces n children, where n is the total number of mothers over as long a period as desired. The second round produces $n/2$ children, the third round produces $n/4$, and so on. The total number of children is $n(1 + 1/2 + 1/4 + 1/8 + \cdots) = 2n$. Dividing this by n gives the average number of children in a family, namely two.

Many readers pointed out that the question could be answered more simply. After showing that the ratio of boys to girls always remains one to one, it follows that in the long run there will be as many boys as there are girls. Since each family has exactly one son, there will be, on the average, also one girl, making an average family size of two children.

ANSWER 2.15—(February 1967)

The best way to make the three coins all heads or all tails is to direct that any coin be turned, then any other coin, then the first coin mentioned. The probability of success on the first move is 1/3. If you fail, the probability is 1/2 that your second move will do the trick. It might be supposed that the sum of those two probabilities is the chance of success in two moves or fewer, but this is incorrect. One must examine the effect of the first two moves on each of the six equally possible initial patterns, *HHT, HTH, HTT, THH, THT, TTH.* The symmetry allows one to pick any two coins for the first two moves. There is success in four cases, so that the chance of success on or before the second move is four out of six, or 2/3.

Seven moves guarantees success if the intent is to make all the coins heads. Of the eight possible starting patterns, only *HHH* is ruled out. You must therefore run through seven pattern variations to make sure you hit *HHH* somewhere along the line. An easily remembered strategy, suggested by Samuel Schwartz, is to label the coins 1, 2, 3 and take them in the order 1, 2, 3, 2, 1, 2, 3. The probability of success on the first move is 1/7, on or before the second move it is 2/7, and so on up to 7/7, or 1, on or before the seventh move.

If the number of coins is n, the required number of moves clearly is $2^n - 1$. The sequence of moves corresponds to the sequence of numbers in a binary Gray code. (On Gray codes, see my *Scientific American* column for August 1972.) Rufus Isaacs and Anthony C. Riddle each analyzed the situation as a competitive game. The player who hides the coins tries to maximize the number of moves, and the player who searches for the pattern tries to minimize the number of moves. The hider's best strategy is to pick randomly one of the seven possible states of the three coins. The searcher's best strategy is to label the corners of a cube with binary numbers 1 through 8, as explained in the August 1972 column, then draw a Hamiltonian path on the cube's edges. The sequence of moves corresponds to the sequence of binary numbers obtained by starting at the corner corresponding to 111 or 000, choosing with equal probability one of the two directions along the path, then traversing the path. If both players play optimally, the expected number of moves is four.

The general problem, given in terms of n switches that turn on a light only when all switches are closed, appeared in *American Mathematical Monthly*, December 1938, page 695, problem E319. For the general theory of search games of this type, see Rufus Isaacs, *Differential Games* (Wiley, 1965), pages 345f.

ANSWER 2.16—(April 1969)

It was stated that if counters are drawn according to a certain procedure from a bag containing an unknown mixture of white and black counters, there is a fixed probability that the last counter will be black. If this is true, it must apply equally to each color. Therefore the probability is 1/2.

Although this answers the problem as posed, there remains the task of proving that the probability is indeed fixed. This can be done by induction, starting with two marbles and then going to three, four, and so on, or it can be done directly. Unfortunately both proofs are too long to give, so that I content myself with referring you to "A Sampling Process," by B. E. Oakley and R. L. Perry, in *The Mathematical Gazette* for February 1965, pages 42–44, where a direct proof is given.

One might hastily conclude that the solution generalizes; that is, if the bag contains a mixture of n colors, the probability that the last counter in the bag is a specified color is $1/n$. Unfortunately, this is not the case. As Perry pointed out in a letter, if there are two red, one white, and one blue counters, the probabilities that the last counter is red, white, or blue are, respectively, 26/72, 23/72, and 23/72.

ANSWER 2.17—(November 1970)

Contrary to most people's intuition, your best bet is that the top card is a black ace.

The situation can be grasped easily by considering simpler cases. In a packet of three cards, including the two black aces and, say, a king, there are three equally probable orderings: *AAK, AKA, KAA.* It is obvious that the probability of the first ace's being on top is 2/3 as against 1/3 that it is the second card. For a full deck of 52 cards the probability of the top card's being the first black ace is 51/1326, the probability that the first black ace is second is 50/1326, that it is third is 49/1326, and so on down to a probability of 1/1326 that it is the 51st card. (It cannot, of course, be the last card.)

In general, in a deck of n cards (n being equal to or greater than 2) the probability that the first of two black aces is on top is $n - 1$ over the sum of the integers from 1 through $n - 1$. The probability that the first black ace is on top in a packet of four cards, for instance, is 1/2.

The problem is given by A. E. Lawrence in "Playing with Probability," in *The Mathematical Gazette*, vol. 53, December 1969, pages 347–54. As David L. Silverman has noticed, by symmetry the most likely position for the *second* black ace is on the bottom. The probability for each position of the second black ace decreases through the same values as before but in reverse order from the last card (51/1326) to the second from the top (1/1326).

Several readers pointed out that the problem of the first black ace is a special case of a problem discussed in *Probability with Statistical Applications*, by Frederick Mosteller, Robert E. K. Rourke, and George B. Thomas, Jr. (Addison-Wesley, 1961). Mosteller likes to call it the "needle in the haystack" problem and give it in the practical form of a manufacturer who has, say, four high-precision widgets randomly mixed in his stock with 200 low-precision ones. An order comes for one high-precision widget. Is it cheaper to search his stock or to tool up and make a new one? His decision depends on how likely he is to find one near the beginning of a search. In the case of the 52-card deck there is a better-than-even chance that an ace will be among the first nine cards at the top of a shuffled deck or—what amounts to the same thing—among the first nine cards picked at random without replacement.

ANSWER 2.18—(April 1972)

Regardless of which coin Smith chooses, the game is fair. The payoff matrices show [see Figure 2.3] that in every case the person least likely

Player with silver dollar

	75	50	25	0	
WIN	75	75	75	75	SUM = 300
LOSS	−100	−100	−100	0	SUM = 300

Player with half-dollar

	125	100	25	0	
WIN	−50	−50	125	125	SUM = 150
LOSS	−50	−50	−50	0	SUM = 150

Player with quarter

	150	100	50	0	
WIN	−25	−25	−25	150	SUM = 75
LOSS	−25	−25	−25	0	SUM = 75

Figure 2.3

Probability

to win (because he has only one coin) wins just enough when he does win to make both his expectation and that of his opponent zero.

As David Silverman suspected when he found this solution, the problem is a special case of the following generalization. If a set of coins has values that are adjacent in the doubling series 1—2—4—8—16 . . . and the game is played as described, it is a fair game regardless of how the coins are divided. We assume, of course, that each player has at least one coin and that each value is represented by only one coin.

Daniel S. Fisher, a high school student in Ithaca, New York, generalized Silverman's generalization. He showed that Silverman's game is fair for any division of ownership of the coins when values of the coins are 1, n, n^2, . . . , n^k and the coins are weighted to fall tails with probability $1/n$ (n equal to or greater than 2).

ANSWER 2.19—(April 1974)

Here is how a coin can be used to decide between alternatives A and B with probabilities expressed by any rational or irrational fraction.

Notational rules:

a. Express A as an endless binary fraction.
b. Number the digits 1, 2, 3, 4, . . . and similarly number the flips of the coin. The nth digit is called the "corresponding digit" of the nth flip.
c. Let the value of each flip be 1 for heads, 0 for tails.

Procedural rules:

a. If the value of a flip equals its corresponding digit, flip again.
b. If the value of a flip is less than its corresponding digit, stop. This decides for A.
c. If the value of a flip is more than its corresponding digit, stop. This decides for B.

Let us see how this works when A is 1/3 and B is 2/3. In binary form A = .01010101 . . . and B = .10101010. . . . The sequence of flips stops with a decision for A if and only if tails (value 0) appears on a flip whose corresponding digit is 1 in the endless binary fraction for A. The 1's are in even positions; therefore the probability of this happening is $1/2^2 + 1/2^4 + 1/2^6 + \cdots$.

The sum of this series is .01010101. . . . This is obvious when we consider its binary fractions:

$$1/4 \ = .01$$
$$1/16 = .0001$$
$$1/64 = .000001$$

$$\cdot \qquad \cdot$$
$$\cdot \qquad \cdot$$
$$\cdot \qquad \cdot$$
$$\overline{\qquad\qquad}$$
$$.010101\ldots$$

Similarly, the sequence of flips stops with a decision for B if and only if heads (value 1) appears on a flip corresponding to 0 in the endless fraction for A. The 0's are in odd positions; therefore the probability of this happening is $1/2 + 1/2^3 + 1/2^5 + \cdots$. This is the same as summing .1 + .001 + .00001 + $\cdots$, a series that just as obviously adds to .10101 . . . = 2/3.

The specific problem given was to decide between A equaling the fractional part of π and B equaling $1 - A$. First express A as a binary fraction:

$$A = .001001000011111101101\ldots$$

As before, the probability of stopping with a decision for A is the probability that you get a tail (0) on a flip whose corresponding digit is 1. This probability is equal to the binary fraction itself, because the fraction is expressing the probability as the sum of an endless series of binary fractions, each a reciprocal of a power of 2. And the probability of stopping with a decision for B is the probability you will get a head (1) on a flip whose corresponding digit is 0.

In the first case the probability is $1/2^3 + 1/2^6 + 1/2^{11} + \cdots$. (The superscripts are the positions of the 1's in the binary fraction for A.) The sum is .00100100001 . . . , the binary fraction for A.

In the second case the probability is $1/2 + 1/2^2 + 1/2^4 + \cdots$. (The superscripts are the positions of the 0's in the binary fraction for A.) The sum is .11011011110 . . . , which is the complement of the previous fraction; that is, 1's have been replaced by 0's, and 0's by 1's. It is the binary fraction for B.

It is not hard to see why the method works. Probability A is expressed as a sum of an endless series of probabilities. Each is a disjoint event, so their sum must equal probability A. There is, of course, a rapidly decreasing probability that the flipping will not stop, but this probability is vanishingly small. The sequence continues only as long as flip values keep matching their corresponding digits. On the nth toss the probability of such a match is $1/2^n$, which has zero measure in the endless series. In other words, the procedure is practically certain to stop, usually quite soon, with a decision.

chapter 3

Numbers

This is not a chapter about number theory, the mathematics of integers. Number theory seems simple but is actually profound and subtle. Instead these puzzles are about numbers as objects of play. The initial problems are catch questions and they get progressively more challenging. Some problems, like "Pied Numbers," will require a calculator, but most will just need a little logic and persistence. The final two problems will appeal to the more theoretical readers.

You will observe a couple of magic tricks here. Martin Gardner has been an amateur magician for 75 years and magic was represented in many of the essays in Scientific American, *as well as in the puzzles collected in this volume.*

Problems

PROBLEM 3.1—Talking Money

If nine thousand nine hundred and nine dollars is properly written $9,909, how should twelve thousand twelve hundred and twelve dollars be written?

PROBLEM 3.2—Many Stories

A six-story house (not counting the basement) has stairs of the same length from floor to floor. How many times as high is a climb from the first to the sixth floor as a climb from the first to the third floor?

PROBLEM 3.3—Base Knowledge

If you think of any base greater than 2 for a number system, I can immediately write down the base without asking you a question. How can I do this? (From Fred Schuh, *The Master Book of Mathematical Recreation*.)

PROBLEM 3.4—Base for a Square
Find a number base other than 10 in which 121 is a perfect square.

PROBLEM 3.5—Missing Number
What is the missing number in the following sequence: 10, 11, 12, 13, 14, 15, 16, 17, 20, 22, 24, 31, 100, ——, 10,000. (Hint: The missing number is in ternary notation.)

PROBLEM 3.6—Curious Set of Integers
The integers 1, 3, 8, and 120 form a set with a remarkable property: the product of any two integers is one less than a perfect square. Find a fifth number that can be added to the set without destroying this property.

PROBLEM 3.7—Matchstick Equation
Move one match in Figure 3.1 to produce a valid equation. Crossing the equality sign with a match to make it a not-equal sign is ruled out.

Figure 3.1

PROBLEM 3.8—Primeval Snake
A "primeval snake" is formed by writing the positive integers consecutively along a snaky path [see Figure 3.2]. If continued upward to infinity, every prime number will fall on the same diagonal line. Explain.

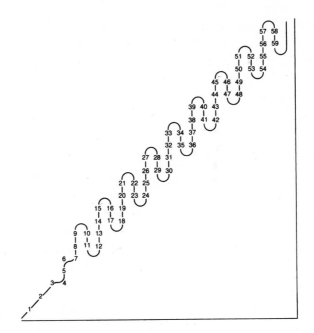

Figure 3.2

PROBLEM 3.9—Christmas and Halloween

Prove (asks Solomon W. Golomb) that Oct. 31 = Dec. 25.

PROBLEM 3.10—Familiar Symbol

Place a familiar mathematical symbol between 2 and 3 to express a number greater than 2 and less than 3.

PROBLEM 3.11—Predicting a Finger Count

On last New Year's day a mathematician was puzzled by the strange way in which his small daughter began to count on the fingers of her left hand. She started by calling the thumb 1, the first finger 2, middle finger 3, ring finger 4, little finger 5, then she reversed direction, calling the ring finger 6, middle finger 7, first finger 8, thumb 9, then back to the first finger for 10, middle finger for 11, and so on. She continued to count back and forth in this peculiar manner until she reached a count of 20 on her ring finger.

"What in the world are you doing?" her father asked.

The girl stamped her foot. "Now you've made me forget where I was. I'll have to start all over again. I'm counting up to 1962 to see what finger I'll end on."

The mathematician closed his eyes while he made a simple mental calculation. "You'll end on your ———," he said.

When the girl finished her count and found that her father was right, she was so impressed by the predictive power of mathematics that she decided to work twice as hard on her arithmetic lessons. How did the father arrive at his prediction and what finger did he predict?

PROBLEM 3.12—Repetitious Number

An unusual parlor trick is performed as follows. Ask spectator A to jot down any three-digit number, and then to repeat the digits in the same order to make a six-digit number (e.g., 394,394). With your back turned so that you cannot see the number, ask A to pass the sheet of paper to spectator B, who is requested to divide the number by 7.

"Don't worry about the remainder," you tell him, "because there won't be any." B is surprised to discover that you are right (e.g., 394,394 divided by 7 is 56,342). Without telling you the result, he passes it on to spectator C, who is told to divide it by 11. Once again you state that there will be no remainder, and this also proves correct (56,342 divided by 11 is 5,122).

With your back still turned, and no knowledge whatever of the figures obtained by these computations, you direct a fourth spectator, D, to divide the last result by 13. Again the division comes out even (5,122 divided by 13 is 394). This final result is written on a slip of paper which is folded and handed to you. Without opening it you pass it on to spectator A.

"Open this," you tell him, "and you will find your original three-digit number."

Prove that the trick cannot fail to work regardless of the digits chosen by the first spectator.

PROBLEM 3.13—Nine to One Equals 100

An old numerical problem that keeps reappearing in puzzle books as though it had never been analyzed before is the problem of inserting mathematical signs wherever one likes between the digits 1, 2, 3, 4, 5, 6, 7, 8, 9 to make the expression equal 100. The digits must remain in the same sequence. There are many hundreds of solutions, the easiest to find perhaps being

$$1 + 2 + 3 + 4 + 5 + 6 + 7 + (8 \times 9) = 100.$$

The problem becomes more of a challenge if the mathematical signs are limited to plus and minus. Here again there are many solutions, for example,

$$1 + 2 + 34 - 5 + 67 - 8 + 9 = 100,$$
$$12 + 3 - 4 + 5 + 67 + 8 + 9 = 100,$$
$$123 - 4 - 5 - 6 - 7 + 8 - 9 = 100,$$
$$123 + 4 - 5 + 67 - 89 = 100,$$
$$123 + 45 - 67 + 8 - 9 = 100,$$
$$123 - 45 - 67 + 89 = 100.$$

"The last solution is singularly simple," writes the English puzzlist Henry Ernest Dudeney in the answer to Problem No. 94 in his *Amusements in Mathematics*, "and I do not think it will ever be beaten."

In view of the popularity of this problem it is surprising that so little effort seems to have been spent on the problem in reverse form. That is, take the digits in descending order, 9 through 1, and form an expression equal to 100 by inserting the smallest possible number of plus or minus signs.

PROBLEM 3.14—Nine-Digit Arrangement

One of the satisfactions of recreational mathematics comes from finding better solutions for problems thought to have been already solved in the best possible way. Consider the following digital problem that appears as Number 81 in Henry Ernest Dudeney's *Amusements in Mathematics*. (There is a Dover reprint of this 1917 book.) Nine digits (0 is excluded) are arranged in two groups. On the left a three-digit number is to be multiplied by a two-digit number. On the right both numbers have two digits each:

$$
\begin{array}{cc}
158 & 79 \\
\underline{23} & \underline{46}
\end{array}
$$

In each case the product is the same: 3,634. How, Dudeney asked, can the same nine digits be arranged in the same pattern to produce as large a product as possible, and a product that is identical in both cases? Dudeney's answer, which he said "is not to be found without the exercise of some judgment and patience," was

$$
\begin{array}{cc}
174 & 96 \\
\underline{32} & \underline{58} \\
5{,}568 & 5{,}568
\end{array}
$$

Victor Meally of Dublin County in Ireland later greatly improved on Dudeney's answer with

$$
\begin{array}{cc}
584 & 96 \\
12 & 73 \\
\hline
7{,}008 & 7{,}008
\end{array}
$$

This remained the record until a Japanese reader found an even better solution. It is believed, although it has not yet been proved, to give the highest possible product. Can you find it without the aid of a computer?

PROBLEM 3.15—Polypowers

By convention, the value of a ladder of exponents such as

$$2^{2^{2^{2^{2}}}}$$

is computed by starting at the top and working down. The highest pair equals 4, then $2^4 = 16$, and $2^{16} = 65{,}536$. How large is $2^{65{,}536}$? Geoffrey W. Hoffmann of West Germany sent me a computer printout of this number. It starts 20035 . . . and has 19,729 digits. Adding another 2 to the ladder gives a number that will never be calculated because the answer, as Hoffmann put it, would require the age of the universe in computer time and the space of the universe to hold the printout.

Even a ladder as short as three 9's is $9^{387{,}420{,}489}$, a number of more than 360 million digits. In 1933 S. Skewes (pronounced skew-ease) published a paper in which he showed that if $\pi(x)$ is the number of primes less than x, and li(x) is the logarithmic integral function, then $\pi(x) - $ li(x) is positive for some x less than

$$10^{10^{10^{34}}}$$

an integer said to be the largest known to play a role in a nontrivial theorem.

In 1971 Aristid V. Grosse, one of the pioneer atomic chemists at Columbia University in 1940 (he later became president of Germantown Laboratories, Inc., affiliated with the Franklin Institute), began an investigation of exponential ladders of identical numbers that are calculated in the opposite direction (up) and their relations to down ladders. He coined the term "polypowers" for ladders of both types.

Ladders of two x's are called "dipowers," of three x's "tripowers" and so on, according to the Greek prefixes. The value of x can be rational or irrational, transcendental, complex, or entirely imaginary. In most cases the polypowers are single-valued, continuous, and differentiable. Since 1 to any polypower of 1 is 1, all these functions and their derivatives, when graphed against x, cross one another at $x = 1$, and their values at 0 are the limits as x approaches 0. Grosse's notes, which fill many volumes, lead into a lush jungle of unusual theorems as well as new classes of numbers.

Up and down dipowers obviously are identical, but for all higher polypowers the two directions give different numbers. The triplet of 9's, for example, when calculated upward is a number of only 77 digits. Except for the triplet of 2's, going "up all the way" on a ladder of identical integers gives the minimum number, and "down all the way" gives the maximum. In what follows, the arrows indicate these maximum and minimum numbers.

What happens when up and down ladders of different lengths are equated? If an up triplet of x's equals a down triplet of x's, $x = 2$. (We exclude $x = 1$ as being trivial.) Each additional x on the up ladder increases the value of x by 1. If three down x's equal four up x's, $x = 3$; if three down equals five up, $x = 4$ and so on.

As an introduction to polypowers, solve the three equations below, which begin a series with down tetrapowers on the left:

$$\overleftarrow{x^{x^{x^x}}} = \overrightarrow{x^{x^{x^x}}}$$

$$\overleftarrow{x^{x^{x^x}}} = \overrightarrow{x^{x^x}}$$

$$\overleftarrow{x^{x^x}} = \overrightarrow{x^{x^x}}$$

You may also enjoy investigating ladders of fractional x's, reciprocals of x, and more exotic forms. Grosse has also developed the concept of a perfect polypower, that is, x to the xth power (up or down) an x number of times. (Example: π to the πth power, π times up, is 588,916,326+.) The reverse operation to polypowers he calls "polyroots." Have these fields been investigated before? In spite of considerable effort, neither he nor I have uncovered references.

PROBLEM 3.16—Figures Never Lie

An old burlesque routine involves two simpleminded men who divide 28 by 13 to get 7, then verify this result by multiplying 13 by 7 to get 28 and finally double-check it by adding 13 seven times to get 28. This is how Irvin S. Cobb told the story in his anthology of 366 jokes (one for leap year), *A Laugh a Day Keeps the Doctor Away* (1923):

"Three patricians of the coal yards fared forth on mercy bent, each in his great black chariot. Their overlord, the yard superintendent, had bade them deliver to seven families a total of twenty-eight tons of coal equally divided.

"Well out of the yards, each with his first load, Kelly and Burke and Shea paused to discuss the problem of equal distribution—how much coal should each family get?

"''Tis this way,' argued Burke. ''Tis but a bit of mathematics. If there are 7 families an' 28 tons o' coal ye divide 28 by 7, which is done as follows: Seven into 8 is 1, 7 into 21 is 3, which makes 13.' He triumphantly exhibited his figures made with a stubby pencil on a bit of grimy paper:

$$
\begin{array}{r}
7/28/13 \\
7 \\
\hline
21 \\
21 \\
\hline
00
\end{array}
$$

"The figures were impressive but Shea was not wholly convinced. 'There's a easy way o' provin' that,' he declared. 'Ye add 13 seven times,' and he made his column of figures according to his own formula. Then, starting from the bottom of the 3 column, he reached the top with a total of 21 and climbed down the column of 1's, thus; '3, 6, 9, 12, 15, 18, 21, 22, 23, 24, 25, 26, 27, 28.' 'Burke is right,' he announced with finality.

"This was Shea's exhibit:

$$
\begin{array}{r}
13 \\
13 \\
13 \\
13 \\
13 \\
13 \\
13 \\
\hline
28
\end{array}
$$

"'There is still some doubt in me mind,' said Kelly. 'Let me demonstrate in me own way. If ye multiply the 13 by 7 and get 28, then 13 is right.' He produced a bit of stubby pencil and a sheet of paper. ''Tis done in this way,' he said. 'Seven times 3 is 21; 7 times 1 is 7, which makes 28. ''Tis thus shown that 13 is the right figure and ye're both right. Would ye see the figures?'

"Kelly's feat in mathematics was displayed as follows:

$$\begin{array}{r} 13 \\ 7 \\ \hline 21 \\ 7 \\ \hline 28 \end{array}$$

"'There is no more argyment,' the three agreed, so they delivered thirteen tons of coal to each family."

The comedian Flournoy Miller made effective use of the routine and published his version of it in his book *Shufflin' Along*. Later, Flip Wilson did the bit on his television show and was sued by Miller's daughter for unauthorized use of the material. The case was apparently settled out of court.

"Is there something special about the numbers 7, 13 and 28?" asked the late William R. Ransom, a mathematician at Tufts University. The answer is no. There are just 22 triplets of numbers—one number is a single digit, the other two are two digits each—that can be substituted for 7, 13, and 28 without changing a single word in the routine. Can you list the 22 triplets?

PROBLEM 3.17—Pied Numbers

The old problem of expressing integers with four 4's (discussed in a *Scientific American* column reprinted as Chapter 5 of *The Magic Numbers of Dr. Matrix*, 1985) has been given many variations. In an intriguing variant proposed by Fitch Cheney one is allowed to use only pi and symbols for addition, subtraction, multiplication, division, square root, and the "round-down function." In the last operation, indicated by brackets, one takes the greatest integer that is equal to or less than the value enclosed by the brackets. Parentheses also may be used, as in algebra, but no other symbols are allowed. Each symbol and pi may be repeated as often as necessary, but the desideratum is to use as few pi symbols as possible. For example, 1 can be written [√π] and 3 even more simply as [π].

Do your best to express the integers from 1 through 20 according to these rules, and to compare them with the best Cheney was able to achieve.

PROBLEM 3.18—Self Numbers

D. R. Kaprekar was a mathematician, diminutive in body but large in brain and heart, who lived in India. For more than 70 years he did highly original work in recreational number theory, at times aided by grants from Indian universities. He contributed frequently to Indian mathematics journals, spoke at conferences, and published some two dozen booklets written in broken English.

Kaprekar is best known outside India for his discovery in 1955 of "Kaprekar's constant." Start with any four-digit number in which not all the digits are alike. Arrange the digits in descending order, reverse them to make a new number, and subtract the new number from the first number. If you keep repeating this process with the remainders, you will (in eight steps or fewer) arrive at Kaprekar's constant, 6174, which then generates itself. Zeros must be preserved. Thus, if you start with 2111 and subtract 1112, you get 0999. Rearranging the digits gives 9990 from which 0999 is taken, and so on.

Here we concern ourselves with a remarkable class of numbers called self-numbers, discovered by Kaprekar in 1949. He wrote many pamphlets about them. They are virtually unknown outside India, although they turned up briefly (under another name) in an article in *The American Mathematical Monthly* (April 1974, page 407). The article contains a proof that there is an infinity of self-numbers.

In explaining self-numbers, it is best to start with a basic procedure that Kaprekar calls digitadition. Select any positive integer and add to it the sum of its digits. Take 47, for example. The sum of 4 and 7 is 11, and 47 and 11 is 58. The new number, 58, is called a generated number. The original number, 47, is the generator. The process can be repeated endlessly, forming a digitadition series: 47, 58, 71, 79, 95, ...

No one has yet found a nonrecursive formula for the partial sum of a digitadition series, given its first and last terms, but there is a simple formula for the sum of all the digits in a digitadition series. Simply subtract the first number from the last and add the sum of the digits in the last number. "Is this not a wonderful new result?" Kaprekar asks in one of his booklets. "The Proof of all this rule is very easy and I have completely written it with me. But as soon as the proof is seen the charm of the whole process is lost, and so I do not wish to give it just now."

Can a generated number have more than one generator? Yes, but not until the number exceeds 100. The smallest such number (Kaprekar calls it a junction number) is 101. It has two generators: 91 and 100. The smallest junction number with three generators is 10,000,000,000,001. It is generated by 10,000,000,000,000, 9,999,999,999,901, and 9,999,999,999,892. The smallest number with four generators, discovered by Kaprekar on June 7, 1961, has twenty-five digits. It is 1 followed by twenty-one zeros and 102. Since then he has found what he conjectures to be the smallest numbers with five and six generators.

A self-number is simply a number that has no generator. In Kaprekar's words, "It is self born." There is an infinity of such numbers, but they are much scarcer than generated numbers. Below 100 there are thirteen: 1, 3, 5, 7, 9, 20, 31, 42, 53, 64, 75, 86, and 97. Self-numbers that are prime are called self-primes. The familiar cyclic number 142,857 is a self-number (multiply it by the digits 1 through 6 and you always get the same six digits in the same cyclic order). The numbers 11,111,111,111,111,111 and 3,333,333,333 are self-numbers. Years of the last century that are self-numbers include: 1906, 1917, 1919, 1930, 1941, 1952, 1963, and 1974.

Consider the powers of 10. The number 10 is generated by 5, 100 by 86, 1,000 by 977, 10,000 by 9968, and 100,000 by 99,959. Why is a millionaire such an important man? Because, answers Kaprekar, 1,000,000 is a self-number! The next power of 10 that is a self-number is 10^{16}.

No one has yet discovered a nonrecursive formula that generates all self-numbers, but Kaprekar has a simple algorithm by which any number can be tested to determine whether it is self-born or generated. Can you discover the procedure? If so, it should be easy to answer this question: What is the next year after 1974 that is a self-number?

PROBLEM 3.19—Persistences of Numbers

N.J.A. Sloane of Bell Laboratories, the author of the valuable reference work *A Handbook of Integer Sequences*, introduced into number theory the concept of the "persistence" of a number. A number's persistence is the number of steps required to reduce it to a single digit by multiplying all its digits to obtain a second number, then multiplying all the digits of that number to obtain a third number, and so on until a one-digit number is obtained. For example, 77 has a persistence of four because it requires four steps to reduce it to one digit: 77→49→36→18→8. The smallest number of persistence one is 10, the smallest of persistence two is 25, the smallest of persistence three is 39, and the smaller of persistence four is 77. What is the smallest number of persistence five?

Sloane determined by computer that no number less than 10^{50} has a persistence greater than 11. He conjectures that there is a number c such that no number has a persistence greater than c. Little is known about persistences in base notations other than 10. In base 2 the maximum persistence is obviously one. In base 3 the second term of the persistence sequence for any number is either zero or a power of 2. Sloane conjectures that in base 3 all powers of 2 greater than 2^{15} include a zero. Calculations show that Sloane's conjecture is true up to 2^{500}, but there is no formal proof of it. Of course, any number with zero as one of its digits is reduced to zero on the next step. Hence if the conjecture is true, it follows that the maximum persistence in base 3 is three, as is illustrated by the following sequence: 222,222,222,222,222→2^{15} (which in base 3 equals 1,122,221,122→1,012→0). Sloane also conjectures that there is a number c for any base notation b such that no number in base b has a persistence greater than c.

Let us call Sloane's persistence a multiplicative persistence to distinguish it from additive persistence, a term introduced by Harvey J. Hindin, then a chemist at Hunter College, after he had learned of Sloane's work. The additive persistence of a number is the number of steps required to reduce it to one digit by successive additions. Recreational mathematicians and accountants know this process as "casting out nines" or obtaining a number's "digital root," procedures that are equivalent to reducing the number modulo 9. For example, 123,456,789 has an additive persistence of two: 123,456,789→45→9.

Unlike multiplicative persistence, additive persistence is relatively trivial and almost everything about it is known. For example, in base 2 the smallest number of additive persistence four is 1,111,111. In base 3 the number is 122,222,222. What is the smallest number of additive persistence four in base 10?

Answers

ANSWER 3.1—(August 1968)

Twelve thousand twelve hundred and twelve dollars is written $13,212.

ANSWER 3.2—(July 1971)

The climb is two-and-a-half times as high.

ANSWER 3.3—(July 1971)

I write "10." This is *any* base written in that base system's notation.

ANSWER 3.4—(August 1969)

121 is a perfect square in any number notation with a base greater than 2. A quick proof is to observe that 11 times 11, in any system, has a product (in the same system) of 121. Craige Schensted showed that, with suitable definitions of "perfect square," 121 is a square even in systems based on negative numbers, fractions, irrational numbers, and complex numbers. "Although the bases may not be exhausted, I am and I assume you are, so I will stop here," he concluded.

ANSWER 3.5—(August 1968)

Each number is 16 in a number system with a different base, starting with base 16 and continuing with bases in descending order, ending with base 2. The missing number, 16 in the ternary system, is 121.

ANSWER 3.6—(February 1967)

The fifth number is 0. The answer is, of course, trivial and intended as a joke. However, a difficult question now arises: Is there a fifth *positive* integer (other than 1, 3, 8, 120) which can be added to the set so that the set retains the property that the product of any two members is one less than a perfect square?

This unusually difficult Diophantine problem goes all the way back to Fermat and Euler. (See L. E. Dickson, *History of the Theory of Numbers*, vol. 2, pages 517 f.) The problem has had an interesting history and was not finally settled until 1968. A student of C. J. Bouwkamp, at the Technological University, Eindhoven, Holland, saw the problem in *Scientific American*, mentioned it to Bouwkamp, who in turn gave it to his colleague J. H. van Lint. In 1968 van Lint showed that if 120 could be replaced by a positive integer, without destroying the set's property, the number would have to be more than 1,700,000 digits. Alan Baker, of Cambridge University, then combined van Lint's results with a very deep number theorem of his own, and finally laid the problem to rest. In a paper by Baker and D. Davenport, *Quarterly Journal of Mathematics*, second series, vol. 78, 1969, pages 129–38, it is proved that there is no replacement for 120, and of course it follows that there can be no fifth member of the set. The proof is complicated, involving the calculation of several numbers to 1,040 decimal places.

It is known that there is an infinity of sets of four positive integers with the desired property, of which 1, 3, 8, and 120 has the smallest sum. You will find 20 other solutions listed in a discussion of the problem by Underwood Dudley and J. H. Hunter in *Journal of Recreational Mathematics*, vol. 4, April 1971, pages 145–46. A simpler proof that there is no fifth number for the 1, 3, 8, 120 set is given by P. Kanagasabapathy and Th. Ponnudurai in a paper published in *The Quarterly Journal of Mathematics*, vol. 3, no. 26, 1975, pages 275–78. The problem is also the topic of "A Problem of Fermat and the Fibonacci Sequence," by V. E. Hoggatt, Jr., and G. E. Bergum, in *The Fibonacci Quarterly*, vol. 15, December 1977, pages 323–30.

Is there a set of five positive integers with the desired property? As far as I know, this remains unanswered.

ANSWER 3.7—(July 1969)

Several readers found an alternate solution to the one in Figure 3.3. The VI on the left is changed to XI, the Roman equivalent of the Arabic 11 on the other side.

Figure 3.3

ANSWER 3.8—(July 1971)

It is well known that every prime greater than 3 is one more or one less than a multiple of 6. It is easy to see that every number of the form $6n \pm 1$ must fall on the same diagonal, therefore the diagonal is certain to catch every prime.

A. P. Evans, William B. Friedman, and others wrote to say that you don't need to know that all primes greater than 3 have the form of $6n \pm 1$. Only four parallel diagonals can be drawn through the snake. Call them, top to bottom, A,B,C,D. All numbers on A are divisible by 3 and therefore cannot be prime. All on B and D are divisible by 2, and hence cannot be prime. Therefore all the primes must fall on C. "I don't suppose it would be sporting," Evans adds, "to ask readers to come up with a diagram on which a straight line can be drawn that contacts all prime numbers and *only* prime numbers."

ANSWER 3.9—(November 1967)

If "Oct." is taken as an abbreviation for "octal" and "Dec." as an abbreviation for "decimal," then 31 (in base-8 notation) is equal to 25 (in base-10 notation). This remarkable coincidence is the basic clue in "A Curious Case of Income Tax Fraud," one of Isaac Asimov's tales about a club called The Black Widowers. (See *Ellery Queen's Mystery Magazine*, November 1976.)

John Friedlein observed that not only does Christmas equal Halloween, each also equals Thanksgiving whenever it falls, as it sometimes does, on Nov. 27. (27 in 9-base notation is 25 in decimal).

Suzanne L. Hanauer established the equivalence of Christmas and Halloween by modulo arithmetic. Oct. 31 can be written 10/31 or 1,031. Dec. 25 is 12/25 or 1,225. And $1,031 = 1,225$ (modulo 194).

David K. Scott and Jay Beattie independently established equality in an even more surprising way. Let the five letters in Oct. and Dec. stand for digits as follows: $O = 6$, $C = 7$, $T = 5$, $D = 8$, $E = 3$. We then decode Oct. 31 = Dec. 25 as:

$$675 \times 31 = 837 \times 25 = 20{,}925$$

Assuming that two different letters cannot have the same digit, that O and D (standing for initial digits of numbers) may be assigned any digit except zero, and that the other three letters may be assigned any digit including zero, there are 24,192 different ways to assign digits. Beattie actually programmed a computer to test all 24,192 possibilities. The program proved (what Scott had conjectured) that the above equation is unique!

ANSWER 3.10—(July 1971)

Just use a decimal point: 2.3.

Larry S. Liebovitch, instead of using a decimal point, solved this problem by using "ln," the symbol of "natural log of." Thus 2 ln 3 = 2.19 +.

ANSWER 3.11—(February 1962)

When the mathematician's little girl counted to 1,962 on her fingers, counting back and forth in the manner described, the count ended on her index finger. The fingers are counted in repetitions of a cycle of eight counts as shown in Figure 3.4. It is a simple matter to apply the concept of numerical congruence, modulo 8, to calculate where the count will fall for any given number. We have only to divide the number by 8, note the remainder, then check to see which finger is so labeled. The number 1,962 divided by 8 has a remainder of 2, so the count falls on the index finger.

In mentally dividing 1,962 by 8 the mathematician recalled the rule that any number is evenly divisible by 8 if its last three digits are evenly divisible by 8, so he had only to divide 962 by 8 to determine the remainder.

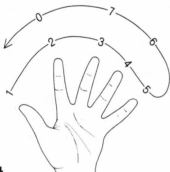

Figure 3.4

ANSWER 3.12—(August 1958)

Writing a three-digit number twice is the same as multiplying it by 1,001. This number has the factors 7, 11, and 13, so writing the chosen number twice is equivalent to multiplying it by 7, 11, and 13. Naturally when the product is successively divided by these same three numbers, the final remainder will be the original number. (This problem is given by Yakov Perelman in his book *Figures for Fun*, Moscow, 1957.)

ANSWER 3.13—(October 1962)

To form an expression equal to 100, four plus and minus signs can be inserted between the digits, taken in reverse order, as follows:

$$98 - 76 + 54 + 3 + 21 = 100$$

There is no other solution with as few as four signs. For a complete tabulation of all solutions for both the ascending and descending sequence see my *Magic Numbers of Dr. Matrix* (Prometheus, 1985, Chapter 6).

ANSWER 3.14—(April 1972)

The solution with the largest product is:

532	98
14	76
7,448	7,448

The problem has 11 basic solutions:

$$532 \times 14 = 98 \times 76 = 7,448$$
$$584 \times 12 = 96 \times 73 = 7,008$$
$$174 \times 32 = 96 \times 58 = 5,568$$
$$158 \times 32 = 79 \times 64 = 5,056$$
$$186 \times 27 = 93 \times 54 = 5,022$$
$$259 \times 18 = 74 \times 63 = 4,662$$
$$146 \times 29 = 73 \times 58 = 4,234$$
$$174 \times 23 = 69 \times 58 = 4,002$$
$$134 \times 29 = 67 \times 58 = 3,886$$
$$138 \times 27 = 69 \times 54 = 3,726$$
$$158 \times 23 = 79 \times 46 = 3,634$$

Many readers found all 11 by hand, others found them with computer programs. Allan L. Sluizer pointed out that the maximum answer has digits 1 through 5 in one of the multiplications, and digits 6 through 9 in the other.

If 0 is included among the digits (though not as an initial digit of a number), we may ask for solutions of the expression $abc \times de = fgh \times ij$. There are 64 solutions, all independently found by Richard Hendrickson, R. F. Forker, and Sluizer. The one with the smallest product is $306 \times 27 = 459 \times 18 = 8,262$. The one with the largest product is $915 \times 64 = 732 \times 80 = 58,560$. The maximum solution is given by Dudeney, in his answer to problem 82 of *Amusements in Mathematics*, as the maximum product obtainable if the 10 digits are divided in any manner whatever to form a pair of multiplications, each of which gives the product. As in all such problems, 0 may not be an initial digit. The lowest product is given by 3,485 × 2 = 6,970 × 1 = 6,970.

"It is extraordinary," Dudeney once declared, "what a large number of good puzzles can be made out of the ten digits." Here are some examples similar to our original problem. How many solutions are there to *ab* × *cde* = *fghi*, using the nine positive digits? And how many to *a* × *bcde* = *fghi*? The seven solutions to the first problem, and the two to the second, are given by Dudeney in his answer to problem 80, *Amusements in Mathematics*.

Using all 10 digits, how many solutions are there for *ab* × *cde* = *fghij*? I have not seen this answered in print, but Y. K. Bhat, a correspondent in New Delhi, found nine:

$$39 \times 402 = 15{,}678$$
$$27 \times 594 = 16{,}038$$
$$54 \times 297 = 16{,}038$$
$$36 \times 495 = 17{,}820$$
$$45 \times 396 = 17{,}820$$
$$52 \times 367 = 19{,}084$$
$$78 \times 345 = 26{,}910$$
$$46 \times 715 = 32{,}890$$
$$63 \times 927 = 58{,}401$$

How about *ab* × *c* = *de* + *fg* = *hi*, excluding 0? In *Modern Puzzles*, problem 73, Dudeney gives the only answer: 17 × 4 = 93 + 25 = 68.

Clement Wood, in his rare *Book of Mathematical Oddities* (Little Blue Book No. 1210), asserts that *ab* × *c* = *de* × *f* = *ghi* (0 excluded) has only two solutions: 38 × 4 = 78 × 2 = 156 and 58 × 3 = 29 × 6 = 174.

One final problem that I leave unanswered: Find the only solution (excluding 0) to *a* × *bc* = *d* × *ef* = *g* × *hi*. This was sent to me in 1972 by Guy J. Crocker, who discovered it. I cannot recall having seen it before.

ANSWER 3.15—(May 1973)

The key to simplifying the three polypower equations is the basic law

$$(a^b)^c = a^{(b \times c)}$$

Applying this to the first equation gives

$$x^{(x^{x^x})} = (x^x)x$$
$$x^{(x^{x^x})} = x^{(x^2)}$$

The two bottom x's are equal; therefore their parenthetical exponents are equal. Cancel the bottom x's and repeat the procedure:

$$x^{(x^x)} = x^2$$

The bottom x's again drop out, leaving $x^x = 2$, which gives x the value $1.55961+$.

The same procedure simplifies the second equation (down-4 equals up-4) to $x^x = 3$, and $x = 1.82545+$. Each succeeding equation increases the value of x^x by 1. The third equation reduces to $x^x = 4$, or $x = 2$.

The general procedure is to replace the up ladder by a number one less than the number of its x's and remove two x's from the down ladder. (Example: Down-5 = up-5 reduces to down-3 = 4.)

Correspondence about polypowers was unusually heavy, and readers raised many interesting questions. Several readers pointed out that parentheses could be placed on a ladder in a variety of ways. In fact, the number of ways is given by the sequence known as the Catalan numbers. However, not all ways of parenthesizing give distinct values for the ladder. Determining the number of such values is a difficult problem, and I do not know the solution.

Many readers called attention to unusual, little-known theorems about infinite ladders of exponents. Consider, for example, a ladder of x's that grows steadily upward to infinity. I would have thought that if x is greater than 1 the ladder's value (working from top down) would diverge as the ladder grows. This is not true. If x is an integer, the value diverges only if x exceeds $e^{1/e} = 1.4446 \ldots$. If x is a real number, it converges only if it is equal to or greater than $e^{-e} = 0.0659 \ldots$ and equal to or less than $e^{1/e}$. I found this amazing.

A delightful paradox is related to the above theorem. Assume that an infinite ladder of x's has a value of 2. What is the value of x? Because all the x's above the bottom x form an infinite chain, we can assume that the value of this chain is also 2. Substituting 2 for this chain gives the equation $x^2 = 2$, for which $x = \sqrt{2}$.

All well and good. Now apply the same dodge to an infinite ladder of x's that equals 4. This leads to $x^4 = 4$, so again $x = \sqrt{2}$. How can an infinite ladder converge to two different numbers? Actually, an infinite ladder of square roots of 2 cannot converge to 4, and in this case the dodge is not applicable. To show this exactly is complicated. You will find it explained in "A Matter of Definition," by M. C. Mitchelmore in *American Mathematical Monthly*, vol. 81, 1974, pages 643–47.

For general discussions of infinite ladders see "Infinite Exponentials," by D. F. Barrow in *American Mathematical Monthly*, vol. 43, 1936, pages 150–60; "Exponentials Reiterated," by R. A. Knoebel, ibid., vol. 88, 1981, pages 235–52; and "Infinite Exponentials," by P. J. Rippon in *Mathematical Gazette*, vol. 67, 1983, pages 189–96. Knoebel gives a long bibliography of earlier references.

Several readers sent references relevant to Grosse's labors, but unfortunately they were all in German or French. I still know of no good references in English to the sort of problems Grosse has been investigating.

Some comments on big numbers may be of interest. I mentioned that the largest number that can be written in conventional notation with no symbols other than three digits is 9^{9^9}. In the next-to-last chapter of Ulysses, Joyce reveals that Leopold Bloom was once fascinated by this number, and a paragraph is devoted to describing how big it is.

The large number that bears Skewes's name was based on the assumption that the Riemann hypothesis is true. What if it isn't? In 1955 Skewes published a proof that the number would then be the much larger

$$10^{10^{10^{10^3}}}$$

For an entertaining account of all this see "Skewered!" by Isaac Asimov in *Fantasy and Science Fiction*, November 1974. Skewes made his calculations at the request of J. E. Littlewood, who tells about it in the chapter titled "Large Numbers" in *A Mathematician's Miscellany* (Methuen, 1953).

Even Skewes's second number is very tiny and no longer the largest ever involved in a legitimate proof. The record is now held by Ronald L. Graham, of Bell Laboratories. Graham's number arose in connection with a problem in a branch of graph theory called Ramsey theory. (See my *Colossal Book of Mathematics*, Chapter 33.) The number can be expressed compactly only in a special notation devised by Donald E. Knuth for handling numbers of such unimaginable magnitude.

ANSWER 3.16—(April 1974)

The 22 triplets that can be substituted for 7, 13, and 28 in Irvin S. Cobb's story are

$$12 \div 2 = 15, \qquad 15 \div 3 = 14, \qquad 16 \div 4 = 13,$$
$$14 \div 2 = 25, \qquad 18 \div 3 = 24, \qquad 24 \div 4 = 15,$$
$$16 \div 2 = 35, \qquad 24 \div 3 = 17, \qquad 28 \div 4 = 25,$$
$$18 \div 2 = 45, \qquad 27 \div 3 = 27, \qquad 36 \div 4 = 18,$$

$$15 \div 5 = 12, \qquad 18 \div 6 = 12, \qquad 28 \div 7 = 13,$$
$$25 \div 5 = 14, \qquad 36 \div 6 = 15, \qquad 49 \div 7 = 16,$$
$$35 \div 5 = 16, \qquad 48 \div 6 = 17,$$
$$45 \div 5 = 18, \qquad\qquad\qquad\qquad 48 \div 8 = 15$$

Readers interested in how William R. Ransom solved this problem will find it explained in his delightful but little-known book, *One Hundred Mathematical Curiosities* (J. Weston Walch, 1955). Using more liberal interpretations of the dialogue in the old burlesque routine, Joseph H. Engel, Sumner Shapiro, and Alan Wayne found other triplets of figures that could be added to the 22 that satisfy a strict interpretation of the dialogue. Wayne recalled having seen the routine performed several times on stage by the Abbott and Costello comedy team.

ANSWER 3.17—(November 1970)

Figure 3.5 gives Fitch Cheney's answers to the problem of expressing the integers 1 through 20 by using pi, as few times as possible, and the symbols specified. He was able to express all integers from 1 through 100 without using more than four pi's in each expression.

$$1 = [\sqrt{\pi}] \qquad\qquad 11 = [(\pi \times \pi) + \sqrt{\pi}]$$
$$2 = [\sqrt{\pi} \ \sqrt{\pi}] \qquad\quad 12 = [\pi \times \pi] + [\pi]$$
$$3 = [\pi] \qquad\qquad\quad 13 = [(\pi \times \pi) + \pi]$$
$$4 = [\pi + \sqrt{\pi}] \qquad\quad 14 = [(\pi \times \pi) + \pi + \sqrt{\pi}]$$
$$5 = [\pi \sqrt{\pi}] \qquad\qquad 15 = [\pi \times \pi] + [\pi + \pi]$$
$$6 = [\pi + \pi] \qquad\qquad 16 = [(\pi \times \pi) + \pi + \pi]$$
$$7 = [\pi^{\sqrt{\pi}}] \qquad\qquad 17 = [\pi \times \pi \times \sqrt{\pi}]$$
$$8 = [(\pi \times \pi) - \sqrt{\pi}] \quad 18 = [\pi \times \pi] + [\pi \times \pi]$$
$$9 = [\pi \times \pi] \qquad\qquad 19 = [(\pi \times \pi) + (\pi \times \pi)]$$
$$10 = [\pi \times \pi] + [\sqrt{\pi}] \quad 20 = [\pi\sqrt{\pi}] \ [\pi + \sqrt{\pi}]$$

Figure 3.5

Hundreds of readers improved on Cheney's answers. Here are some typical ways of shortening six of the expressions:

$$14 = [[\pi] \times (\pi + \sqrt{\pi})]$$
$$15 = [\pi] \times [\pi\sqrt{\pi}]$$
$$16 = [\pi\sqrt{\pi} \times [\pi]]$$
$$18 = [\pi] \times [\pi + \pi]$$
$$19 = [\pi(\pi + \pi)]$$
$$20 = [\pi^\pi/\sqrt{\pi}] \text{ or } [(\pi\sqrt{\pi})^{\sqrt{\pi}}]$$

These improvements reduce the total number of pi's to 50. John W. Gosling was the first of many readers to achieve 50, but it is only fair to add that the problem did not specifically allow exponention and that many who wrote earlier than Gosling would probably have achieved 50 had they used exponents for integers 7 and 20. (Without exponents, 7 requires three pi's and 20 requires four.) Numerous readers lowered the number of pi's below 50 by adding other symbols, such as the factorial sign or the "unary negative operator," which has the effect of rounding up instead of down. Bernard Wilde and Carl Thune, Mark T. Longley-Cook, V. E. Hoggatt, Jr., Robert L. Caswell, and others conjectured that by using nested radical signs to reduce a divisor, any positive integer can be expressed with three pi's.

Cheney and John Leech each pointed out that if $-[-\pi]$ is interpreted in a standard way to mean 4, then further reductions are possible:

$$2 = -[-\sqrt{\pi}]$$
$$4 = -[-\pi]$$
$$8 = -[-\pi] - [-\pi]$$
$$10 = -[-\pi \times \pi]$$
$$11 = [(-[-\pi])^{\sqrt{\pi}}]$$
$$12 = [-\pi \times (-\pi)]$$
$$13 = -[\pi \times (-\pi)]$$
$$16 = [-\pi] \times [-\pi]$$

Answer 3.18—(May 1975)

D. R. Kaprekar's method of testing a number, N, to see if it is a self-number is as follows. Obtain N's digital root by adding its digits, then adding the digits of the result, and so on, until only one digit remains. If the digital root is odd, add 9 to it and divide by 2. If it is even, simply divide by 2. In either case call the result C.

Subtract C from N. Check the remainder to see if it generates N. If it does not, subtract 9 from the last result and check again. Continue subtracting 9's, each time checking the result to see if it generates N. If this fails to produce a generator of N in k steps, where k is the number of digits in N, then N is a self-number.

For example, we want to test the year 1975. Its digital root, 4, is even, so that we divide 4 by 2 to obtain $C = 2$. 1975 minus 2 is 1973, which fails to generate 1975. 1973 minus 9 is 1964. This also fails. But 1964 minus 9 is 1955, and 1955 plus the sum of its digits, 20, is 1975; therefore 1975 is a generated number. Since 1975 has four digits, we would have had only one more step to go to settle the matter. With this simple procedure, it does not take long to determine that the next self-year after 1974 is 1985. There is only one other self-year in the last century: 1996.

For progress on the problem of finding a nonrecursive formula for the sum of a digitadition series, see "The Sum of a Digitadition Series," by Kenneth B. Stolarsky (*Proceedings of the American Mathematical Society* 59, August 1976, pp. 1–5). Among his references on digitations, the earliest is a 1906 French article.

ANSWER 3.19—(February 1979)

The smallest number in base 10 of multiplicative persistence 5 is 679. The chart in Figure 3.6 gives the smallest numbers with multiplicative persistence of 11 or less. It is taken from N.J.A. Sloane's paper "The Persistence of a Number," in *Journal of Recreational Mathematics* (6, 1973, pp. 97–98).

The smallest number in base 10 of additive persistence 4 is

Persistence	Number
1	10
2	25
3	39
4	77
5	679
6	6,788
7	68,889
8	2,677,889
9	26,888,999
10	3,778,888,999
11	277,777,788,888,899

Figure 3.6

Numbers

19,999,999,999,999,999,999,999. The sum of the digits in this number is 199, the smallest number of additive persistence 3. The sum of the digits in 199 is 19, the smallest number of additive persistence 2. The sum of the digits in 19 is 10, the smallest number of additive persistence 1. More generally, the second step in the sequence for the smallest number of additive persistence k gives the smallest number of additive persistence $k - 1$. All such numbers start with 1 and are followed by 9's. Therefore the smallest number of additive persistence 5 is the number consisting of 1 followed by 2,222,222,222,222,222, 222,222 9's. For further discussion of the problem see Harvey J. Hinden's "The Additive Persistence of a Number" (*Journal of Recreational Mathematics*, 7, 1974, pp. 134–35).

chapter 4

Algebra

Algebraic puzzles are certainly the oldest type of puzzle. In fact these problems predate "algebra" and were fundamentally involved with the invention of algebra. What were truly mystifying to the ancient Babylonians, Chinese, and Arabs, as well as the medieval scholars, today are regarded as rote "word problems." While they may still be found in puzzle books, they are generally no longer considered entertaining.

There are some new twists, however. Some problems are entertaining simply because they appear unsolvable, as if there was not enough information provided. The initial, easier, puzzles are "algebraic" only in appearance, merely requiring a little insight. Some of the problems involve Diophantine equations, which are equations that cannot be solved without additional work, typically some case analysis.

Problems

PROBLEM 4.1—Apple Picking
If you took 3 apples from a basket that held 13 apples, how many apples would you have?

PROBLEM 4.2—Greek Lifespan
A Greek was born on the seventh day of 40 B.C. and died on the seventh day of A.D. 40. How many years did he live?

PROBLEM 4.3—Fixed-Point Theorem
One morning, exactly at sunrise, a Buddhist monk began to climb a

tall mountain. The narrow path, no more than a foot or two wide, spiraled around the mountain to a glittering temple at the summit.

The monk ascended the path at varying rates of speed, stopping many times along the way to rest and to eat the dried fruit he carried with him. He reached the temple shortly before sunset. After several days of fasting and meditation he began his journey back along the same path, starting at sunrise and again walking at variable speeds with many pauses along the way. His average speed descending was, of course, greater than his average climbing speed.

Prove that there is a spot along the path that the monk will occupy on both trips at precisely the same time of day.

PROBLEM 4.4—GCD Times LCM

Find two positive integers, x and y, such that the product of their greatest common divisor and their lowest common multiple is xy.

PROBLEM 4.5—Naming Animals

A farmer has 20 pigs, 40 cows, 60 horses. How many horses does he have if you call the cows horses? (From T. H. O'Beirne)

PROBLEM 4.6—The Sum Is the Product

What three positive integers have a sum equal to their product?

PROBLEM 4.7—Trotting Dog

A boy, a girl, and a dog are at the same spot on a straight road. The boy and the girl walk forward—the boy at four miles per hour, the girl at three miles per hour. As they proceed the dog trots back and forth between them at 10 miles per hour. Assume that each reversal of its direction is instantaneous. An hour later, where is the dog and which way is it facing? (From A. K. Austin)

PROBLEM 4.8—Was Fermat Wrong?

Time, March 7, 1938, reported that one Samuel Isaac Krieger claimed to have found a counterexample to Fermat's unproved last theorem. Krieger announced that it was $1,324^n + 731^n = 1,961^n$, where n is a certain positive integer greater than 2, and which Krieger refused to disclose. A reporter on the *New York Times*, said *Time*, easily proved that Krieger was mistaken. How?

PROBLEM 4.9—Counting Earrings

In a certain African village there live 800 women. Three percent of them are wearing one earring. Of the other 97 percent, half are wearing two earrings, half are wearing none. How many earrings all together are being worn by the women?

PROBLEM 4.10—Square Year

"I was n years old in the year n^2," said Smith in 1971. When was he born?

PROBLEM 4.11—Colliding Missiles

Two missiles speed directly toward each other, one at 9,000 miles per hour and the other at 21,000 miles per hour. They start 1,317 miles apart. Without using pencil and paper, calculate how far apart they are one minute before they collide.

PROBLEM 4.12—Coleridge's Apples

Who would have thought that the poet Samuel Taylor Coleridge would have been interested in recreational mathematics? Yet the first entry in the first volume of his private notebooks (published in 1957 by Pantheon Books) reads: "Think any number you like—double—add 12 to it—halve it—take away the original number—and there remains six." Several years later, in a newspaper article, Coleridge spoke of the value of this simple trick in teaching principles of arithmetic to the "very young."

The notebook's second entry is: "Go into an Orchard—in which there are three gates—thro' all of which you must pass—Take a certain number of apples—to the first man [presumably a man stands by each gate] I give half of that number & half an apple—to the 2nd [man I give] half of what remain & half an apple—to the third [man] half of what remain & half an apple—and yet I never cut one Apple."

Determine the smallest number of apples Coleridge could start with and fulfill all the stated conditions.

PROBLEM 4.13—How Old Is the Rose-Red City?

Two professors, one of English and one of mathematics, were having drinks in the faculty club bar.

"It is curious," said the English professor, "how some poets can write one immortal line and nothing else of lasting value. John

William Burgon, for example. His poems are so mediocre that no one reads them now, yet he wrote one of the most marvelous lines in English poetry: 'A rose-red city half as old as Time.'"

The mathematician, who liked to annoy his friends with improvised brainteasers, thought for a moment or two, then raised his glass and recited:

> *A rose-red city half as old as Time.*
> *One billion years ago the city's age*
> *Was just two-fifths of what Time's age will be*
> *A billion years from now. Can you compute*
> *How old the crimson city is today?*

The English professor had long ago forgotten his algebra, so he quickly shifted the conversation to another topic, but you should have no difficulty with the problem.

PROBLEM 4.14—Absent-Minded Teller

An absent-minded bank teller switched the dollars and cents when he cashed a check for Mr. Brown, giving him dollars instead of cents, and cents instead of dollars. After buying a five-cent newspaper, Brown discovered that he had left exactly twice as much as his original check. What was the amount of the check?

PROBLEM 4.15—What Price Pets?

The owner of a pet shop bought a certain number of hamsters and half that many pairs of parakeets. He paid $2 each for the hamsters and $1 for each parakeet. On every pet he placed a retail price that was an advance of 10 percent over what he paid for it.

After all but seven of the creatures had been sold, the owner found that he had taken in for them an amount of money exactly equal to what he had originally paid for all of them. His potential profit, therefore, was represented by the combined retail value of the seven remaining animals. What was this value?

PROBLEM 4.16—Plane in the Wind

An airplane flies in a straight line from airport A to airport B, then back in a straight line from B to A. It travels with a constant engine speed and there is no wind. Will its travel time for the same round trip be greater, less, or the same if, throughout both flights, at the same engine speed, a constant wind blows from A to B?

Problem 4.17—Three Watch Hands

Assume an idealized, perfectly running watch with a sweep second hand. At noon all three hands point to exactly the same spot on the dial. What is the next time at which the three hands will be in line again, all pointing in the same direction? The answer is: Midnight.

The first part of this problem—much the easiest—is to prove that the three hands are together only when they point straight up. The second part, calling for more ingenuity, is to find the exact time or times, between noon and midnight, when the three hands come *closest* to pointing in the same direction. "Closest" is defined as follows: two hands point to the same spot on the dial, with the third hand a minimum distance away. When does this occur? How far away is the third hand?

It is assumed (as is customary in problems of this type) that all three hands move at a steady rate, so that time can be registered to any desired degree of accuracy.

Problem 4.18—How Far Did the Smiths Travel?

At ten o'clock one morning Mr. Smith and his wife left their house in Connecticut to drive to the home of Mrs. Smith's parents in Pennsylvania. They planned to stop once along the way for lunch at Patricia Murphy's Candlelight Restaurant in Westchester.

The prospective visit with his in-laws, combined with business worries, put Mr. Smith in a sullen, uncommunicative mood. It was not until eleven o'clock that Mrs. Smith ventured to ask: "How far have we gone, dear?"

Mr. Smith glanced at the mileage meter. "Half as far as the distance from here to Patricia Murphy's," he snapped.

They arrived at the restaurant at noon, enjoyed a leisurely lunch, then continued on their way. Not until five o'clock, when they were 200 miles from the place where Mrs. Smith had asked her first question, did she ask a second one. "How much farther do we have to go, dear?"

"Half as far," he grunted, "as the distance from here to Patricia Murphy's."

They arrived at their destination at seven that evening. Because of traffic conditions Mr. Smith had driven at widely varying speeds.

Nevertheless, it is quite simple to determine (and this is the problem) exactly how far the Smiths traveled from one house to the other.

PROBLEM 4.19—Beer Signs on the Highway

Smith drove at a steady clip along the highway, his wife beside him. "Have you noticed," he said, "that those annoying signs for Flatz beer seem to be regularly spaced along the road? I wonder how far apart they are."

Mrs. Smith glanced at her wristwatch, then counted the number of Flatz beer signs they passed in one minute.

"What an odd coincidence!" exclaimed Smith. "When you multiply that number by 10, it exactly equals the speed of our car in miles per hour."

Assuming that the car's speed is constant, that the signs are equally spaced, and that Mrs. Smith's minute began and ended with the car midway between two signs, how far is it between one sign and the next?

PROBLEM 4.20—Twenty Bank Deposits

A Texas oilman who was an amateur number theorist opened a new bank account by depositing a certain integral number of dollars, which we shall call x. His second deposit, y, also was an integral number of dollars. Thereafter each deposit was the sum of the two previous deposits. (In other words, his deposits formed a generalized Fibonacci series.) His 20th deposit was exactly a million dollars. What are the values of x and y, his first two deposits? (I am indebted to Leonard A. Monzert of West Newton, Massachusetts, for sending the problem of which this is a version.)

The problem reduces to a Diophantine equation that is somewhat tedious to solve, but a delightful shortcut using the golden ratio becomes available if I add that x and y are the two positive integers that begin the longest possible generalized Fibonacci chain ending in a term of 1,000,000.

PROBLEM 4.21—Fifty Miles an Hour

A train goes 500 miles along a straight track, without stopping, completing the trip with an average speed of exactly 50 miles per hour. It travels, however, at different speeds along the way. It seems plausible that nowhere along the 500 miles of track is there a segment of 50 miles that the train traverses in precisely one hour.

Prove that this is not the case.

PROBLEM 4.22—Early Commuter

A commuter is in the habit of arriving at his suburban station each evening exactly at five o'clock. His wife always meets the train and drives him home. One day he takes an earlier train, arriving at the station at four. The weather is pleasant, so instead of telephoning home he starts walking along the route always taken by his wife. They meet somewhere on the way. He gets into the car and they drive home, arriving at their house ten minutes earlier than usual. Assuming that the wife always drives at a constant speed, and that on this occasion she left just in time to meet the five o'clock train, can you determine how long the husband walked before he was picked up?

PROBLEM 4.23—Two Ferryboats

Two ferryboats start at the same instant from opposite sides of a river, traveling across the water on routes at right angles to the shores. Each travels at a constant speed, but one is faster than the other. They pass at a point 720 yards from the nearest shore. Both boats remain in their slips for 10 minutes before starting back. On the return trips they meet 400 yards from the other shore.

How wide is the river?

PROBLEM 4.24—Professor on the Escalator

When Professor Stanislaw Slapenarski, the Polish mathematician, walked very slowly down the down-moving escalator, he reached the bottom after taking 50 steps. As an experiment, he then ran up the same escalator, one step at a time, reaching the top after taking 125 steps.

Assuming that the professor went up five times as fast as he went down (that is, took five steps to every one step before), and that he made each trip at a constant speed, how many steps would be visible if the escalator stopped running?

PROBLEM 4.25—Marching Cadets and a Trotting Dog

A square formation of Army cadets, 50 feet on the side, is marching forward at a constant pace [see Figure 4.1]. The company mascot, a small terrier, starts at the center of the rear rank [position A in the illustration], trots forward in a straight line to the center of the front rank [position B], then trots back again in a straight line to the center of the rear. At the instant he returns to position A, the cadets have advanced exactly 50 feet. Assuming that the dog trots at a constant speed and loses no time in turning, how many feet does he travel?

If you solve this problem, which calls for no more than a knowledge of elementary algebra, you may wish to tackle a much more difficult version proposed by the famous puzzlist Sam Loyd (see *Mathematical Puzzles of Sam Loyd*, vol. 2, Dover paperback, 1960, page 103). Instead of moving forward and back through the marching cadets, the mascot trots with constant speed around the *outside* of the square, keeping as close as possible to the square at all times. (For the problem we assume that he trots along the perimeter of the square.) As before, the formation has marched 50 feet by the time the dog returns to point A. How long is the dog's path?

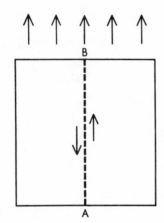

Figure 4.1

Answers

Answer 4.1—(August 1968)

You would have three apples.

Answer 4.2—(August 1969)

The Greek lived 79 years. There was no year 0.

Answer 4.3—(June 1961)

A man goes up a mountain one day, down it another day. Is there a spot along the path that he occupies at the same time of day on both trips? This problem was called to my attention by psychologist Ray Hyman, of the University of Oregon, who in turn found it in a monograph entitled "On Problem-Solving," by the German Gestalt psychologist Karl Duncker. Duncker writes of being unable to solve it and of observing with satisfaction that others to whom he put the problem had the same difficulty. There are several ways to go about it, he continues, "but probably none is . . . more drastically evident than the following. Let ascent and descent be divided between *two* persons on the same day. They must *meet*. Ergo. . . . With this, from an unclear dim condition not easily surveyable, the situation has suddenly been brought into full daylight."

ANSWER 4.4—(July 1971)

This is true for any two positive integers.

ANSWER 4.5—(August 1964)

The farmer has 60 horses. Calling a cow a horse doesn't make it a horse.

John Appel and Daniel Rosenblum were the first to tell me that this is a version of a joke attributed to Abraham Lincoln. He once asked a man who had been arguing that slavery was not slavery but a form of protection how many legs a dog would have if you called its tail a leg. The answer, said Lincoln, is four because calling a tail a leg does not make it a leg.

ANSWER 4.6—(July 1971)

$$1 + 2 + 3 = 1 \times 2 \times 3$$

Problem E 2262, in *The American Mathematical Monthly* (November 1971, pages 1021–23), by G. J. Simmons and D. E. Rawlinson, generalized this question by asking for all other sets of k positive integers of which the same statement could be made. It turns out it can be made for all positive integers, but only a very small set have unique answers. When $k = 2$, the only answer is $2 + 2 = 2 \times 2$. Our problem provided the only answer for $k = 3$. For $k = 4$ it is $2 + 4 + 1 + 1 = 2 \times 4 \times 1 \times 1$.

Readers of the periodical showed that for all values of k not exceeding 1,000, the only values with unique solutions are 2, 3, 4, 6, 24, 114,

174, and 444. It is possible, the editor comments, that no other values have unique answers other than the eight listed.

Answer 4.7—(July 1971)

The dog can be at any point between the boy and the girl, facing either way. Proof: At the end of one hour, place the dog anywhere between the boy and the girl, facing in either direction. Time-reverse all motions and the three will return at the same instant to the starting point.

This question (about the boy, the girl, and the dog) stirred up a hornet's nest. Some mathematicians defended the answer as being valid, others insisted the problem has *no* answer because it is logically contradictory. There is no way the three can start moving, it was argued, because the instant they do the dog will no longer be between the boy and the girl. This plunges us into deep waters off the coast of Zeno.

Answer 4.8—(July 1971)

The first number, 1,324, raised to any power must end in 6 or 4. The other two numbers, 731 and 1,961, raised to any power must end in 1. Since no number ending in 6 or 4, added to a number ending in 1, can produce a number ending in 1, the equation has no solution.

Martin Kruskal provided a photocopy of the *New York Times* account (February 22, 1938) of Samuel Isaac Krieger's preposterous claim to have disproved Fermat's last theorem. He had saved the clipping since he had seen it as a small boy.

Answer 4.9—(August 1968)

Among the 97 percent of the women, if half wear two earrings and half none, this is the same as if each wore one. Assuming, then, that each of the 800 women is wearing one earring, there are 800 earrings in all.

Answer 4.10—(July 1971)

Smith was born in 1892. He was 44 in $44^2 = 1936$.

Answer 4.11—(August 1958)

The two missiles approach each other with combined speeds of 30,000 miles per hour, or 500 miles per minute. By running the scene backward in time we see that one minute before the collision the missiles would have to be 500 miles apart.

Answer 4.12—(March 1965)

Seven is the smallest number of apples (we rule out "negative apples") that satisfies the conditions of Coleridge's problem.

Answer 4.13—(October 1960)

The rose-red city's age is 7 billion years. Let x be the city's present age; y, the present age of Time. A billion years ago the city would have been $x - 1$ billion years old and a billion years from now Time's age will be $y + 1$. The data in the problem permit two simple equations:

$$2x = y$$
$$x - 1 = \frac{2}{5}(y + 1)$$

These equations give x, the city's present age, a value of 7 billion years; and y, Time's present age, a value of 14 billion years. The problem presupposes a Big Bang theory of the creation of the cosmos.

Answer 4.14—(May 1959)

To determine the value of Brown's check, let x stand for the dollars and y for the cents. The problem can now be expressed by the following equation: $100y + x - 5 = 2(100x + y)$. This reduces to $98y - 199x = 5$, a Diophantine equation with an infinite number of integral solutions. A solution by the standard method of continued fractions gives as the lowest values in positive integers: $x = 31$ and $y = 63$, making Brown's check $31.63. This is a unique answer to the problem because the next lowest values are: $x = 129$, $y = 262$, which fails to meet the requirement that y be less than 100.

There is a much simpler approach to the problem and many readers wrote to tell me about it. As before, let x stand for the dollars on the check, y for the cents. After buying his newspaper, Brown has left $2x + 2y$. The change that he has left, from the x cents given him by the cashier, will be $x - 5$.

We know that y is less than 100, but we don't know yet whether it is less than 50 cents. If it is less than 50 cents, we can write the following equations:

$$2x = y$$
$$2y = x - 5$$

If y is 50 cents or more, then Brown will be left with an amount of cents ($2y$) that is a dollar or more. We therefore have to modify the above equations by taking 100 from $2y$ and adding 1 to $2x$. The equations become:

$$2x + 1 = y$$
$$2y - 100 = x - 5$$

Each set of simultaneous equations is easily solved. The first set gives x a minus value, which is ruled out. The second set gives the correct values.

ANSWERS 4.15—(February 1960)

Let x be the number of hamsters originally purchased and also the number of parakeets. Let y be the number of hamsters among the seven unsold pets. The number of parakeets among the seven will then be $7 - y$. The number of hamsters sold (at a price of $2.20 each, which is a markup of 10 percent over cost) will be $x - y$, and the number of parakeets sold (at $1.10 each) will be $x - 7 + y$.

The cost of the pets is therefore $2x$ dollars for the hamsters and x dollars for the parakeets—a total of $3x$ dollars. The hamsters that were sold brought $2.2(x - y)$ dollars and the parakeets sold brought $1.1(x - 7 + y)$ dollars—a total of $3.3x - 1.1y - 7.7$ dollars.

We are told that these two totals are equal, so we equate them and simplify to obtain the following Diophantine equation with two integral unknowns:

$$3x = 11y + 77$$

Since x and y are positive integers and y is not more than 7, it is a simple matter to try each of the eight possible values (including zero) for y to determine which of them makes x also integral. There are only two such values: 5 and 2. Each would lead to a solution of the problem were it not for the fact that the parakeets were bought in pairs. This eliminates 2 as a value for y because it would give x (the number of parakeets purchased) the odd value of 33. We conclude therefore that y is 5.

A complete picture can now be drawn. The shop owner bought 44 hamsters and 22 pairs of parakeets, paying altogether $132 for them.

He sold 39 hamsters and 21 pairs of parakeets for a total of $132. There remained five hamsters worth $11 retail and two parakeets worth $2.20 retail—a combined value of $13.20, which is the answer to the problem.

ANSWER 4.16—(February 1960)

Since the wind boosts the plane's speed from A to B and retards it from B to A, one is tempted to suppose that these forces balance each other so that total travel time for the combined flights will remain the same. This is not the case, because the time during which the plane's speed is boosted is shorter than the time during which it is retarded, so the overall effect is one of retardation. The total travel time in a wind of constant speed and direction, regardless of the speed or direction, is always greater than if there were no wind.

ANSWER 4.17—(November 1963)

A quick way to prove that all three hands of a watch with a sweep second hand are together only when they point to 12 is to apply elementary Diophantine analysis. When the hour hand coincides with one of the other hands, the difference between the distances traveled by each must be an integral number of hours. During the 12-hour period the hour hand makes one circuit around the dial. Assume that it travels a distance x, less than one complete circuit, to arrive at a position with all three hands together. After the hour hand has gone a distance of x, the minute hand will have gone a distance of $12x$, making the difference $11x$. In the same period of time the second hand will have gone a distance of $720x$, making the difference $719x$. All three hands can be together only when x has a value that makes both $719x$ and $11x$ integral. But 719 and 11 are both prime numbers, therefore x can take only the values of 0 and 1, which it has at 0 and 12 o'clock respectively.

Aside from the case in which all three hands point straight up, the closest the hands come to pointing in the same direction (defining "closest" as the minimum deviation of one hand when the other two coincide) is at 16 minutes 16 and 256/719 seconds past 3, and again at 43 minutes 43 and 463/719 seconds after 8.

The two times are mirror images in the sense that if a watch showing one time is held up to a mirror and the image is read as though it were an unreversed clock, the image would indicate the other time.

The sum of the two times is 12 hours. In both instances the second and hour hands coincide, with the minute hand separated from them by a distance of 360/719 of one degree of arc. (The distance is 5 and 5/719 seconds if we define a second as a sixtieth of the distance of a clock minute.) In the first instance the minute hand is behind the other two by this distance, in the second instance it leads by the same distance.

Another simple proof that the three hands are never together except at 12 was found by Henry D. Friedman, of Sylvania Electronic Systems. The hour and minute hands meet 11 times, with periods of 12/11 hours that divide the clock's circumference into 11 equal parts. The minute and second hands similarly divide the circumference into 59 equal parts. All three hands can meet only at a point where $r/11 = s/59$, r and s being positive integers with r less than 11 and s less than 59. Since 11/59 cannot be reduced to a lower fraction, r/s, there can be no meeting of the three hands except at 12 o'clock.

ANSWER 4.18—(February 1962)

To find the mileage covered by the Smiths on their trip from Connecticut to Pennsylvania, the various times of day that are given are irrelevant, since Smith drove at varying speeds. At two points along the way Mrs. Smith asked a question. Smith's answers indicate that the distance from the first point to Patricia Murphy's Candlelight Restaurant is two thirds of the distance from the start of the trip to the restaurant, and the distance from the restaurant to the second point is two thirds of the distance from the restaurant to the end of the trip. It is obvious, therefore, that the distance from point to point (which we are told is 200 miles) is two thirds of the total distance. This makes the total distance 300 miles. Figure 4.2 should make it all clear.

Figure 4.2

ANSWER 4.19—(October 1960)

The curious thing about the problem of the Flatz beer signs is that it is not necessary to know the car's speed to determine the spacing of the signs. Let x be the number of signs passed in one minute. In an hour the car will pass $60x$ signs. The speed of the car, we are told, is

10x miles per hour. In 10x miles it will pass 60x signs, so in one mile it will pass 60x/10x, or 6, signs. The signs therefore are 1/6 mile, or 880 feet, apart.

ANSWER 4.20—(November 1970)

The Texas oilman's bank deposit problem reduces to the Diophantine equation $2,584x + 4,181y = 1,000,000$. It can be solved by Diophantine techniques such as the continued-fraction method. The first two deposits are $154 and $144.

The shortcut, given that x and y start the longest possible Fibonacci chain terminating in 1,000,000, rests on the fact that the longer a generalized Fibonacci series continues, the closer the ratio of two adjacent terms approaches the golden ratio. To find the longest generalized Fibonacci chain that ends with a given number, place the number over x and let it equal the golden ratio. In this case the equation is

$$\frac{1,000,000}{x} = \frac{1 + \sqrt{5}}{2}$$

Solve for x and change the result to the nearest integer. It is 618,034. Because no other integer, when related to 1,000,000, gives a closer approximation of the golden ratio, 618,034 is the next-to-last term of the longest possible chain of positive integers in a generalized Fibonacci series ending in 1,000,000. One can now easily work backward along the chain to the first two terms. (This method is explained in Litton Industries' *Problematical Recreations*, edited by Angela Dunn, Booklet 10, Problem 41.)

ANSWER 4.21—(December 1979)

Divide the 500-mile track into 10 segments of 50 miles each. If any segment is traversed in one hour, the problem is solved, and so it must be assumed that traversing each segment takes either less than an hour or more than an hour. It then follows that somewhere along the track there will be at least one pair of adjacent segments, one (call it A) traversed in less than an hour and the other (call it B) traversed in more than an hour.

Imagine an enormous measuring rod 50 miles long that is placed over segment A. In your mind slide the rod slowly in the direction of segment B until it coincides with B. As you slide the rod, the average time taken by the train to go the 50 miles covered by the rod varies continuously from less than an hour (for *A*) to more than an hour (for *B*). Therefore there must be at least one position where the rod covers a 50-mile length of track that was traversed by the train in exactly one hour.

For a more technical analysis of the problem, in terms of a jogger who averages a mile in eight minutes and runs an integral number of miles, see "Comments on 'Kinematics Problem for Joggers,'" by R. P. Boas, in the *American Journal of Physics* (vol. 42, August 1974, page 695).

ANSWER 4.22—(February 1957)

The commuter has walked for 55 minutes before his wife picks him up. Since they arrive home 10 minutes earlier than usual, this means that the wife has chopped 10 minutes from her usual travel time to and from the station, or five minutes from her travel time to the station. It follows that she met her husband five minutes before his usual pick-up time of five o'clock, or at 4:55. He started walking at four, therefore he walked for 55 minutes. The man's speed of walking, the wife's speed of driving, and the distance between home and station are not needed for solving the problem. If you tried to solve it by juggling figures for these variables, you probably found the problem exasperating.

When this problem was presented in *Scientific American* it was unfortunately worded, suggesting that the wife habitually arrived early at the station and waited for the five-o'clock train. If this is the case, the husband's walking time lies within a range of 50 to 55 minutes.

A number of readers pointed out that the problem yields readily to solution by what Army logisticians call a "march graph" [see Figure 4.3]. Time is plotted on the horizontal axis, distance on the vertical. The graph shows clearly that the wife could leave home up to 10 minutes earlier than the leaving time required to just meet the train. The lower limit (50 minutes) of her husband's walking time can occur only when the wife leaves a full 10 minutes earlier and either drives habitually at infinite speed (in which case her husband arrives home at the same moment she leaves), or the husband walks at an infinitesimal speed (in which case she meets him at the station after he has walked 50 minutes and gotten nowhere). "Neither image rings false," wrote David W. Weiser, assistant professor of natural science at the University of Chicago, in one of the clearest analyses I received of the problem, "considering the way of a wife with a car, or of a husband walking past a tavern."

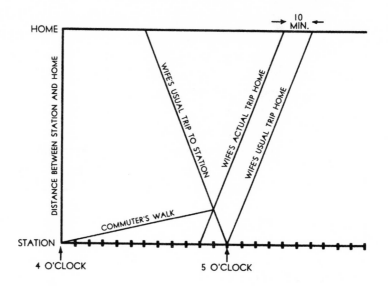

Figure 4.3

ANSWER 4.23—(November 1957)

When the ferryboats meet for the first time [top of Figure 4.4], the combined distance traveled by the boats is equal to the width of the river. When they reach the opposite shore, the combined distance is twice the width of the river; and when they meet the second time [bottom of Figure 4.4], the total distance is three times the river's width. Since the boats have been moving at a constant speed for the same period of time, it follows that each boat has gone three times as far as when they first met and had traveled a combined distance of one river-width. Since the white boat had traveled 720 yards when the first meeting occurred, its total distance at the time of the second meeting must be 3 × 720, or 2,160 yards. The bottom illustration shows clearly that this distance is 400 yards more than the river's width, so we subtract 400 from 2,160 to obtain 1,760 yards, or one mile, as the width of the river. The time the boats remained at their landings does not enter into the problem.

The problem can be approached in other ways. Many readers solved it as follows. Let x equal the river-width. On the first trip the ratio of distances traveled by the two boats is $x - 720 : 720$. On the second trip it is $2x - 400 : x + 400$. These ratios are equal, so it is easy to solve for x. (The problem appears in Sam Loyd's *Cyclopedia of Puzzles*, 1914, page 80.)

Algebra

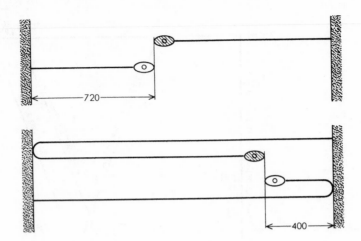

Figure 4.4

ANSWER 4.24—(May 1959)

Let n be the number of steps visible when the escalator is not moving, and let a unit of time be the time it takes Professor Slapenarski to walk down one step. If he walks down the down-moving escalator in 50 steps, then $n - 50$ steps have gone out of sight in 50 units of time. It takes him 125 steps to run up the same escalator, taking five steps to every one step before. In this trip, $125 - n$ steps have gone out of sight in 125/5, or 25, units of time. Since the escalator can be presumed to run at constant speed, we have the following linear equation that readily yields a value for n of 100 steps:

$$\frac{n - 50}{50} = \frac{125 - n}{25}$$

ANSWER 4.25—(February 1960)

Let 1 be the length of the square of cadets and also the time it takes them to march this length. Their speed will also be 1. Let x be the total distance traveled by the dog and also its speed. On the dog's forward trip his speed relative to the cadets will be $x - 1$. On the return trip his speed relative to the cadets will be $x + 1$. Each trip is a distance of 1 (relative to the cadets), and the two trips are completed in unit time, so the following equation can be written:

$$\frac{1}{x - 1} + \frac{1}{x - 1} = 1$$

COMBINATORIAL & NUMERICAL PROBLEMS

This can be expressed as the quadratic: $x^2 - 2x - 1 = 0$, for which x has the positive value of $1 + \sqrt{2}$. Multiply this by 50 to get the final answer: $120.7 +$ feet. In other words, the dog travels a total distance equal to the length of the square of cadets plus that same length times the square root of 2.

Loyd's version of the problem, in which the dog trots *around* the moving square, can be approached in exactly the same way. I paraphrase a clear, brief solution sent by Robert F. Jackson of the Computing Center at the University of Delaware.

As before, let 1 be the side of the square and also the time it takes the cadets to go 50 feet. Their speed will then also be 1. Let x be the distance traveled by the dog and also his speed. The dog's speed with respect to the speed of the square will be $x - 1$ on his forward trip, $\sqrt{x^2 - 1}$ on each of his two transverse trips, and $x + 1$ on his backward trip. The circuit is completed in unit time, so we can write this equation:

$$\frac{1}{x - 1} + \frac{2}{\sqrt{x^2 - 1}} + \frac{1}{x + 1} = 1$$

This can be expressed as the quartic equation: $x^4 - 4x^3 - 2x^2 + 4x + 5 = 0$. Only one positive real root is not extraneous: $4.18112+$. We multiply this by 50 to get the desired answer: $209.056+$ feet.

Theodore W. Gibson, of the University of Virginia, found that the first form of the above equation can be written as follows, simply by taking the square root of each side:

$$\frac{1}{\sqrt{x - 1}} + \frac{1}{\sqrt{x + 1}} = 1$$

which is remarkably similar to the equation for the first version of the problem.

Many readers sent analyses of variations of this problem: a square formation marching in a direction parallel to the square's diagonal, formations of regular polygons with more than four sides, circular formations, rotating formations, and so on. Thomas J. Meehan and David Salsburg each pointed out that the problem is the same as that of a destroyer making a square search pattern around a moving ship and showed how easily it could be solved by vector diagrams on what the Navy calls a "maneuvering board."

part two

_____

Geometric
Puzzles

chapter 5

Plane Geometry

The geometry of Euclid continues to surprise and entertain us. The elements of planar geometry—points, lines, and circles—are the building blocks of many wonderful structures. These puzzles do not require knowledge of obscure theorems. They do require you to "prove" their answer, often by introducing new lines to the diagrams, rotating objects, or restating what is given.

The matchstick puzzle is a genre that is centuries old, and Gardner included several new examples in his Scientific American *column. They are not all together within this volume, as some are geometric and some emphasize other aspects.*

Problems

PROBLEM 5.1—Largest Triangle

Each of the two equal sides of an isosceles triangle is one unit long. Without using calculus, find the length of the third side that maximizes the triangle's area.

PROBLEM 5.2—Lines and Triangles

Draw six line segments of equal length to form eight equilateral triangles.

PROBLEM 5.3—Leave Two Triangles

It is easy to see how to remove four matches and leave two equilateral triangles, or to remove three matches and leave two equilateral triangles, but can you remove just *two* matches from Figure 5.1 and leave two equilateral triangles? There must be no "loose ends."

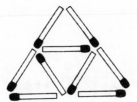

Figure 5.1

PROBLEM 5.4—Hexagon's Area

An equilateral triangle and a regular hexagon have perimeters of the same length. If the triangle has an area of two square units, what is the area of the hexagon?

PROBLEM 5.5—Trisecting a Square

From one corner of a square extend two lines that exactly trisect the square's area [see Figure 5.2]. Into what ratios do these trisecting lines cut the two sides of the square?

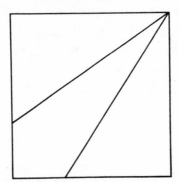

Figure 5.2

PROBLEM 5.6—Circumscribed Hexagon

Regular hexagons are inscribed in and circumscribed outside a circle [see Figure 5.3]. If the smaller hexagon has an area of three square

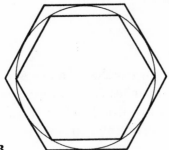

Figure 5.3

units, what is the area of the larger hexagon? (From Charles W. Trigg, *Mathematical Quickies*, reprinted by Dover in 1985.)

PROBLEM 5.7—Leave Four Squares

Change the positions of two matches in Figure 5.4 to reduce the number of unit squares from five to four. "Loose ends"—matches not used as sides of unit squares—are not allowed. An amusing feature of this classic is that, even if someone solves it, you can set up the pattern again in mirror-reflected form or upside down (or both) and the solution will be as difficult as before.

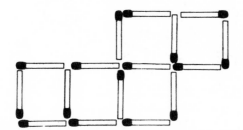

Figure 5.4

PROBLEM 5.8—Geometry and Divisibility

Assuming that the angle cannot be trisected by compass and straightedge, prove that no number in the doubling series 1, 2, 4, 8, 16, 32, . . . is a multiple of 3. (From Robert A. Weeks.)

PROBLEM 5.9—Hole-in-Metal Problem

A metal sheet has the shape of a two-foot square with semicircles on opposite sides [see Figure 5.5]. If a disk with a diameter of two feet is removed from the center as shown, what is the area of the remaining metal?

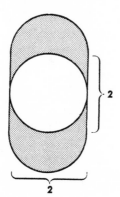

Figure 5.5   **2**

Plane Geometry

109

PROBLEM 5.10—Triangle of Tangents

Two tangents to a circle are drawn [see Figure 5.6] from a point, *C*. Tangent line segments *YC* and *XC* are necessarily equal. Each has a length of 10 units. Point *P*, on the circle's circumference, is randomly chosen between *X* and *Y*. Line *AB* is then drawn tangent to the circle at *P*. What length is the perimeter of triangle *ABC*?

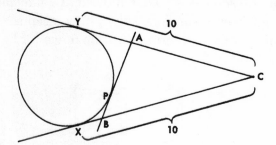

Figure 5.6

PROBLEM 5.11—Hidden Cross

The puzzle shown in Figure 5.7 is reproduced from the September—October 1978 issue of the magazine *Games*. The task is to trace in the larger figure a shape geometrically similar to the smaller one shown beside it.

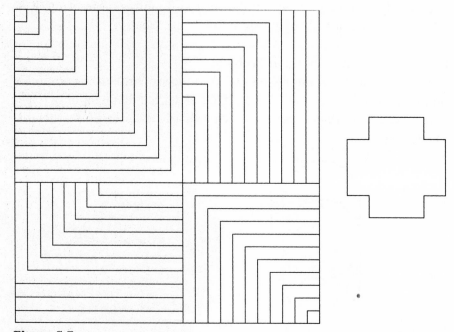

Figure 5.7

PROBLEM 5.12—Guess the Diagonal

A rectangle is inscribed in the quadrant of a circle as shown in Figure 5.8. Given the unit distances indicated, can you accurately determine the length of the diagonal $\overline{AC}$?

Time limit: one minute!

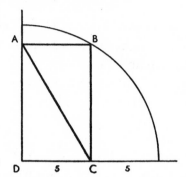

Figure 5.8

PROBLEM 5.13—Circle on the Chessboard

A chessboard has squares that are two inches on the side. What is the radius of the largest circle that can be drawn on the board in such a way that the circle's circumference is entirely on black squares?

PROBLEM 5.14—Bisecting Yin and Yang

Two mathematicians were dining at the Ying and Yang, a Chinese restaurant on West Third Street in Manhattan. They chatted about the symbol on the restaurant's menu [see Figure 5.9].

"I suppose it's one of the world's oldest religious symbols," one of them said. "It would be hard to find a more attractive way to symbolize the great polarities of nature: good and evil, male and female, inflation and deflation, integration and differentiation."

"Isn't it also the symbol of the Northern Pacific Railway?"

Figure 5.9

Plane Geometry

"Yes. I understand that one of the chief engineers of the railroad saw the emblem on a Korean flag at the Chicago World's Fair in 1893 and urged his company to adopt it. He said it symbolized the extremes of fire and water that drove the steam engine."

"Do you suppose it inspired the construction of the modern baseball?"

"I wouldn't be surprised. By the way, did you know that there is an elegant method of drawing one straight line across the circle so that it exactly bisects the areas of the Yin and Yang?"

Assuming that the Yin and Yang are separated by two semi-circles, show how each can be simultaneously bisected by the same straight line.

PROBLEM 5.15—Cocircular Points

Five paper rectangles (one with a corner torn off) and six paper disks have been tossed on a table. They fall as shown in Figure 5.10. Each corner of a rectangle and each spot where edges are seen to intersect marks a point. The problem is to find four sets of four "cocircular" points: four points that can be shown to lie on a circle.

For example, the corners of the isolated rectangle [bottom right in

Figure 5.10

Figure 5.10] constitute such a set, because the corners of any rectangle obviously lie on a circle. What are the other three sets?

PROBLEM 5.16—Twelve Matches

Assuming that a match is a unit of length, it is possible to place 12 matches on a plane in various ways to form polygons with integral areas. Figure 5.11 shows two such polygons: a square with an area of nine square units, and a cross with an area of five.

The problem is this: Use all 12 matches (the entire length of each match must be used) to form in similar fashion the perimeter of a polygon with an area of exactly four square units.

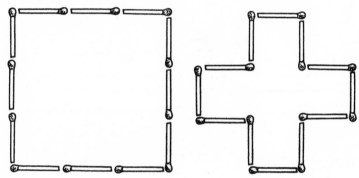

Figure 5.11

PROBLEM 5.17—Steiner Ellipses

This oldie goes back to Jakob Steiner, a noted Swiss geometer of the nineteenth century. My excuse for reviving it is that it is one of the best examples I know of a problem that is difficult to solve by calculus or analytic geometry but is ridiculously easy if approached with the right turn of mind and some knowledge of elementary plane and projective geometry.

We are given a 3, 4, 5 triangle. Its area is six square units. We wish to calculate both the smallest area of an ellipse that can be circumscribed around it and the largest area of an ellipse that can be inscribed inside it.

PROBLEM 5.18—Overlap Squares

In 1950, when Charles W. Trigg, dean emeritus of Los Angeles City College, was editing the problems department of *Mathematics Magazine*, he introduced a popular section headed "Quickies." A quickie, Trigg then explained, is a problem "which may be solved by laborious methods, but which with proper insight may be disposed of with dispatch." His 1967 book, *Mathematical Quickies* (reprinted by Dover in 1985), is a splendid collection of 270 of the best quickies he encountered or invented in his distinguished career as a problem expert.

In one quickie from Trigg's book [see Figure 5.12] the smaller square has a side of three inches and the larger square a side of four. Corner *D* is at the center of the small square. The large square is rotated around *D* until the intersection of two sides at point *B* exactly trisects $\overline{AC}$. How quickly can you compute the area of overlap (shown shaded) of the two squares?

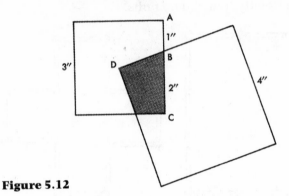

Figure 5.12

PROBLEM 5.19—Triple Beer Rings

In *Problematical Recreations*, No. 7, a series of puzzle booklets once issued annually by Litton Industries in Beverly Hills, California, the following problem appeared. A man places his beer glass on the bar three times to produce the set of triple rings shown in Figure 5.13. He does it carefully, so that each circle passes through the center of the other two. The bartender thinks the area of mutual overlap

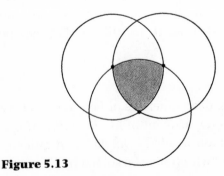

Figure 5.13

GEOMETRIC PUZZLES

[shaded] is less than one-fourth of the area of a circle, but the customer says it is more than one-fourth. Who is right?

The solution can be obtained the hard way by finding the area of an equilateral triangle inscribed in the shaded section and then adding the areas of the three segments of circles on each side of the triangle. A reader of this column, Tad Dunne of Willowdale in Ontario, sent me a beautiful graphical "look and see" solution that involves no geometrical formulas and almost no arithmetic, although it does make use of a repeating wallpaper pattern. Can you rediscover it?

PROBLEM 5.20—Rolling Pennies

If one penny rolls around another penny without slipping, how many times will it rotate in making one revolution? One might guess the answer to be one, since the moving penny rolls along an edge equal to its own circumference, but a quick experiment shows that the answer is two; apparently the complete revolution of the moving penny adds an extra rotation. Suppose we roll the penny, without slipping, from the top of a six-penny triangle [see Figure 5.14] all the way around the sides and back to the starting position. How many times will it rotate? It is easy to see from the picture that the penny rolls along arcs with a total distance (expressed in fractions of full circles) of 12/6, or two full circles. Therefore it must rotate at least twice. Because it makes a complete revolution, shall we add one more rotation and say it rotates three times? No, a test discloses that it makes four rotations! The truth is that for every degree of arc along which it rolls, it rotates two degrees. We must double the length of the path along which it rolls to get the correct answer: four rotations. With this in mind it is easy to solve other puzzles of this type, which are often found in puzzle books. Simply calculate the length of the path in degrees, multiply by two, and you have the number of degrees of rotation.

All of this is fairly obvious, but there is a beautiful theorem lurking here that has not been noticed before to my knowledge. Instead of close-packing the coins around which a coin is rolled, join them to form an irregular closed chain; Figure 5.15 shows a random chain of nine pennies. (The only proviso is that as a penny rolls around it without slipping, the rolling coin must touch every coin in the chain.) It turns out, surprisingly, that regardless of the shape of a chain of a given length, the number of rotations made by the rolling penny by

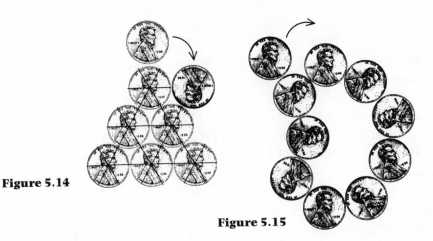

Figure 5.14

Figure 5.15

the time it returns to its starting position is always the same! In the nine-penny case the penny makes exactly five rotations. If the penny is rolled *inside* the chain, it will make exactly one rotation. This is also a constant, unaffected by the shape of the chain.

Can you prove (only the most elementary geometry need be used) that for any closed chain of n pennies ($n > 2$) the number of rotations of the moving penny, as it rolls once around the outside of the chain, is constant? If so, you will see immediately how to apply the same proof to a penny rolling inside a chain of n coins ($n > 6$) as well as how to derive a simple formula, for each case, that expresses the number of rotations as a function of n.

Problem 5.21—How Many Spots?

This double problem was given by D. Mollison of Trinity College, Cambridge, in a 1966 problems contest for members of the Archimedeans, a Cambridge student mathematics society. The first question: What is the maximum number of points that can be placed on or within the figure shown in Figure 5.16, provided that no two points are separated by a distance that is less than the square root of 2?

The second question: In how many different patterns can this maximum number of points be placed, not counting rotations and reflections of patterns as being different? The dotted lines were added to the figure to show that it is formed by a unit square surrounded by four half-squares.

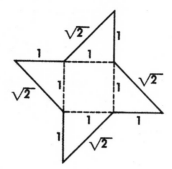

Figure 5.16

Problem 5.22—Amorous Bugs

Four bugs—A, B, C, and D—occupy the corners of a square 10 inches on a side [Figure 5.17]. A and C are male, B and D are female. Simultaneously A crawls directly toward B, B toward C, C toward D, and D toward A. If all four bugs crawl at the same constant rate, they will describe four congruent logarithmic spirals which meet at the center of the square.

How far does each bug travel before they meet? The problem can be solved without calculus.

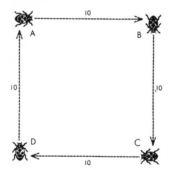

Figure 5.17

Problem 5.23—Three Squares

Using only elementary geometry (not even trigonometry), prove that angle *C* in Figure 5.18 equals the sum of angles *A* and *B*.

I am grateful to Lyber Katz for this charmingly simple problem. He writes that as a child he went to school in Moscow, where the problem was given to his fourth-grade geometry class for extra credit to those who solved it. "The number of blind alleys the problem leads to," he adds, "is extraordinary."

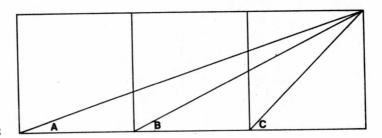

Figure 5.18

PROBLEM 5.24—Intersecting Circles

This is one of those elegant theorems in old-fashioned plane geometry that seem at first to be exceedingly difficult to establish but that yield readily to the right insight. Three circles of unit radius, with centers at X, Y, Z, intersect at a common point, O [see Figure 5.19]. The problem is to prove that the other three intersection points, A, B, C, lie on a circle that also has a unit radius. The problem comes by way of Frank R. Bernhart.

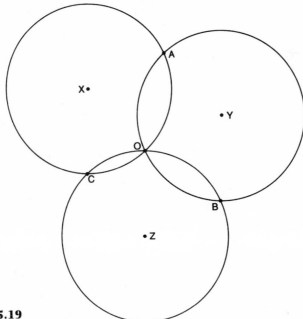

Figure 5.19

PROBLEM 5.25—Three Circles

Draw three nonoverlapping circles of three different sizes anywhere on a piece of paper. For each pair of circles draw their two common tangents. If you have not encountered this beautiful theorem before, you may be surprised to find that the intersections of the three pairs of tangents lie on a straight line [see Figure 5.20].

As one would expect, there are many ways the theorem can be proved by adding construction lines to the figure. *Popular Computing* reported in its issue of December 1974, however, that the theorem lends itself to an elegant solution if one leaves the two-dimensional plane for a three-dimensional extension. Quoting from an earlier book in which they found the problem, the editors of the magazine report that when the theorem was shown to John Edson Sweet, a professor of engineering at Cornell University who died in 1916, he studied the picture for a moment and then said, "Yes, that's perfectly self-evident."

What was Professor Sweet's sweet solution?

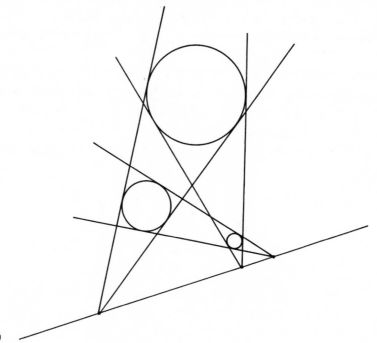

Figure 5.20

PROBLEM 5.26—Different Distances

It is easy to place three counters on the cells of a 3 × 3 checkerboard so no two pairs of counters are the same distance apart. We assume that each counter marks the exact center of a cell and that distances are measured on a straight line joining the centers. Discounting rotations and reflections, there are five solutions, which appear in the top part of Figure 5.21.

It also is easy to put four counters on a 4 × 4 board so that all distances between pairs are different. There are 16 ways of doing it. On the 5 × 5 board the number jumps to 28.

In the January 1972 issue of the *Journal of Recreational Mathematics* Sidney Kravitz asked for solutions to the order-5 and the order-6 squares. The solution for the order-6 square proved to be difficult because for the first time the 3, 4, 5 right triangle (the smallest Pythagorean triangle) enters the picture. The fact that an integral distance of 5 is now possible both orthogonally and diagonally severely limits the patterns. As readers of the journal discovered, there are only two solutions. They appear in the bottom part of Figure 5.21.

For what squares of side *n* is it possible to place *n* counters so that all distances are different? As reported in the fall 1976 issue of the journal, John H. Muson proved (using a pigeonhole argument) that no solutions are possible on squares of order 16 and higher. Harry L. Nelson lowered the limit to 15. Milton W. Green, with an exhaustive-search computer program, established impossibility for orders 8 and 9 and found a unique solution for order 7. David Babcock, an editor of *Popular Computing*, confirmed these results through order 8. The problem was finally laid to rest in 1976 by Michael Beeler. His com-

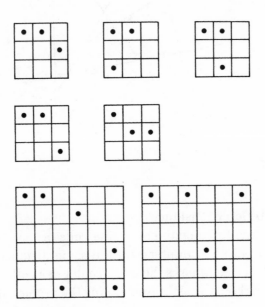

Figure 5.21

puter program confirmed the results indicated above and proved impossibility for orders 10 through 14.

The order-7 board, therefore, is the largest for which there is a solution. The solution is unique and extremely hard to find without a computer. You may nonetheless enjoy searching for it.

PROBLEM 5.27—Rigid Square

Raphael M. Robinson, a mathematician at the University of California at Berkeley, is known throughout the world for his solution of a famous minimum problem in set theory. In 1924 Stefan Banach and Alfred Tarski dumbfounded their colleagues by showing that a solid ball can be cut into a finite number of point sets that can then be rearranged (without altering their rigid shape) to make two solid balls each the same size as the original. The minimum number of sets required for the "Banach-Tarski paradox" was not established until 20 years later, when Robinson came up with an elegant proof that it was five. (Four are sufficient if one neglects the single point in the center of the ball!)

Here, on a less significant but more recreational level, is an unusual minima problem devised by Robinson for which the minimum is not yet known. Imagine that you have before you an unlimited supply of rods all the same length. They can be connected only at their ends. A triangle formed by joining three rods will be rigid but a four-rod square will not: it is easily distorted into other shapes without bending or breaking a rod or detaching the ends. The simplest way to brace the square so that it cannot be deformed is to attach eight more rods [see Figure 5.22] to form the rigid skeleton of a regular octahedron.

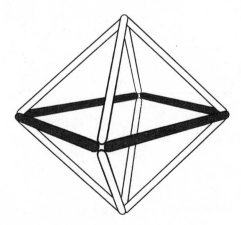

Figure 5.22

Suppose, however, you are confined to the plane. Is there a way to add rods to the square, joining them only at the ends, so that the square is made absolutely rigid? All rods must, of course, lie perfectly flat on the plane. They may not go over or under one another or be bent or broken in any way. The answer is: Yes, the square *can* be made rigid. But what is the smallest number of rods required?

Answers

ANSWER 5.1—(July 1971)

With one unit side as base and the other unit side free to rotate [see Figure 5.23], the triangle's area is greatest when the altitude is maximum. The third side will then be the square root of 2.

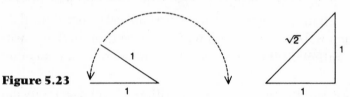

Figure 5.23

ANSWER 5.2—(April 1969)

Figure 5.24 left shows how I answered the question. On the right is a second solution found independently by readers Harry Kemmerer and Gary Rieveschl.

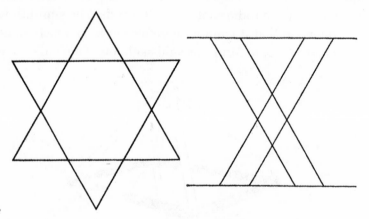

Figure 5.24

ANSWER 5.3—(July 1969)

Two equilateral triangles can be obtained as shown in Figure 5.25.

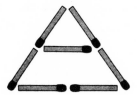

Figure 5.25

ANSWER 5.4—(August 1968)

Three square units, as shown in Figure 5.26.

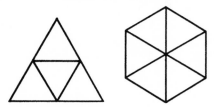

Figure 5.26

ANSWER 5.5—(August 1968)

The trisecting lines also trisect each side of the square. As Piet Hein, who sent me this problem, points out, this is easily seen by dividing any rectangle into halves by drawing the main diagonal from the corner where the trisecting lines originate. Each half of the rectangle obviously must be divided by a trisecting line into two triangles such that the smaller is half the area of the larger. Since the two triangles share a common altitude, this is done by making the base of the smaller triangle half the base of the larger.

ANSWER 5.6—(July 1971)

Instead of inscribing the hexagon as shown, turn it to the position shown in Figure 5.27. The lines divide the larger hexagon into 24 congruent triangles, 18 of which form the smaller hexagon. The ratio of

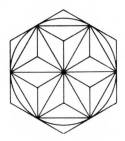

Figure 5.27

Plane Geometry

areas is 18 : 24 = 3 : 4, and so if the smaller hexagon has an area of three, the larger one has an area of four.

ANSWER 5.7—(July 1969)

Four squares can be obtained as shown in Figure 5.28.

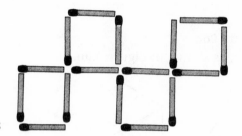

Figure 5.28

ANSWER 5.8—(August 1968)

Any angle can be bisected with a compass and straightedge. By repeated bisections we can divide any angle into 2, 4, 8, 16, . . . equal parts. If any number in this series is a multiple of 3, then repeated bisection obviously would allow trisection of the angle with compass and straightedge. Since this has been proved impossible, no number in the doubling series is evenly divisible by 3.

ANSWER 5.9—(August 1968)

The two semicircles together form a circle that fits the hole. The remaining metal therefore has a total area of four square feet.

ANSWER 5.10—(August 1968)

The triangle's perimeter is 20 units. Lines tangent to a circle from an exterior point are equal, therefore $\overline{YA} = \overline{AP}$ and $\overline{BP} = \overline{XB}$. Since $\overline{AP} + \overline{BP}$ is a side of triangle ABC, it is easy to see that the triangle's perimeter is 10 + 10 = 20. This is one of those curious problems that can be solved in a different way on the assumption that they have answers. Since P can be anywhere on the circle from X to Y, we move P to a limit (either Y or X). In both cases one side of triangle ABC shrinks to zero as side $\overline{AB}$ expands to 10, producing a degenerate straight-line "triangle" with sides of 10, 10, and 0 and a perimeter of 20. (Thanks to Philip C. Smith, Jr.)

ANSWER 5.11—(February 1979)

The solution is shown in Figure 5.29. The puzzle is adapted from an optical illusion in Charles H. Paraquin's *Eye Teasers; Optical Illusion Puzzles* (Sterling Publishing Co., Inc., 1978).

Figure 5.29

ANSWER 5.12—(November 1957)

Line $\overline{AC}$ is one diagonal of the rectangle [Figure 5.30]. The other diagonal is clearly the 10-unit radius of the circle. Since the diagonals are equal, line $\overline{AC}$ is 10 units long.

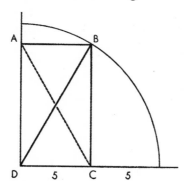

Figure 5.30

ANSWER 5.13—(August 1958)

If you place the point of a compass at the center of a black square on a chessboard with two-inch squares, and extend the arms of the compass a distance equal to the square root of 10 inches, the pencil will trace the largest possible circle that touches only black squares.

ANSWER 5.14—(October 1960)

Figure 5.31 shows how to construct a straight line that bisects both the Yin and the Yang. A simple proof is obtained by drawing the two broken semicircles. Circle K's diameter is half that of the monad; therefore its area is one-fourth that of the monad. Take region G from this circle, add H, and the resulting region is also one-fourth the monad's area. It follows that area G equals area H, and of course half of G must equal half of H. The bisecting line takes half of G away from circle K, but restores the same area (half of H) to the circle, so the black area below the bisecting line must have the same area as circle K. The small circle's area is one-fourth the large circle's area, therefore the Yin is bisected. The same argument applies to the Yang.

The foregoing proof was given by Henry Dudeney in his answer to problem 158, *Amusements in Mathematics*. After it appeared in *Scientific American*, four readers (A. E. Decae, F. J. Hooven, Charles W. Trigg, and B.H.K. Willoughby) sent the following alternative proof, which is much simpler. In Figure 5.31, draw a horizontal diameter of the small circle K. The semicircle below this line has an area that is clearly 1/8 that of the large circle. Above the diameter is a 45-degree sector of the large circle (bounded by the small circle's horizontal diameter and the diagonal line) which also is obviously 1/8 the area of the large circle. Together, the semicircle and sector have an area of 1/4 that of the large circle, therefore the diagonal line must bisect both Yin and Yang. For ways of bisecting the Yin and Yang with curved lines, see Dudeney's problem, cited above, and Trigg's article, "Bisec-

Figure 5.31

tion of Yin and of Yang," in *Mathematics Magazine*, vol. 34, No. 2, November-December 1960, pages 107–8.

The Yin-Yang symbol (called the *T'ai-chi-t'u* in China and the *Tomoye* in Japan) is usually drawn with a small spot of Yin inside the Yang and a small spot of Yang inside the Yin. This symbolizes the fact that the great dualities of life are seldom pure; each usually contains a bit of the other. There is an extensive Oriental literature on the symbol. Sam Loyd, who bases several puzzles on the figure (*Sam Loyd's Cyclopedia of Puzzles*, page 26), calls it the Great Monad. The term "monad" is repeated by Dudeney, and also used by Olin D. Wheeler in a booklet entitled *Wonderland*, published in 1901 by the Northern Pacific Railway. Wheeler's first chapter is devoted to a history of the trademark, and is filled with curious information and color reproductions from Oriental sources.

For more on the symbol, see Schuyler Cammann, "The Magic Square of Three in Old Chinese Philosophy and Religion," *History of Religions*, vol. 1, no. 1, Summer 1961, pages 37–80; my *New Ambidextrous Universe* (Freeman, 1990); and George Sarton, *A History of Science*, vol. 1 (Harvard University Press, 1952), page 11. Carl Gustav Jung cites some English references on the symbol in his introduction to the book of *I Ching* (1929), and there is a book called *The Chinese Monad: Its History and Meaning*, by Wilhelm von Hohenzollern, the date and publisher of which I do not know.

ANSWER 5.15—(November 1965)

Three sets of four cocircular points on the picture of randomly dropped rectangles and disks are shown as black spots in Figure 5.32. The four corners of the rectangle were mentioned in the statement of the problem. The four points on the small circle are obviously cocircular. The third set consists of points A, B, C, D. To see this, draw dotted line $\overline{BD}$ and think of it as the diameter of a circle. Since the angles at A and C are right angles, we know (from a familiar theorem of plane geometry) that A and C must lie on the circle of which $\overline{BD}$ is the diameter.

When this problem was first published, I asked only for three sets of four cocircular points. The problem proved to be better than either its originator or I realized. Many readers were quick to call attention to the fourth set. Its four points are: A, the unmarked intersection immediately below A on the right, B, and the unmarked corner just

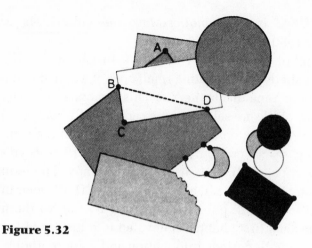

Figure 5.32

above *B*. The line segment joining *B* to the point below *A* is the diameter of a circle on which the other two points lie, since each is the vertex of a right angle subtending the diameter.

The problem, modified to exclude this fourth set of points, appears in Stephen Barr's *Second Miscellany of Puzzles* (Macmillan, 1969).

ANSWER 5.16—(November 1957)

Twelve matches can be used to form a right triangle with sides of three, four, and five units, as shown at left in Figure 5.33. This triangle will have an area of six square units. By altering the position of three matches as shown at right in the illustration, we remove two square units, leaving a polygon with an area of four.

The above solution is the one to be found in many puzzle books. There are hundreds of other solutions. Elton M. Palmer, Oakmont, Pennsylvania, correlated this problem with the polyominoes, pointing out that each of the five tetrominoes (figures made with four squares) can provide the base for a large number of solutions. We simply add and subtract the same amount in triangular areas to accom-

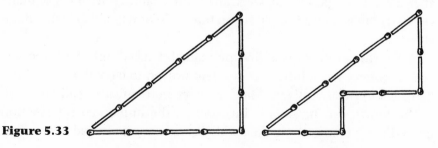

Figure 5.33

GEOMETRIC PUZZLES

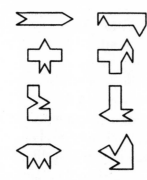

Figure 5.34

modate all 12 matches. Figure 5.34 depicts some representative samples, each row based on a different tetromino.

Eugene J. Putzer, staff scientist with the General Dynamics Corporation; Charles Shapiro, Oswego, New York; and Hugh J. Metz, Oak Ridge, Tennessee, suggested the star solution shown in Figure 5.35. By adjusting the width of the star's points you can produce any desired area between 0 and 11.196, the area of a regular dodecagon, the largest area possible with the 12 matches.

Figure 5.35

ANSWER 5.17—(April 1977)

The ellipse problem was given for a 3, 4, 5 triangle, but we shall solve it for any triangle.

The largest ellipse that can be inscribed in an equilateral triangle is a circle, and the smallest ellipse that can be circumscribed around an equilateral triangle is also a circle. By parallel projection we can transform an equilateral triangle into a triangle of any shape. When that is done, the inscribed and circumscribed circles become noncircular ellipses.

Parallel projection does not alter the ratios of the areas of the triangle and the two closed curves; therefore the ellipses that result will have the maximum and minimum areas for any triangle produced by the projection. In other words, the ratio of the area of the smallest ellipse that can be circumscribed around any triangle to the area of the triangle is the same as the ratio of a circle to an inscribed equilateral triangle. Similarly, the ratio of the area of the largest ellipse that can be inscribed in any triangle to the area of the triangle is the same as the ratio of a circle to a circumscribed equilateral triangle.

It is easy to show that the ratio of the inside circle to the triangle is $\pi/3\sqrt{3}$ and that the ratio of the outside circle to the triangle is four times that number. Applying this result to the 3, 4, 5 triangle, the area of the largest inside ellipse is $2\pi/\sqrt{3}$, and the area of the smallest outside ellipse is $8\pi/\sqrt{3}$.

For a more formal proof see Heinrich Dörrie's *100 Great Problems of Elementary Mathematics* (Dover, 1965), page 378 ff.

ANSWER 5.18—(November 1967)

To solve the problem of the overlapping squares, extend two sides of the large square as shown by the dotted lines in Figure 5.36. This obviously divides the small square into four congruent parts. Since the small square has an area of nine inches, the overlap [shaded] must have an area of 9/4, or 2¼ inches. The amusing thing about the problem is that the area of overlap is constant regardless of the large square's position as it rotates around D. The fact that B trisects $\overline{AC}$ is irrelevant information, designed to mislead.

The problem appears as No. 52 in *Mathematical Challenges: Selected Problems from the Mathematics Student Journal*, edited by Mannis Charosh (National Council of Teachers of Mathematics, 1965). A second solution is given, and the problem is generalized for any regular polygon.

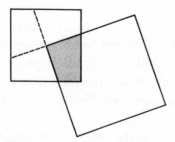

Figure 5.36

ANSWER 5.19—(April 1969)

Three intersecting circles, each passing through the centers of the other two, can be repeated on the plane to form the wallpaper pattern shown in Figure 5.37. Each circle is made up of six delta-shaped figures (labeled D) and 12 "bananas" (labeled B). One-fourth of a circle's area must therefore equal the sum of one-and-a-half deltas plus three bananas. The area common to three mutually intersecting circles [shown shaded in the illustration] consists of three bananas and one delta, and therefore it is smaller than one-fourth of a circle by an amount equal to half a delta. Computation shows that the mutual overlap is a little more than .22 of the circle's area.

Figure 5.37

ANSWER 5.20—(February 1966)

You were asked to prove that if a penny is rolled once around a closed chain of pennies, touching every penny, the number of rotations made by the penny is a constant regardless of the chain's shape. We shall prove this first for a chain of nine pennies.

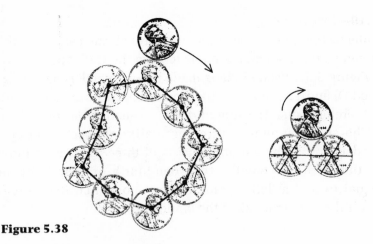

Figure 5.38

Join the centers of the coins by straight lines as shown at the left of Figure 5.38 to form a nine-sided polygon. The total length (in degrees of arc) of the perimeters of the pennies, outside the polygon, is the same as the total of the polygon's conjugate angles. (The conjugate of an angle is the difference between that angle and 360 degrees.) The sum of the conjugate angles of a polygon of n sides is always $\frac{1}{2}n + 1$ perigons (a perigon is an angle of 360 degrees).

As the penny rolls around the chain, however, for every pair of pennies it touches, it fails to touch two arcs of $\frac{1}{6}$ perigon each, which together are $\frac{1}{3}$ perigon [see right of Figure 5.38]. For n pennies, it will fail to touch $\frac{1}{3}n$ perigons. We subtract this form $\frac{1}{2}n + 1$ to obtain $\frac{1}{6}n + 1$ as the total perimeter over which the penny rolls in making one circuit around the chain.

As explained, the penny rotates two degrees for each degree of arc over which it rolls, therefore the penny must make a total of $\frac{1}{3}n + 2$ rotations. This obviously is a constant, regardless of the chain's shape, because the centers of any chain of pennies must mark the vertices of an n-sided polygon. (The formula also applies to a degenerate chain of two pennies, whose centers can be taken as the corners of a degenerate polygon of two sides.)

A similar argument establishes $(\frac{1}{3}n) - 2$ as the number of rotations made by a penny rolling once around the inside of a closed chain of six or more pennies and touching every penny. The formula gives zero rotations for the chain of six, inside which it fits snugly, touching all six coins. For an open chain of n pennies it is easy to show that the rolling penny makes $\frac{1}{3}(2n + 4)$ rotations in a complete circuit.

Several readers pointed out that in experimenting with rolling coin problems, quarters and dimes are preferable to pennies because their milled edges prevent slippage. Here is an intriguing variation to explore. What sort of theorems emerge in calculating the number of rotations of a coin when rolled once around a closed chain of coins, allowing each coin to be of arbitrary size?

ANSWER 5.21—(February 1967)

Five spots can be placed on the figure as shown in Figure 5.39 so that each pair is separated by a distance equal to the square root of 2 or more. There is enough leeway to allow each dot to be shifted slightly and therefore the number of different patterns is infinite. Did you fall into the carefully planned trap of thinking each spot had to fall on a vertex?

This problem appeared earlier in *Eureka*, the journal of the Archimedeans, October 1966, page 19.

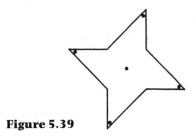

Figure 5.39

ANSWER 5.22—(November 1957)

At any given instant the four bugs form the corners of a square which shrinks and rotates as the bugs move closer together. The path of each pursuer will therefore at all times be perpendicular to the path of the pursued. This tells us that as A, for example, approaches B, there is no component in B's motion which carries B toward or away from A. Consequently A will capture B in the same time that it would take if B had remained stationary. The length of each spiral path will be the same as the side of the square: 10 inches.

If three bugs start from the corners of an equilateral triangle, each bug's motion will have a component of 1/2 (the cosine of a 60-degree angle is 1/2) its velocity that will carry it toward its pursuer. Two bugs will therefore have a mutual approach speed of 3/2 velocity. The bugs meet at the center of the triangle after a time interval equal to twice the side of the triangle divided by three times the velocity, each tracing a path that is 2/3 the length of the triangle's side.

ANSWER 5.23—(February 1970)

There are many ways to prove that angle C in the figure is the sum of angles A and B. Here is one [see Figure 5.40]. Construct the squares as indicated. Angle B equals angle D because they are corresponding angles of similar right triangles. Since angles A and D add to angle C, B can be substituted for D, and it follows immediately that C is the sum of A and B.

This little problem produced a flood of letters from readers who sent dozens of other proofs. Scores of correspondents avoided construction lines by making the diagonals equal to the square roots of 2, 5, and 10, then using ratios to find two similar triangles from which the desired proof would follow. Others generalized the problem in unusual ways.

Charles Trigg published 54 different proofs in the *Journal of Recreational Mathematics*, vol. 4, April 1971, pages 90–99. A proof using paper cutting, by Ali R. Amir-Moéz, appeared in the same journal, vol. 5, Winter 1973, pages 8–9. For other proofs, see Roger North's contribution to *The Mathematical Gazette*, December 1973, pages 334–36, and its continuation in the same journal, October 1974, pages 212–15. For a generalization of the problem to a row of n squares, see Trigg's "Geometrical Proof of a Result of Lehmer's," in *The Fibonacci Quarterly*, vol. 11, December 1973, pages 539–40.

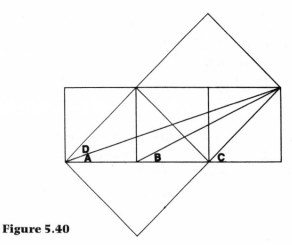

Figure 5.40

Answer 5.24—(April 1974)

Draw three line segments connecting O to each of the centers, X, Y and Z, of the three circles. Add six line segments to connect each center to the two nearest of the intersection points, A, B, and C. The nine line segments are shown as solid lines in Figure 5.41. Each line segment is a unit in length; therefore the lines form three rhombuses. Now through each of the intersection points A, B, and C draw another line [dotted], making each line parallel to a radius line segment of one of the circles. This forms three additional rhombuses. Because opposite sides of parallelograms are equal, we know that the three dotted line segments are equal and that each is one unit in length. Consequently they meet at a point. Q, that is the center of a circle with a radius of one unit. The intersection points A, B, and C lie on this circle, which is the assertion we were asked to prove.

Many readers sent other ways of proving the theorem. For discussions of the problem see George Polya, *Mathematical Discovery*, vol. 2, 1965, pages 53–58; and Ross Honsberger, *Mathematical Gems II*, the Mathematical Association of America, 1976, page 18.

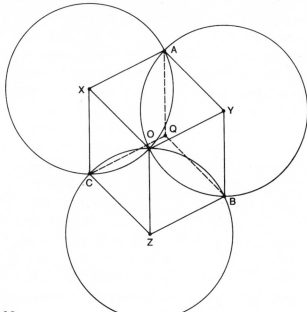

Figure 5.41

ANSWER 5.25—(May 1975)

John Edson Sweet's solution to the three-circle theorem is given in the answer to Problem 62 in L. A. Graham's *Ingenious Mathematical Problems and Methods* (Dover, 1959). Instead of circles, suppose you are looking down on three unequal spheres. The tangent lines for each pair of balls are the edges of three cones into which the two balls fit snugly. The cones rest on the plane that supports the balls, and the apexes of the cones therefore lie on the plane.

Now imagine that a flat plate is placed on top of all three balls. Its underside is a second plane, tangent to all three balls, and tangent to all three cones. This second plane also will contain the three apexes of the cones. Because the apexes lie on both planes, they must lie on the intersection of the two planes, which of course is a straight line.

C. Stanley Ogilvy wrote to say he had included the three-circle problem in his book *Excursions in Geometry* (Oxford University Press, 1969). He too assumed that the proof by way of the three cones took care of all cases; but one day, after giving the problem to his class at Hamilton College, a student pointed out that the proof does not apply when a small sphere is between two larger ones. In such cases it is not possible for the two intersecting planes to be mutually tangent to all three spheres.

Many readers sent in other ways of proving the theorem. Bernard F. Burke, Richard I. Felver, Clyde E. Holvenstot, David B. Shear, and Radu Vero each proposed turning the drawing so that the line (on which the intersections of the tangent pairs lie) is horizontal and above the circles. The circles can now be viewed as being equal spheres inside three mutually intersecting pipes of identical circular cross section, seen in perspective. The tangent lines become the parallel sides of the three pipes. Since the pipes all must rest on a plane, their parallel sides seen in perspective will all have vanishing points on the horizon line.

It is not necessary that the circles be nonintersecting; indeed, the theorem can be stated in a more general way, in terms of "centers of similitude" instead of tangents, to hold for circles that lie entirely within one another. I am indebted to Donald Keeler for explaining this, as well as for pointing out that the theorem is known as "Monge's theorem" after the French mathematician and friend of Napoleon, Gaspard Monge, who gave it in a 1798 treatise. R. C. Archibald, in *The American Mathematical Monthly* (vol. 22, 1915, p. 65) traced the theorem back to the ancient Greeks (writes Keeler).

Daniel Sleator found that the theorem has an analog with four spheres in space. Each of the four triplets will have the apexes of its three cones on a straight line. Because these four lines intersect each other at six points, the four lines must be coplanar. Therefore, the vertexes of the six cones all lie on a plane. The theorem generalizes to all higher Euclidian dimensions. (*See* "Monge's Theorem in Many Dimensions," by Richard Walker, in *The Mathematical Gazette*, vol. 60, October 1976, pp. 185–88.)

Monge's theorem, for three circles on the plane, is mentioned in Herbert Spencer's autobiography. It is, writes Spencer, "a truth which I never contemplate without being struck by its beauty at the same time that it excites feelings of wonder and of awe: the fact that apparently unrelated circles should in every case be held together by this plexus of relations, seeming so utterly incomprehensible."

ANSWER 5.26—(April 1977)

Figure 5.42 shows the only way (not counting rotations and reflections) to place seven counters on an order-7 matrix so that all distances between pairs of counters are different.

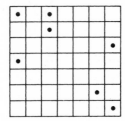

Figure 5.42

Paul Erdös called my attention to a paper he wrote with Richard Guy, "Distinct Distances Between Lattice Points," *Elemente der Mathematik* (vol. 25, 1970, pages 121–33), in which they give the solution for the order-7 matrix and prove that no solutions exist for higher orders. They raise several related open questions, such as: What is the minimum number of lattice points that give distinct distances and are placed so that no point can be added without duplicating a distance?

When points are not confined to lattice points, we have a famous unsolved problem in combinatorial geometry. What is the smallest number of distinct distances determined by n points on the Euclidean plane? Fan Chung, in her paper "The Number of Different Distances Determined by n Points in the plane," *Journal of Combinatorial The-*

ory (series A, vol. 36, 1984, pages 342–54), cites earlier references going back to Erdös's proposal of the problem in 1946. A note added in proof gives the best-known lower bound as $n^{4/5}$. The bound is discussed in a paper by Chung, E. Szemerédi, and W. T. Trotter, "On the Number of Different Distances," *Discrete and Computational Geometry,* vol. 7, 1992, pages 1–11.

ANSWER 5.27—(November 1963)

Raphael M. Robinson's best solution to his problem of bracing a square on the plane with the minimum number of rods, all equal in length to the square's side, calls for 31 rods in addition to the 4 used for the square. Figure 5.43 shows two of several equally good patterns.

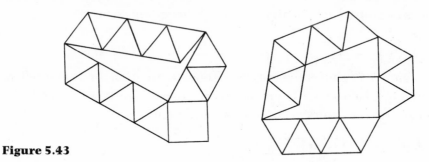

Figure 5.43

This answer was reduced to 25 rods [see Figure 5.44] by 57 *Scientific American* readers. As I was recovering from the shock of this elegant improvement seven readers—G. C. Baker, Joseph H. Engel, Kenneth J. Fawcett, Richard Jenney, Frederick R. Kling, Bernard M. Schwartz, and Glenwood Weinert—staggered me with the 23-rod solution shown in Figure 5.45. Later, about a dozen more readers sent

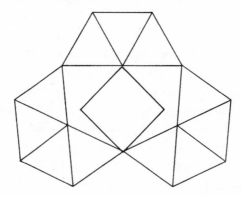

Figure 5.44

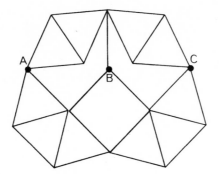

Figure 5.45

the same solution. The rigidity of the structure becomes apparent when one realizes that points *A*, *B*, and *C* must be collinear.

All solutions with fewer than 23 rods proved to be either nonrigid or geometrically inexact. For example, many readers sent the pattern shown in Figure 5.46, left, which is not rigid, or the pattern shown at the right, which, although rigid, unfortunately includes line A, a trifle longer than one unit.

The problem obviously can be extended to other regular polygons. The hexagon is simply solved with internal braces (how many?), but the pentagon is a tough one. T. H. O'Beirne managed to rigidify a regular pentagon with 64 additional rods, but it is not known if this number is minimal.

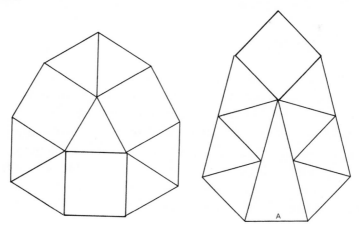

Figure 5.46

chapter 6

Solid Geometry

The problems in this chapter involve cubes, spheres, and polyhedra as objects to be played with. They do not require much mathematical knowledge beyond, say, the volume of a sphere. Two well-known problems are found here: Archimede's calculation of the volume of the intersection of two cylinders and the "hole in the sphere."

The last, and hardest, problem originally appeared as a "quickie." Gardner devised this problem and was startled when readers discovered that it was far subtler than he ever imagined. His readership provided a series of ever-improving solutions; just coming close to the right answer should be reason for celebration. Of course, many of the results that Gardner quoted over the years were improved upon by his industrious subscribers.

Problems

PROBLEM 6.1—Unstable Polyhedra

Each face of a convex polyhedron can serve as a base when the solid is placed on a horizontal plane. The center of gravity of a regular polyhedron is at the center, therefore it is stable on any face. Irregular polyhedrons are easily constructed that are unstable on certain faces; that is, when placed on a table with an unstable face as the base, they topple over. Is it possible to make a model of an irregular convex polyhedron that is unstable on *every* face?

PROBLEM 6.2—Rigid Cube

You want to construct a rigid wire skeleton of a one-inch cube by using 12 one-inch wire segments for the cube's 12 edges. These you intend to solder together at the cube's eight corners.

"Why not cut down the number of soldering points," a friend suggests, "by using one or more longer wires that you can bend at sharp right angles at various corners?"

Adopting your friend's suggestion, what is the smallest number of corners where soldering will be necessary to make the cube's skeleton rigid? (Thanks to Philip C. Smith, Jr.)

PROBLEM 6.3—Coplanar Points

Consider these three points: the center of a regular tetrahedron and any two of its corner points. The three points are coplanar (lie on the same plane). Is this also true of all irregular tetrahedrons?

PROBLEM 6.4—Bricks and Cubes

Can a 6 × 6 × 6 cube be made with 27 bricks that are each 1 × 2 × 4 units [see Figure 6.1]?

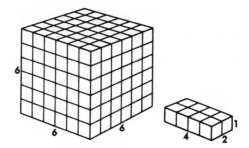

Figure 6.1

PROBLEM 6.5—Five-Match Cube

How can you make a cube with five paper matches? No bending or splitting of matches is allowed.

PROBLEM 6.6—Returning Explorer

An old riddle runs as follows. An explorer walks one mile due south, turns and walks one mile due east, turns again and walks one mile due north. He finds himself back where he started. He shoots a bear. What color is the bear? The time-honored answer is: "White," because the explorer must have started at the North Pole. But not long ago someone made the discovery that the North Pole is not the only starting point that satisfies the given conditions! Can you think of any other spot on the globe from which one could walk a mile south, a mile east, a mile north, and find himself back at his original location?

Solid Geometry

PROBLEM 6.7—Cutting the Cube

A carpenter, working with a buzz saw, wishes to cut a wooden cube, three inches on a side, into 27 one-inch cubes. He can do this easily by making six cuts through the cube, keeping the pieces together in the cube shape [see Figure 6.2]. Can he reduce the number of necessary cuts by rearranging the pieces after each cut?

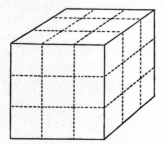

Figure 6.2

PROBLEM 6.8—Cork Plug

Many old puzzle books explain how a cork can be carved to fit snugly into square, circular, and triangular holes [Figure 6.3]. An inter-

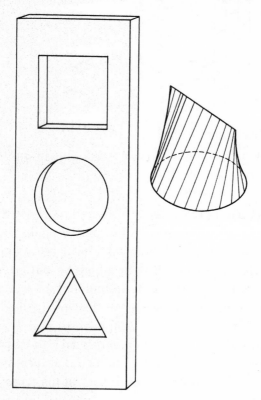

Figure 6.3

esting problem is to find the volume of the cork plug. Assume that it has a circular base with a radius of one unit, a height of two units, and a straight top edge of two units that is directly above and parallel to a diameter of the base. The surface is such that all vertical cross sections which are made perpendicular to the top edge are triangles.

The surface may also be thought of as generated by a straight line connecting the sharp edge with the circular edge and moving so that it is at all times parallel to a plane that is perpendicular to the sharp edge. The plug's volume can of course be determined by calculus, but there is a simple way to find it with little more information than knowing that the volume of a right circular cylinder is the area of its base times its altitude.

PROBLEM 6.9—How Long Is a "Lunar"?

In H. G. Wells's novel *The First Men in the Moon* our natural satellite is found to be inhabited by intelligent insect creatures who live in caverns below the surface. These creatures, let us assume, have a unit of distance that we shall call a "lunar." It was adopted because the moon's surface area, if expressed in square lunars, exactly equals the moon's volume in cubic lunars. The moon's diameter is 2,160 miles. How many miles long is a lunar?

PROBLEM 6.10—Sliced Cube and Sliced Doughnut

An engineer noted for his ability to visualize three-dimensional structure, was having coffee and doughnuts. Before he dropped a sugar cube into his cup, he placed the cube on the table and thought: If I pass a horizontal plane through the cube's center, the cross section will of course be a square. If I pass it vertically through the center and four corners of the cube, the cross section will be an oblong rectangle. Now suppose I cut the cube this way with the plane. . . . To his surprise, his mental image of the cross section was a regular hexagon.

How was the slice made? If the cube's side is half an inch, what is the side of the hexagon?

After dropping the cube into his coffee, the engineer turned his attention to a doughnut lying flat on a plate. "If I pass a plane horizontally through the center," he said to himself, "the cross section will be two concentric circles. If I pass the plane vertically through the center, the section will be two circles separated by the width of the hole.

But if I turn the plane so. . . ." He whistled with astonishment. The section consisted of two perfect circles that intersected!

How was this slice made? If the doughnut is a perfect torus, three inches in outside diameter and with a hole one inch across, what are the diameters of the intersecting circles?

PROBLEM 6.11—Find the Hexahedrons

A polyhedron is a solid bounded by plane polygons known as the faces of the solid. The simplest polyhedron is the tetrahedron, consisting of four faces, each a triangle [Figure 6.4 top]. A tetrahedron can have an endless variety of shapes, but if we regard its network of edges as a topological invariant (that is, we may alter the length of any edge and the angles at which edges meet but we must preserve the structure of the network), then there is only one basic type of tetrahedron. It is not possible, in other words, for a tetrahedron to have sides that are anything but triangles.

The five-sided polyhedron has two basic varieties [Figure 6.4 middle and bottom]. One is represented by the Great Pyramid of Egypt (four triangles on a quadrilateral base). The other is represented by a tetrahedron with one corner sliced off; three of its faces are quadrilaterals, two are triangles.

John McClellan, an artist in Woodstock, New York, asks this question: How many basic varieties of convex hexahedrons, or six-sided solids, are there altogether? (A solid is convex if each of its sides can

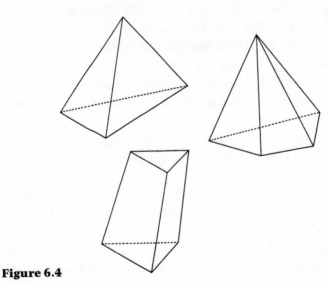

Figure 6.4

be placed flat against a table top.) The cube is, of course, the most familiar example.

If you search for hexahedrons by chopping corners from simpler solids, you must be careful to avoid duplication. For example, the Great Pyramid, with its apex sliced off, has a skeleton that is topologically equivalent to that of the cube. Be careful also to avoid models that cannot exist without warped faces.

PROBLEM 6.12—Papered Cube

What is the largest cube that can be completely covered on all six sides by folding around it a pattern cut from a square sheet of paper with a side of three inches? (The pattern must, of course, be all in one piece.)

PROBLEM 6.13—Crossed Cylinders

One of Archimedes' greatest achievements was his anticipation of some of the fundamental ideas of calculus. The problem illustrated in Figure 6.5 is a classic example of a problem that most mathematicians today would regard as unsolvable without a knowledge of calculus (indeed, it is found in many calculus textbooks) but that yielded readily to Archimedes' ingenious methods. The two circular cylinders intersect at right angles. If each cylinder has a radius of one unit, what is the volume of the shaded solid figure that is common to both cylinders?

No surviving record shows exactly how Archimedes solved this problem. There is, however, a startlingly simple way to obtain the answer; indeed, one need know little more than the formula for the area of a circle (π times the square of the radius) and the formula for the volume of a sphere (four-thirds π times the cube of the radius). It

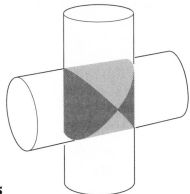

Figure 6.5

Solid Geometry

may have been the method Archimedes used. In any case, it has become a famous illustration of how calculus often can be side-stepped by finding a simple approach to a problem.

PROBLEM 6.14—Hole in the Sphere

This incredible problem—incredible because it seems to lack sufficient data for a solution—appeared in an issue of *The Graham Dial*, a publication of Graham Transmissions Inc. A cylindrical hole six inches long has been drilled straight through the center of a solid sphere. What is the volume remaining in the sphere?

PROBLEM 6.15—Mirror-Symmetric Solids

On plane figures an axis of symmetry is a straight line that divides the figure into congruent halves that are mirror images of each other. The hearts on playing cards, for example, have one axis of symmetry. So do spades and clubs, but the diamond has two such axes. A square has four axes of symmetry. A regular five-pointed star has five, and a circle has an infinite number. A swastika or a yin-yang symbol has no axis of symmetry.

If a plane figure has at least one axis of symmetry, it is said to be superposable on its mirror image in the following sense. If you view the figure in a vertical mirror, with the edge of the mirror resting on a horizontal plane, you can imagine sliding the figure into the mirror and, if necessary, rotating it on the plane so that it coincides with its mirror reflection. You are not allowed to turn the figure over on the plane because that would require rotating it through a third dimension.

A plane of symmetry is a plane that slices a solid figure into congruent halves, one half a mirror reflection of the other. A coffee cup has a single plane of symmetry. The Great Pyramid of Egypt has four such planes. A cube has nine: three are parallel to a pair of opposite faces and six pass through corresponding diagonals of opposite faces. A cylinder and a sphere each have an infinite number of planes of symmetry.

Think of a solid object bisected by a plane of symmetry. If you place either half against a mirror, with the sliced cross section pressing against the glass, the mirror reflection, together with the bisected half, will restore the shape of the original solid. Any solid with at least one plane of symmetry can be superposed on its mirror image by making, if necessary, a suitable rotation in space.

Discussing this in my book The *Ambidextrous Universe* (Charles Scribner's Sons, 1979), I stated on page 19 that if a three-dimensional object had no plane of symmetry (such as a helix, a Möbius strip, or an overhand knot in a closed loop of rope), it could not be superposed on its mirror image without imagining its making an impossible rotation that "turned it over" through a fourth dimension.

The statement is false! As many readers of the book pointed out, there are solid figures totally lacking a plane of symmetry that nonetheless can, by a suitable rotation in ordinary space, be superposed on their mirror images. In fact, one is so simple that you can fold it in a trice from a square sheet of paper. How is it done?

PROBLEM 6.16—Rectangle and the Oil Well

An oil well being drilled in flat prairie country struck pay sand at an underground spot exactly 21,000 feet from one corner of a rectangular plot of farmland, 18,000 feet from the opposite corner, and 6,000 feet from a third corner. How far is the underground spot from the fourth corner? By solving the problem you will discover a useful formula of great generality and delightful simplicity.

PROBLEM 6.17—Touching Cubes

You have one red cube and a supply of white cubes all the same size as the red one. What is the largest number of white cubes that can be placed so that they all abut the red cube, that is, a positive-area portion of a face of each white cube is pressed flat against a positive-area portion of a face of the red cube. Touching at corner points or along edges does not count.

Answers

ANSWER 6.1—(August 1968)

No. If a convex polyhedron were unstable on every face, a perpetual motion machine could be built. Each time the solid toppled over to a new base it would be unstable and would topple over again.

ANSWER 6.2—(July 1971)

The smallest number of soldering points remains eight no matter how wires are bent. Because an odd number of edges meet at each corner of a cube, every point requires soldering.

ANSWER 6.3—(August 1968)

Yes. *Any* three points in space are coplanar.

ANSWER 6.4—(August 1968)

No. Think of the order-6 cube as made up of 27 cubes with sides of two units and alternately colored black and white. Since 27 is an odd number there will be 13 cubes of one color and 14 of another. No matter how a brick is placed within this cube, half of its unit cubes will be black and half white, so that if the cube can be formed, it must contain as many black unit cubes as white. This contradicts the fact that the large cube has more unit cubes of one color than of the other, therefore there is no way to build the order-6 cube with the 27 bricks.

ANSWER 6.5—(August 1968)

If "cube" is taken in the numerical sense, the five matches can be used to form 1 or 27, or VIII, or I with an exponent of 3. If the bottoms of the matches are straight, the arrangement shown in Figure 6.6 produces a tiny cube at the center.

Figure 6.6

ANSWER 6.6—(February 1957)

Is there any other point on the globe, besides the North Pole, from which you could walk a mile south, a mile east, and a mile north and find yourself back at the starting point? Yes indeed; not just one point but an infinite number of them! You could start from any point on a circle drawn around the South Pole at a distance slightly more than 1 + 1/2π miles (about 1.16 miles) from the Pole—the distance is "slightly more" to take into account the curvature of the earth. After walking a mile south, your next walk of one mile east will take you on a complete circle around the Pole, and the walk one mile north from there will then return you to the starting point. Thus your starting point could be any one of the infinite number of points on the circle with a radius of about 1.16 miles from the South Pole. But this is not all. You could also start at points closer to the Pole, so that the walk east would carry you just twice around the Pole, or three times, and so on.

ANSWER 6.7—(February 1957)

There is no way to reduce the cuts to fewer than six. This is at once apparent when you focus on the fact that a cube has six sides. The saw cuts straight—one side at a time. To cut the one-inch cube at the center (the one which has no exposed surfaces to start with) must take six passes of the saw.

This problem was originated by Frank Hawthorne, supervisor of mathematics education, State Department of Education, Albany, New York, and first published in *Mathematics Magazine*, September–October 1950 (Problem Q-12).

Cubes of 2 × 2 × 2 and 3 × 3 × 3 are unique in the sense that regardless of how the pieces are rearranged before each cut (provided each piece is cut somewhere), the former will always require three cuts and the latter six to slice into unit cubes.

The 4 × 4 × 4 cube requires nine cuts if the pieces are kept together as a cube, but by proper piling before each cut, the number of cuts can be reduced to six. If at each piling you see that every piece is cut as nearly in half as possible, the minimum number of cuts will be achieved. In general, for an $n \times n \times n$ cube, the minimum number of cuts is $3k$ where k is defined by

$$2^k \geq n > 2^{k-1}$$

This general problem was posed by L. R. Ford, Jr., and D. R. Fulkerson, both of The Rand Corporation, in the *American Mathematical Monthly*, August–September 1957 (Problem E1279), and answered in the March 1958 issue. The problem is a special case of a more general problem (the minimum cuts for slicing an $a \times b \times c$ block into unit cubes) contributed by Leo Moser, of the University of Alberta, to *Mathematics Magazine*, vol. 25, March–April 1952, p. 219.

Eugene J. Putzer and R. W. Lowen generalized the problem still further in a research memorandum, "On the Optimum Method of Cutting a Rectangular Box into Unit Cubes," issued in 1958 by Convair Scientific Research Laboratory, San Diego. The authors considered blocks of *n* dimensions, with integral sides, which are to be sliced by a minimum number of planar cuts into unit hypercubes. In three dimensions the problem is one which the authors feel might "have important applications in the cheese and sugarloaf industries."

ANSWER 6.8—(August 1958)

Any vertical cross section of the cork plug at right angles to the top edge and perpendicular to the base will be a triangle. If the cork were a cylinder of the same height, corresponding cross sections would be rectangles. Each triangular cross section is obviously one-half the area of the corresponding rectangular cross section. Since all the triangular sections combine to make up the cylinder, the plug must be one-half the volume of the cylinder. The cylinder's volume is 2π, so our answer is simply π. (This solution is given in "No Calculus, Please," by J. H. Butchart and Leo Moser in *Scripta Mathematica*, September-December 1952.)

Actually, the cork can have an infinite number of shapes and still fit the three holes. The shape described in the problem has the least volume of any convex solid that will fit the holes. The largest volume is obtained by the simple procedure of slicing the cylinder with two plane cuts as shown in Figure 6.7. This is the shape given in most puzzle books that include the plug problem. Its volume is equal to $2\pi - 8/3$ (I am indebted to J. S. Robertson, East Setauket, New York, for sending this calculation.)

Figure 6.7

ANSWER 6.9—(February 1960)

The volume of a sphere is $4\pi/3$ times the cube of the radius. Its surface is 4π times the square of the radius. If we express the moon's radius in "lunars" and assume that its surface in square lunars equals its volume in cubic lunars, we can determine the length of the radius simply by equating the two formulas and solving for the value of the radius. Pi cancels out on both sides, and we find that the radius is three lunars. The moon's radius is 1,080 miles, so a lunar must be 360 miles.

ANSWER 6.10—(October 1960)

A cube, cut in half by a plane that passes through the midpoints of six sides as shown in Figure 6.8, produces a cross section that is a regular hexagon. If the cube is half an inch on the side, the side of the hexagon is $\sqrt{2}/4$ inch.

To cut a torus so that the cross section consists of two intersecting circles, the plane must pass through the center and be tangent to the torus above and below, as shown in Figure 6.9. If the torus and hole have diameters of three inches and one inch, each circle of the section will clearly have a diameter of two inches.

This way of slicing, and the two ways described earlier, are the only ways to slice a doughnut so that the cross sections are circular. Everett A. Emerson sent a full algebraic proof that there is no fourth way.

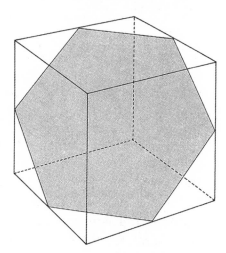

Figure 6.8

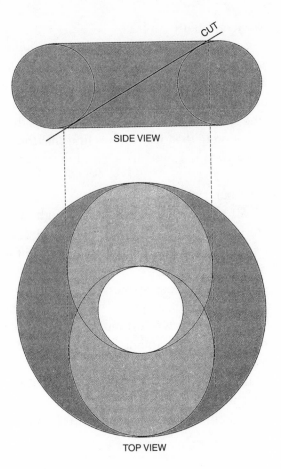

SIDE VIEW

TOP VIEW

Figure 6.9

ANSWER 6.11—(June 1961)

The seven varieties of convex hexahedrons, with topologically distinct skeletons, are shown in Figure 6.10. I know of no simple way to prove that there are no others. An informal proof is given by John McClellan in his article on "The Hexahedra Problem," *Recreational Mathematics Magazine*, no. 4, August 1961, pages 34–40.

There are 34 topologically distinct convex heptahedra, 257 octahedra, and 2,606 9-hedra. The three nonconvex (concave or re-entrant) hexahedra are shown in Figure 6.11. There are 26 nonconvex heptahedra and 277 nonconvex octahedral. See the following papers by P. J. Federico: "Enumeration of Polyhedra: The Number of 9-hedra," *Journal of Combinatorial Theory*, vol. 7, September 1969, pages 155–61; "Polyhedra with 4 to 8 Faces," *Geometria Dedicata*, vol. 3, 1975,

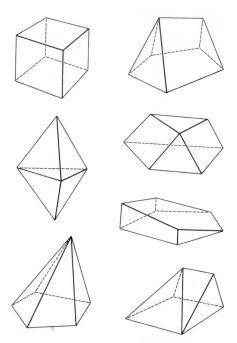

Figure 6.10

pages 469–81; and "The Number of Polyhedra," *Philips Research Reports*, vol. 30, 1975, pages 220–31.

A formula for calculating the number of topologically distinct convex polyhedra, given the number of faces, remains unfound.

Paul R. Burnett called my attention to the Old Testament verse, Zechariah 3:9. In a modern translation by J. M. Powis Smith it reads:

"For behold the stone I have set before Joshua; upon a single stone with seven facets I will engrave its inscription."

An outline of a formal proof that there are just seven distinct convex hexahedra is given in "Euler's Formula for Polyhedra and Related Topics," by Donald Crowe, in *Excursions into Mathematics*, by Anatole Beck, Michael Bleicher, and Donald Crowe (Worth, 1969, pages 29–30).

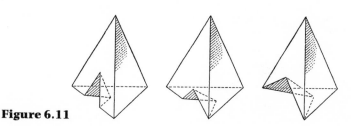

Figure 6.11

ANSWER 6.12—(November 1967)

If overlapping of paper is not allowed, the largest cube that can be folded from a pattern cut from a square sheet of paper with a three-inch side is a cube with a side that is three-fourths of the square root of 2. The pattern, shown in Figure 6.12, is folded along dotted lines.

In stating the problem, however, I did not forbid overlapping. This was not an oversight. It simply did not occur to me that overlapping would permit better solutions than the one above, which was given as the problem's only answer. John H. Halton, a mathematician at the University of Wisconsin, was the first to send a cut-and-fold technique by which one can approach as closely as one wishes to the ultimate-size cube with a surface area equal to the area of the square sheet! (Three readers, David Elwell, James F. Scudder, and Siegfried Spira, each came close to such a discovery by finding ways to cover a cube larger than the one given in the answer, and George D. Parker found a complete solution that was essentially the same as Halton's.)

Halton's technique involves cutting the square so that opposite sides of the cube are covered with solid squares and the rest of the cube is wrapped with a ribbon folded into one straight strip so that the amount of overlap can be as small as one wishes. In wrapping, the overlap can again be made as small as one pleases simply by reducing the width of the ribbon. Assuming infinite patience and paper of infinitely small molecular dimensions, as Halton put it, this procedure will cover a cube approaching as closely as desired to the limit, a cube of side $\sqrt{3}/\sqrt{2}$.

Fitch Cheney, a mathematician at the University of Hartford, found another way to do the same thing by extending the pattern in Figure 6.12. By enlarging the center square as shown in Figure 6.13, he makes the four surrounding squares rectangles and the four corner triangles correspondingly larger. The shaded areas, cut as shown, can be folded as indicated to extend each corner triangle. (*A* turns over once, *B* three

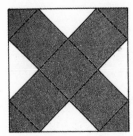

Figure 6.12

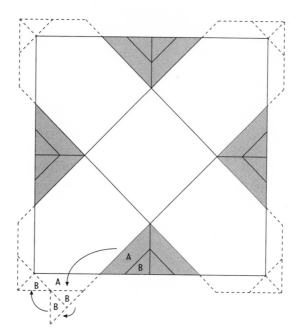

Figure 6.13

times.) The result: a pattern exactly like the one given, except that it is larger. Since the unavoidable overlap can be made as small as one wants, it is clear that this method also approaches the $\sqrt{3}/\sqrt{2}$ cube as a limit.

ANSWER 6.13—(October 1962)

Two circular cylinders of unit radius intersect at right angles. What is the volume common to both cylinders? The problem is solved easily, without the use of calculus, by the following elegant method:

Imagine a sphere of unit radius inside the volume common to the two cylinders and having as its center the point where the axes of the cylinders intersect. Suppose that the cylinders and sphere are sliced in half by a plane through the sphere's center and both axes of the cylinders [at left of Figure 6.14]. The cross section of the volume common to the cylinders will be a square. The cross section of the sphere will be a circle that fills the square.

Now suppose that the cylinders and sphere are sliced by a plane that is parallel to the previous one but that shaves off only a small portion of each cylinder [at right of the illustration]. This will produce parallel tracks on each cylinder, which intersect as before to form a square cross section of the volume common to both cylinders.

Solid Geometry

155

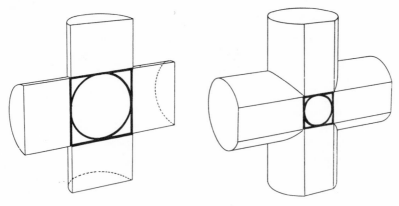

Figure 6.14

Also as before, the cross section of the sphere will be a circle inside the square. It is not hard to see (with a little imagination and pencil doodling) that any plane section through the cylinders, parallel to the cylinders' axes, will always have the same result: a square cross section of the volume common to the cylinders, enclosing a circular cross section of the sphere.

Think of all these plane sections as being packed together like the leaves of a book. Clearly, the volume of the sphere will be the sum of all circular cross sections, and the volume of the solid common to both cylinders will be the sum of all the square cross sections. We conclude, therefore, that the ratio of the volume of the sphere to the volume of the solid common to the cylinders is the same as the ratio of the area of a circle to the area of a circumscribed square. A brief calculation shows that the latter ratio is $\pi/4$. This allows the following equation, in which x is the volume we seek:

$$\frac{4\pi r^3/3}{x} = \frac{\pi}{4}$$

The π's drop out, giving x a value of $16r^3/3$. The radius in this case is 1, so the volume common to both cylinders is 16/3. As Archimedes pointed out, it is exactly 2/3 the volume of a cube that encloses the sphere; that is, a cube with an edge equal to the diameter of each cylinder.

A number of readers pointed out that this solution makes use of what is called "Cavalieri's theorem," after Bonaventura Cavalieri, a seventeenth-century Italian mathematician. "This theorem in its sim-

GEOMETRIC PUZZLES

plest form," wrote Fremont Reizman, "says that two solids are equal in volume if they have equal altitudes and equal cross sections at equal heights above the base. But to prove it, Cavalieri had to anticipate calculus a bit by building his figures up from a stack of laminae and passing to the limit." The principle was known to Archimedes. In a lost book called *The Method* that was not found until 1906 (it is the book in which Archimedes gives the answer to the crossed-cylinders problem), he attributes the principle to Democritus, who used it for obtaining the formula for the volume of a pyramid or cone.

Several readers solved the problem by applying Cavalieri's theorem in a slightly different way. Granville Perkins, for instance, did it by circumscribing (around the solid common to both cylinders) a cube. Using the two faces parallel to both axes as bases, he constructed two pyramids with apices at the cube's center. By slicing laminae parallel to these bases the problem is easily solved.

For those of you who care to tackle the problem of finding the volume of the solid common to *three* orthogonally intersecting cylinders of unit radius, I give only the solution: $8 (2 - \sqrt{2})$. A calculus solution is given in S. I. Jones, *Mathematical Nuts* (privately printed in Tennessee, 1932), pages 83 and 287. Discussions of the two-cylinder case can be found in J. H. Butchart and Leo Moser, "No Calculus Please," *Scripta Mathematica*, vol. 18, September 1952, pages 221–36, and Richard M. Sutton, "The 'Steinmetz Problem' and School Arithmetic," *Mathematics Teacher*, vol. 50, October 1957, pages 434–35.

ANSWER 6.14—(November 1957)

Without resorting to calculus, the problem can be solved as follows. Let R be the radius of the sphere. As Figure 6.15 indicates, the radius of the cylindrical hole will then be the square root of $R^2 - 9$, and the altitude of the spherical caps at each end of the cylinder will be $R - 3$. To determine the residue after the cylinder and caps have been removed, we add the volume of the cylinder, $6\pi (R^2 - 9)$, to twice the volume of the spherical cap, and subtract the total from the volume of the sphere, $4\pi R^3/3$. The volume of the cap is obtained by the following formula, in which A stands for its altitude and r for its radius: $\pi A (3r^2 + A^2)/6$.

When this computation is made, all terms obligingly cancel out except 36π—the volume of the residue in cubic inches. In other

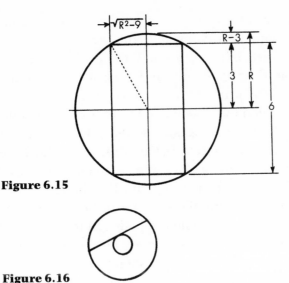

Figure 6.15

Figure 6.16

words, the residue is constant regardless of the hole's diameter or the size of the sphere!

The earliest reference I have found for this beautiful problem is on page 86 of Samuel I. Jones's *Mathematical Nuts*, 1932. A two-dimensional analog of the problem appears on page 93 of the same volume. Given the longest possible straight line that can be drawn on a circular track of any dimensions [see Figure 6.16], the area of the track will equal the area of a circle having the straight line as a diameter.

John W. Campbell, Jr., editor of *Astounding Science Fiction*, was one of several readers who solved the sphere problem quickly by reasoning adroitly as follows: The problem would not be given unless it has a unique solution. If it has a unique solution, the volume must be a constant which would hold even when the hole is reduced to zero radius. Therefore the residue must equal the volume of a sphere with a diameter of six inches, namely 36π.

ANSWER 6.15—(April 1981)

Figure 6.17 shows how to fold a square of paper into a shape that has no plane of symmetry yet can be superposed on its mirror reflection. The figure is said to have "mirror-rotation symmetry," a type of symmetry of great importance in crystallography. I took this example from page 42 of *Symmetry in Science and Art*, by A. V. Shubnikov and V. A. Koptsik (Plenum Press, 1974).

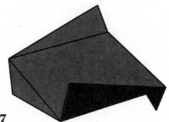

Figure 6.17

Paul Schwink of Carlisle, Iowa, and Piet Hein of Copenhagen pointed out that the solution is unnecessarily complicated. The same result can be obtained by folding just one pair of opposite sides, one up, one down.

ANSWER 6.16—(June 1964)

Consider first a spot p on the surface inside the rectangle shown at the left of Figure 6.18. Adding two broken coordinate lines provides a set of right triangles. Because $e^2 = a^2 + c^2$ and $g^2 = b^2 + d^2$, we can write the equality

$$e^2 + g^2 = a^2 + c^2 + b^2 + d^2$$

And since $f^2 = a^2 + d^2$ and $h^2 = b^2 + c^2$, we can write

$$f^2 + h^2 = a^2 + d^2 + b^2 + c^2$$

The right sides of both equations are the same, therefore,

$$e^2 + g^2 = f^2 + h^2$$

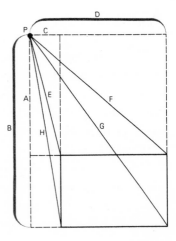

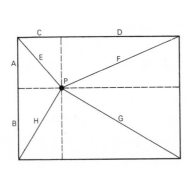

Figure 6.18

Solid Geometry

159

Exactly the same analysis applies to the right diagram, in which point p is outside the rectangle. If you think of p in either diagram as belowground, this will lengthen certain sides of the right triangles involved, but the relations expressed by the equations remain unchanged. In other words, regardless of where point p is located in space—above, below, or even on the edge or corner of the rectangle itself—the sum of the squares of its distances from two opposite corners of the rectangle will equal the sum of the squares of its distances from the other two corners. Applying this simple formula to the three distances given yields 27,000 as the fourth distance. The sides of the rectangle are not, of course, determined by the given data.

ANSWER 6.17—(July 1971)

Twenty white cubes can abut the gray cube. Arrange seven white cubes as shown in Figure 6.19. The gray cube goes on top as indicated. Seven more white cubes, in the same pattern and position as the first layer, form layer 3. Between layer 1 and layer 3 six more white cubes can be placed: two on each of two opposite sides of the gray cube and single cubes on the remaining two sides.

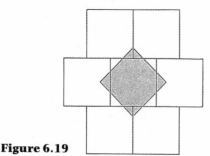

Figure 6.19

The problem of the touching cubes was one I thought of several years ago and had answered with 20 cubes. I was staggered to receive two different solutions, each with 22 cubes. Figure 6.20 shows how five white cubes can abut the top side of the gray cube. Since none extends beyond line AB, this formation can go on four sides of the gray cube [see Figure 6.21].

Two more cubes plug up the holes on face A and its opposite side. This solution was first received from Kenneth J. Fawcett, Jr., and later from Rudolf K. M. Bergan, Michael J. and Alice E. Fischer, Leigh Janes, K. B. Mallory, Allen J. Schwenk, and George Starbuck.

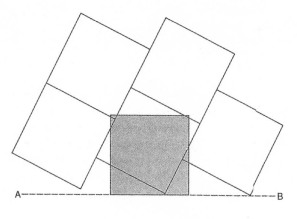

Figure 6.20

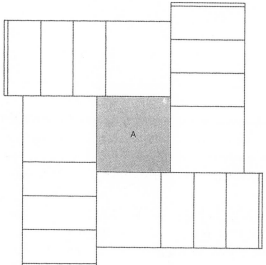

Figure 6.21

The other solution, found independently by Bergan, Rudolph A. Krutar, and Robert S. Holmes, is shown as drawn by Holmes [Figure 6.22]. Eight cubes go on two opposite faces of the gray cube, and six abut the gray cube in the middle layer. Even the fact that as many as eight nonintersecting unit squares can overlap one unit square is, as far as I know, a previously unknown result.

Stanley Ogilvy later pointed out that because the bottom corners of the three lowest squares in Figure 6.20 are not on a horizontal line, there is just enough room below them to permit three more squares, joined face to face, to go beneath the other five squares. This provides another way for eight cubes to abut one face, and leads to another solution with 22 cubes.

Solid Geometry

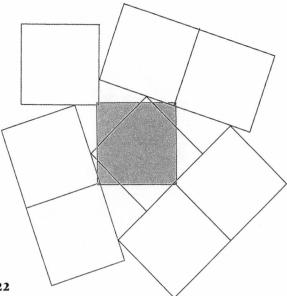

Figure 6.22

While I was still recovering from the 22-cube solution, Holmes (who was working for his doctorate in particle physics at the University of Rochester) delivered the knockout punch: a 24-cube solution! Later Janes, in collaboration with Michael Bradley, reported a 23-cube solution.

It is hard to believe, but as far as I know no one has seriously considered before the simple question of how many unit cubes can share a positive surface area with a given unit cube. My innocent answer of 20 is indeed the best if one adds the condition that the surface of the given unit cube (we shall distinguish it from the others by making it gray and leaving the other cubes white) be completely covered by the touching cubes. This proviso, however, was not part of the original problem.

Holmes's technique begins with placing seven white cubes on one gray face [see Figure 6.23]. Three pairs of cubes (think of each pair as being glued together) are placed around a face of the gray cube so that the midpoint of each pair touches a corner of the gray face. A seventh white cube (P, drawn with broken lines) overlaps the gray face as indicated, the two faces having an axis of symmetry shown as a diagonal line. By rotating the white cubes clockwise, keeping corner A on the left edge of the fixed gray face and preserving the bilateral symmetry,

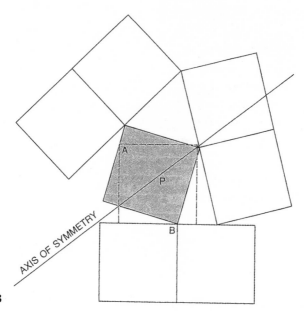

Figure 6.23

the pattern shown in Figure 6.24 is reached. If the broken-line cube P is now moved up a trifle, the two meeting corners of each pair of glued cubes can have a tiny positive-area overlap with a corner of the gray face. These three overlaps can be made arbitrarily small without

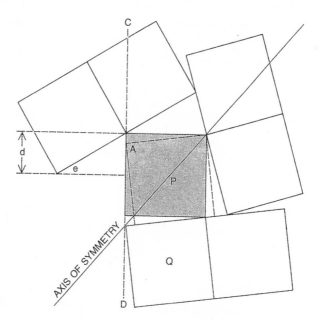

Figure 6.24

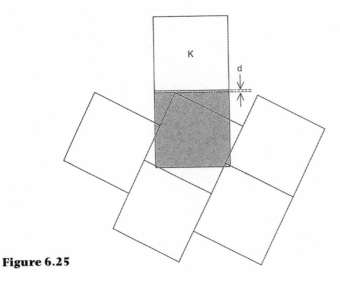

Figure 6.25

allowing the white cubes P and Q to project to the left beyond the vertical line $\overline{CD}$. As a result angle e and distance d can also be made arbitrarily small.

This pattern of seven cubes goes on the front and rear faces of the gray cube. Then one cube goes exactly on top of the gray cube, another goes flush against the gray cube's base and two more cubes abut the right face of the gray cube. Although 18 cubes now abut five faces of the gray one, its sixth face (on the left) remains completely exposed. Figure 6.25 shows the gray cube with the exposed face toward you. On both its left and right sides are seven cubes; they are not shown in the drawing. (Also not shown are the single cubes above and below and the two cubes that about the gray cube's back face.) Cube K is placed so that it overlaps the top of the gray face along a thin horizontal strip of height d that can be arbitrarily small. This allows five more cubes to abut the face, below cube K, as shown, bringing the total number of touching cubes to 24 (7 + 7 + 1 + 1 + 2 + 1 + 5).

A full proof of this construction would be long and tedious, but you should have no difficulty convincing yourself that it can be done in spite of the extremely minute overlaps that are involved. The 24-cube solution is probably maximum, although proving it appears to be formidable. Until that is done there will remain the gnawing suspicion that one or more additional white cube can somehow be squeezed in.

Theodore Katsanis posed an interesting related problem. What is the *minimum* number of unit cubes that can touch another unit cube in such a way that no other cube can be added? If "touch" is defined as in the original problem, the answer is obviously six. Suppose, however, we enlarge the meaning of touch to include contact along edges and corners. The maximum problem is again trivial, answered by 26 cubes, but the minimum problem is not. The lowest Katsanis could get is nine, but perhaps you can do better.

chapter 7

Shapes &
Dissections

Geometric dissections take a figure, such as a square, and cut it into smaller pieces. Several questions can be asked. Can the pieces be reassembled into a given shape? Can the pieces be identical? Can each piece be an acute triangle? Can each piece not fit inside the others?

Sometimes we begin by defining the pieces themselves, with their shapes carefully prescribed. The best-known set of shapes is the pentominoes, formed from five unit squares. While pentominoes have been known for a century, it was not until Gardner wrote about them that they became popular. (Gardner's original December 1957 column reported on Solomon Golomb's ideas.)

Problems

PROBLEM 7.1—Dissecting a Square

If one-fourth of a square is taken from its corner, is it possible to dissect the remaining area into four congruent (same size and shape) parts? Yes, it can be done in the manner shown at the left in Figure 7.1. Similarly, an equilateral triangle with one fourth of its area cut

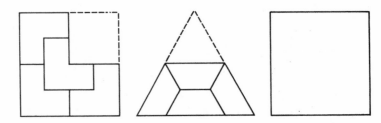

Figure 7.1

from a corner, as in the center figure in the illustration, can also be divided into four congruent parts. These are typical of a large variety of geometric puzzles. Given a certain geometric figure, the task is to cut it into a specified number of identical shapes that completely fill the larger figure.

Can the square at the right in the illustration be dissected into five congruent parts? Yes, and the answer is unique. The pieces can be any shape, however complex or bizarre, provided that they are identical in shape and size. An asymmetric piece may be "turned over"; that is, it is considered identical with its mirror image. The problem is annoyingly intractable until suddenly the solution strikes like lightning.

PROBLEM 7.2—Slicing a Trapezoid

The trapezoid shown in Figure 7.2 is called a triamond, or an order-3 polyiamond, because it can be formed by joining three equilateral triangles. There is an old puzzle that asks how to cut the triamond into four congruent parts. Figure 7.2 gives the traditional solution.

Richard Brady, a mathematics teacher in Washington, D.C., tells me that when Andrew Miller, one of his pupils, encountered the problem in Harold R. Jacobs's *Geometry* (W. H. Freeman and Company, 1974, page 188), he found a different solution. In Miller's new solution all four regions do not have the same shape as the larger figure, but they are identical if one or more may be turned over. What is the new solution?

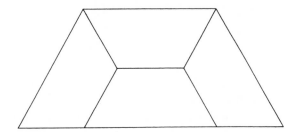

Figure 7.2

PROBLEM 7.3—Five Congruent Polygons

L. Vosburgh Lyons contributed a fiendish dissection problem to a magic magazine in 1969 [see Figure 7.3]. The polygon [at left in illustration] can be dissected into four congruent polygons [at right]. Can you discover the only way in which the same polygon can be cut into *five* congruent polygons?

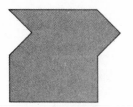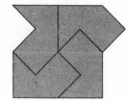

Figure 7.3

Problem 7.4—Damaged Patchwork Quilt

The patchwork quilt in Figure 7.4, of size 9 by 12, was originally made up of 108 unit squares. Parts of the quilt's center became worn, making it necessary to remove eight squares as indicated.

The problem is as follows. Cut the quilt along the lattice lines into just two parts that can be sewn together to make a 10 × 10 square quilt. The new quilt cannot, of course, have any holes. Either part can be rotated, but neither one can be turned over because the quilt's underside does not match its upper side.

The puzzle is an old one, but the solution is beautiful. In fact, the solution is unique even if it is not required that the cutting be along lattice lines.

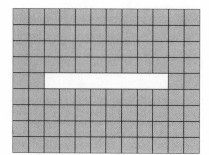

Figure 7.4

Problem 7.5—Seven File Cards

A sheet of legal-sized paper, $8\frac{1}{2} \times 12\frac{1}{2}$ inches, has an area of $106\frac{1}{4}$ square inches. Seven file cards of the 3 × 5-inch size have a combined area of 105 square inches. Obviously it is not possible to cover the large sheet completely with the seven cards, but what is the largest area that *can* be covered? The cards must be placed flat, and they may not be folded or cut in any way. They may overlap the edges of the sheet, however, and it is not necessary for their sides to be parallel with the sides of the sheet. Figure 7.5 shows how the seven cards can be arranged to cover an area of $98\frac{3}{4}$ square inches. This is not the maximum.

Everyone in the family, young and old, will enjoy working on this puzzle. If the required materials are not handy, a sheet of cardboard can be cut to the $8\frac{1}{2} \times 12\frac{1}{2}$-inch size, and the seven 3×5 rectangles can be cut from paper. It is a good plan to rule the large sheet into half-inch squares so that the area left exposed can be computed quickly.

The problem was first posed by Jack Halliburton in *Recreational Mathematics Magazine*, December 1961.

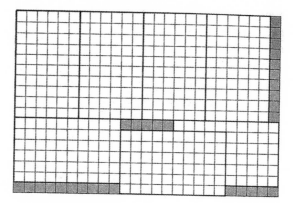

Figure 7.5

PROBLEM 7.6—Acute Dissection

Given a triangle with one obtuse angle, is it possible to cut the triangle into smaller triangles, all of them acute? (An acute triangle is a triangle with three acute angles. A right angle is of course neither acute nor obtuse.) If this cannot be done, give a proof of impossibility. If it can be done, what is the smallest number of acute triangles into which any obtuse triangle can be dissected?

Figure 7.6 shows a typical attempt that leads nowhere. The triangle has been divided into three acute triangles, but the fourth is obtuse, so nothing has been gained by the preceding cuts.

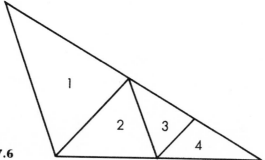

Figure 7.6

Shapes & Dissections

The problem (which came to me by way of Mel Stover of Winnipeg) is amusing because even the best mathematician is likely to be led astray by it and come to a wrong conclusion. My pleasure in working on it led me to ask myself a related question: What is the smallest number of acute triangles into which a *square* can be dissected? For days I was convinced that nine was the answer; then suddenly I saw how to reduce it to eight. I wonder how many of you can discover an eight-triangle solution, or perhaps an even better one. I am unable to prove that eight is the minimum, though I strongly suspect that it is.

PROBLEM 7.7—Gunport Problem

What is the maximum number of 1×1 "holes" that can be obtained by arranging dominoes on an $m \times n$ field? It is assumed that m and n are each greater than 1.

Bill Sands was able to prove that the number of holes cannot exceed the number of dominoes. (See his article, "The Gunport Problem," in *Mathematics Magazine*, vol. 44, 1971, pages 193–96.) He also showed that if either side of the field is a multiple of 3, a repeated pattern provides a simple way of achieving the maximum number of holes [see Figure 7.7]. In other words, if one side of the field is of the form $3k$, the maximum number of holes is $mn/3$. Otherwise the maximum number of holes must be less than this.

Murray R. Pearce of Bismarck, North Dakota, writing in the November 1973 issue of the London monthly *Games & Puzzles*, conjectured that if neither side is $3k$ but both are equal modulo 3 (that is, both are either $3k + 1$ or $3k + 2$), the maximum number of holes is $(mn - 4)/3$, and if one side is $3k + 1$ and the other $3k + 2$, the maximum is $(mn - 2)/3$. Examples of how the predicted maximum can be obtained for the three types of fields that have no side equal to $3k$ are shown in Figure 7.8.

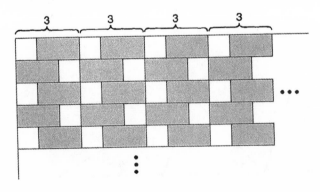

Figure 7.7

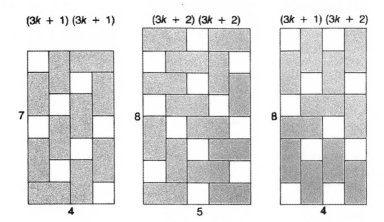

Figure 7.8

Pearce's formulas set a maximum of 26 holes for the 8 × 10 field. Sands confessed he was unable to do better than 24 holes, using 28 dominoes. Can you find a 26-hole solution, using 27 dominoes?

PROBLEM 7.8—Exploring Tetrads

The most sensational news in recreational mathematics in 1976 was surely the announcement by two University of Illinois mathematicians that they had proved the four-color-map conjecture. This famous conjecture is often confused with a simpler theorem in topology, which is easily proved; it states that no more than four regions on the plane can have a mutual border. Michael R. W. Buckley, in the *Journal of Recreational Mathematics*, vol. 8 (1975), proposed the name tetrad for four simply connected planar regions, each pair of which shares a finite portion of a common boundary.

The tetrad at the left in Figure 7.9 has no holes. Note that only three regions are congruent. The tetrad at the right has four congruent regions and a large hole. Is it possible, Buckley asked, to construct a tetrad with four congruent regions and no hole?

This question has been answered affirmatively by Scott Kim, and I am grateful to him for his permission to give some of his results here.

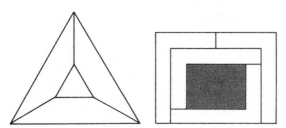

Figure 7.9

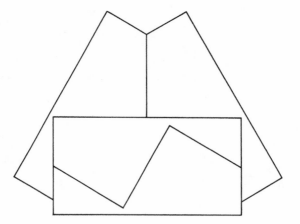

Figure 7.10

Figure 7.10 shows a solution with four congruent hexagons. It is not known if a solution can be achieved with a polygon of fewer than six sides or if there is a solution with an outside border that is convex.

Part A of Figure 7.11 shows a solution with congruent polyhexes of order 4. (A polyhex is a union of congruent regular hexagons.) It is easy to show that no solution with lower-order polyhexes is possible.

Part B of the illustration shows a solution with congruent polyamonds of order 10. (A polyamond is a union of congruent equilateral triangles.) It is not known whether there is a solution with lower-order polyiamonds.

Part C of the illustration shows a solution with congruent polyominoes of order 12. (A polyomino is a union of congruent squares.) It is not known whether there is a solution with lower-order polyominoes.

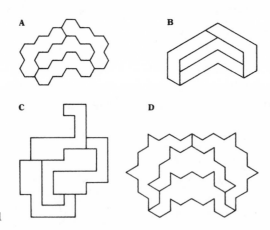

Figure 7.11

GEOMETRIC PUZZLES

Part D of the illustration shows a solution with congruent pieces that have bilateral symmetry and a bilaterally symmetric border. Is there such a solution with polygons of fewer sides?

It has been known since the 1870s that in three dimensions an infinite number of congruent solids can be put together so that every pair shares a common portion of surface. For those of you who are unfamiliar with this result, here is the problem in terms of polycubes. (A polycube is a union of identical cubes.) Show how an infinite number of congruent polycubes can be put together, with no interior hole, so that every pair shares part of a surface. Such a structure proves that an unlimited number of colors are required for coloring any three-dimensional "map."

PROBLEM 7.9—Square-Triangle Polygons

An unlimited number of cardboard squares and equilateral triangles, each with unit sides, are assumed to be available. With these pieces it is easy to form convex polygons with from 3 to 10 sides [see Figure 7.12]. Can you make an 11-sided convex polygon with the pieces? And what is the largest number of sides a convex polygon formed by the pieces can have?

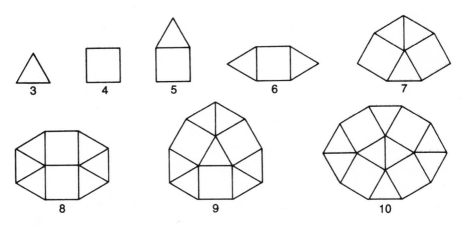

Figure 7.12

PROBLEM 7.10—Two Pentomino Posers

For pentomino buffs, here are two problems, the first one easy and the second difficult.

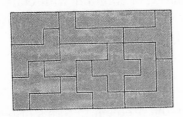 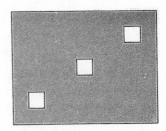

Figure 7.13

A. At the left of Figure 7.13 the 12 pentominoes are arranged to form a 6 × 10 rectangle. Divide the rectangle, along the black lines only, into two parts that can be fitted together again to make the three-holed pattern at the right of the illustration.

B. Arrange the 12 pentominoes to form a 6 × 10 rectangle but in such a way that each pentomino touches the border of the rectangle. Of several thousand fundamentally different ways of making the 6 × 10 rectangle (rotations and reflections are not considered different), only two are known to meet the condition of this problem. Asymmetrical pieces may be turned over and placed with either side against the table.

PROBLEM 7.11—Pentomino Farms

Victor G. Feser of Saint Louis University has proposed four maximum-area problems, each using the full set of 12 pentominoes. Three have been solved, and the fourth is probably solved.

1. Form a rectangular "fence" around the largest rectangular field. The 4 × 7 has been proved maximum [see Figure 7.14a].
2. Form a rectangular fence around the largest field of any shape. The maximum is 61 unit squares [see Figure 7.14b].
3. Form a fence of any shape around the largest rectangular field. The 9 × 10 is maximum [see Figure 7.14c].
4. Form a fence of any shape around the largest field of any shape. (As in the preceding problems, the fence must be at least one unit thick at all points.) This is the most difficult of the four. In Figure 7.14d you see a solution of 127 squares. This was believed to be maximum until Donald E. Knuth, the Stanford computer scientist, raised it to 128. Knuth has an informal proof that 128 cannot be exceeded. You will find it a pleasant and difficult task to find a 128 solution.

GEOMETRIC PUZZLES

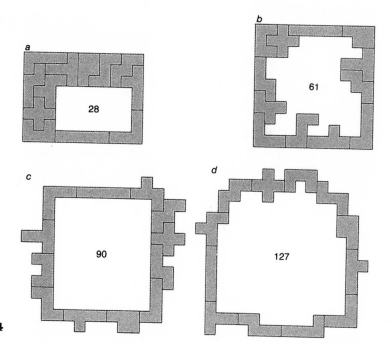

Figure 7.14

PROBLEM 7.12—Kobon Triangles

Kobon Fujimura, a Japanese puzzle expert, invented a problem in combinatorial geometry. It is simple to state, but no general solution has yet been found. What is the largest number of nonoverlapping triangles that can be produced by n straight line segments?

It is not hard to discover by trial and error that for $n = 3, 4, 5$, and 6 the maximum number of triangles is respectively one, two, five, and seven [see Figure 7.15]. For seven lines the problem is no longer easy.

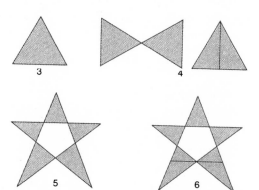

Figure 7.15

Search for the maximum number of nonoverlapping triangles that can be produced by seven, eight, and nine lines.

The problem of finding a formula for the maximum number of triangles as a function of the number of lines appears to be extremely difficult.

Problem 7.13—Acute and Isosceles Triangles

An acute triangle is one with each interior angle less than 90 degrees. What is the smallest number of nonoverlapping acute triangles into which a square can be divided? I asked myself that question and I solved it by showing how to cut a square into eight acute triangles as is indicated in Figure 7.16, top (for details, see problem 7.6).

Since then I have received many letters from readers who were unable to find a solution with nine acute triangles but who pointed out that solutions are possible for 10 or any higher number. The middle illustration in Figure 7.16 shows how it is done with 10. Note that obtuse triangle ABC is cut into seven acute triangles by a pentagon of five acute triangles. If ABC is now divided into an acute and an obtuse triangle by BD, as is indicated in Figure 7.16 bottom, we can use the same pentagonal method for cutting the obtuse triangle BCD into seven acute triangles, thereby producing 11 acute triangles for the entire square. A repetition of the procedure will produce 12, 13, 14, . . . acute triangles.

Apparently the hardest dissection to find is the one with nine acute triangles. Nevertheless, it can be done.

There are many comparable problems about cutting figures into nonoverlapping triangles, of which I shall mention only two. It is easy to divide a square into any even number of triangles of equal area, but can a square be cut into an odd number of such triangles? The surprising answer is no. As far as I know, this was first proved by Paul Monsky in *American Mathematical Monthly* (vol. 77, no. 2, February 1970, pages 161–64).

Another curious theorem is that any triangle can be cut into n isosceles triangles provided n is greater than 3. A proof by Gali Salvatore appeared in *Crux Mathematicorum* (vol. 3, no. 5, May 1977, pages 134–35). Another proof, by N. J. Lord, is in *The Mathematical Gazette* (vol. 66, June 1982, pages 136–37).

The case of the equilateral triangle is of particular interest. It is easy

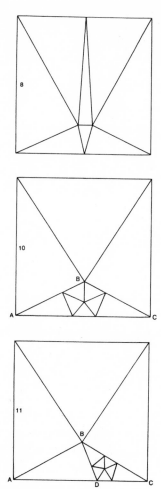

Figure 7.16

to cut it into four isosceles triangles (all equilateral) or into three isosceles triangles. (Some triangles cannot be cut into three or two isosceles triangles, which is why the theorem requires that n be 4 or more.) Can you cut an equilateral triangle into five isosceles triangles? I shall show how it can be done with none of the five triangles equilateral, with just one equilateral, and with just two equilateral. It is not possible for more than two of the isosceles triangles to be equilateral.

Problem 7.14—Rectangling the Rectangle

There is a classic dissection problem known as squaring the square: Can a square be cut into a finite number of smaller squares no two of

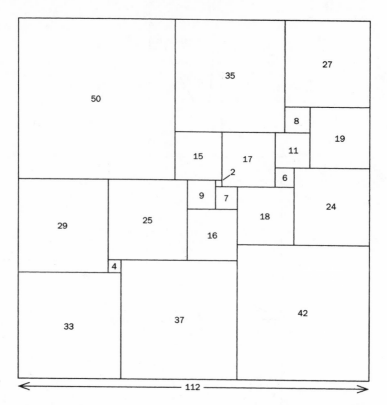

Figure 7.17

which are the same size? The solution to this problem was discussed in an article by graph theorist William T. Tutte that is reprinted in Chapter 17 of my *2nd Scientific American Book of Mathematical Puzzles & Diversions.* At the time Tutte wrote, the best solution known ("best" meaning with a minimum number of different squares) required 24 squares. In 1978 this figure was lowered to 21 squares by A.J.W. Duijvestijn, a Dutch mathematician, as was reported in *Scientific American* (June 1978, pages 86–88; see Figure 7.17). It had been known that 21 was the smallest number possible, and Duijvestijn was also able to show that his pattern for that number of squares is unique. More recently he has found two squared squares containing 22 squares and the best solution, also with 22 squares, for squaring the domino (a rectangle with one side twice the length of the other). See Duijvestijn's paper "A Simple Perfect Squared of Lowest Order," in *Journal of Combinatorial Theory*, Series B, vol. 25, 1978, pp. 240–43.

A somewhat analogous problem is to divide a nonsquare rectangle into the minimum number of smaller rectangles in such a way that no

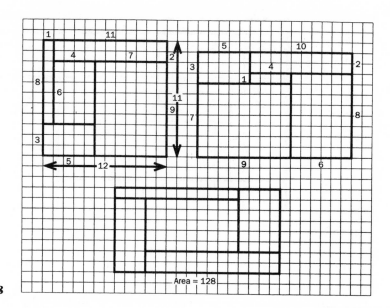

Figure 7.18

two sides of two different rectangles have the same length. It is not hard to show that the minimum number of internal rectangles is five. Now add the condition that all sides of all six rectangles (including the outer one) are integers. Scott Kim has proved that no solution is possible in which the integers 1 through 12 are used for the 12 different edge lengths, although one can come close. Figure 7.18, top left, shows a solution by Kim that includes 11 twice and omits 10. If the sides of the outside rectangle are ignored, the consecutive integers from 1 through 10 will do the trick, as Ronald L. Graham has proved with the rectangle shown in Figure 7.18, top right.

Figure 7.18, bottom, is reproduced from *Mathematical Puzzles*, a challenging collection by Stephen Ainley (G. Bell and Sons, Ltd., 1977, page 59). It displays one of the two solutions in which all 12 edges are different, although not consecutive, and the area of the outside rectangle is reduced to 128. Ainley calls this a minimal area but gives no proof.

Now relax the conditions slightly so that only the 10 different edges of the five internal rectangles must be distinct and integral. What solution has an outer rectangle of the smallest area? This problem was first solved by C.R.J. Singleton, who sent it to me in 1972. There are two solutions. Can you find them?

The more difficult but closely related problem of finding the

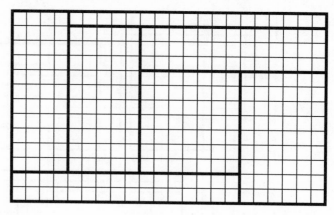

Figure 7.19   22 x 13

smallest rectangle that can be divided into incomparable rectangles was posed by Edward M. Reingold as Problem E2422 in *The American Mathematical Monthly* (vol. 80, 1973, p. 69); a solution was published in the same journal the following year (vol. 81, no. 6, June-July 1974, pages 664–66). Two rectangles are called incomparable if neither one can be placed inside the other and aligned so that corresponding sides are parallel. Reingold, Andrew C. C. Yao, and Bill Sands, in "Tiling with Incomparable Rectangles" (*Journal of Recreational Mathematics,* vol. 8, 1975–76, pages 112–19), prove many theorems about this problem.

The minimum number of incomparable rectangles needed to tile a larger rectangle is seven. If all sides are integral, the outside rectangle with both the smallest area and the smallest perimeter is the 22 × 13 rectangle shown in Figure 7.19. It was found by Sands. A square can be tiled with seven incomparable rectangles having integral sides if and only if its side is 34 or larger, but eight rectangles can tile a square of side 27. This square is the smallest one known that can be tiled with incomparable rectangles, but it has not been proved to be the smallest one possible. For details consult the paper by Reingold, Yao, and Sands.

In 1975 the first problem, of rectangling the rectangle, was generalized to three dimensions by Kim. There is an elegant proof that a cube cannot be cut into smaller cubes no two of which are alike. (See my *2nd Scientific American Book of Mathematical Puzzles & Diversions*, page 208.) Can a cube be "boxed" by cutting it into smaller boxes (rectangular parallelepipeds) so that no two boxes share a common edge length? The answer is yes, and Kim was able to show that the minimum number of interior boxes is 23. Later William H. Cutler devised a second proof that 23 is minimal. Cutler found 56 essentially different ways to box the cube. If all the edges of such a cube are integral, the smallest cube that can be boxed is not known. (See "Subdividing a Box Into Completely Incongruent Boxes," by William Cutler, in *Journal of Recreational Mathematics*, 12, 1979–80, pages 104–11.)

Another unsolved problem is to determine the noncubical box of smallest volume that can be sliced into 23 (or possibly more) boxes with no edge in common and all edges integral. Cutler found a box that is $147 \times 157 \times 175$ and splits into the following 23 boxes:

$13 \times 112 \times 141$	$27 \times 36 \times 48$
$18 \times 72 \times 82$	$34 \times 110 \times 135$
$23 \times 41 \times 73$	$57 \times 87 \times 97$
$31 \times 69 \times 78$	$16 \times 74 \times 140$
$38 \times 42 \times 90$	$21 \times 52 \times 65$
$14 \times 70 \times 75$	$28 \times 55 \times 123$
$19 \times 53 \times 86$	$35 \times 62 \times 127$
$26 \times 49 \times 56$	$17 \times 24 \times 67$
$33 \times 46 \times 60$	$22 \times 107 \times 131$
$45 \times 68 \times 85$	$30 \times 54 \times 134$
$15 \times 44 \times 50$	$37 \times 83 \times 121$
$20 \times 40 \times 92$	

It is not easy to fit these boxes together to make the large box. As far as I know no one has worked on a three-dimensional version of Reingold's incomparable rectangles.

Answers

ANSWER 7.1—(October 1962)

A square can be dissected into five congruent parts only in the manner shown in Figure 7.20. The consternation of those who find themselves unable to solve this problem is equaled only by their feeling of foolishness when shown the answer.

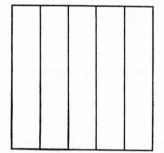

Figure 7.20

ANSWER 7.2—(February 1979)

Figure 7.21 displays the only other way the triamond can be cut into four identical regions.

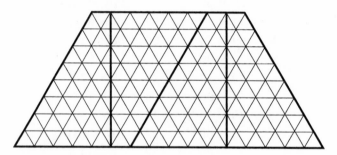

Figure 7.21

ANSWER 7.3—(November 1978)

The large polygon in Figure 7.22 can be cut into five congruent polygons as shown. The method obviously enables one to dissect the polygon into any desired number of congruent shapes. L. Vosburgh Lyons first published this in *The Pallbearers Review,* July 1969, page 268.

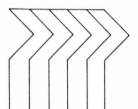

Figure 7.22

ANSWER 7.4—(April 1981)

Figure 7.23 shows how the damaged patchwork quilt is cut into two parts that can be sewn together to make a square quilt without a hole. The problem is No. 215 in Henry Ernest Dudeney's *Puzzles* and *Curious Problems* (Thomas Nelson and Sons, Ltd., 1931).

The problem of the mutilated quilt leads to a variety of generalizations. In the problem given, if the rectangle is chessboard colored, the 10×10 square will not preserve such coloring. I spent several days exploring squares of side n, with an associated rectangle of $(n - 1) \times (n + 2)$ sides, with a hole of $1 \times (n - 2)$ sides parallel to the rectangle's long side and as centered as possible. Assuming the rectangle to be chessboard colored, reader Eric Stott asked, "Into how few pieces can it be sliced along lattice lines to form a properly colored square?"

Consider the rectangle shown in Figure 7.24. I found it could be sliced into as few as three pieces that will form a standard chessboard as shown at the top of Figure 7.25. The generalization to squares of even-numbered sides is obvious. If the square's side is odd, one must distinguish between $n = 1$ (mod 4) and $n = 3$ (mod 4). Examples of each are shown in the middle and bottom illustrations. Again it is easy to see how they generalize.

Two-piece dissections are impossible; four-piece dissections are plentiful and easy to find. In all such solutions with three pieces I believe it is impossible to avoid mirror reflecting the asymmetric piece.

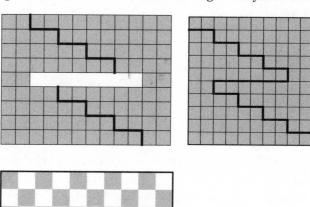

Figure 7.23

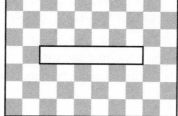

Figure 7.24

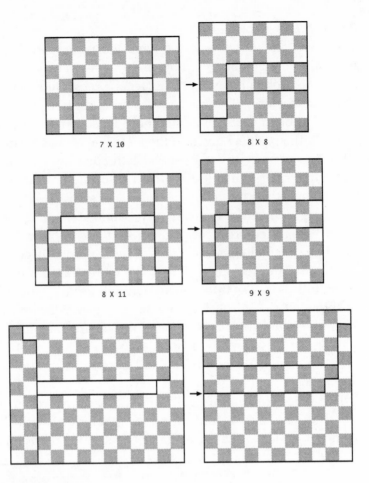

7 X 10   8 X 8

8 X 11   9 X 9

Figure 7.25

ANSWER 7.5—(October 1962)

If it is required that the cards be placed with their edges parallel to the edges of the sheet, a maximum of 100 square inches can be covered. Figure 7.26 shows one of many different ways in which the cards can be placed.

Stephen Barr was the first to point out that if the central card is tilted as shown in Figure 7.27 the covered area can be increased to 100.059+ square inches. Then Donald Vanderpool wrote to say that if the card is rotated a bit more than this (keeping it centered on the exposed strip) the covered area can be increased even more. The angle at which maximum coverage is achieved must be obtained by calculus.

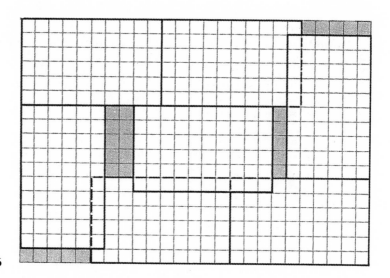

Figure 7.26

James A. Block found that the angle could be varied from 6° 12′ to 6° 13′ without altering the covered area expressed to five decimal places: 100.06583+ square inches. The problem's history has been sketched in Joseph S. Madachy, *Mathematics on Vacation* (New York: Scribner's, 1966), pages 133–35. He gives a calculation by R. Robinson Rowe for an angle of 6° 12′ 37.8973″ that gives a covering of 100.065834498+ square inches.

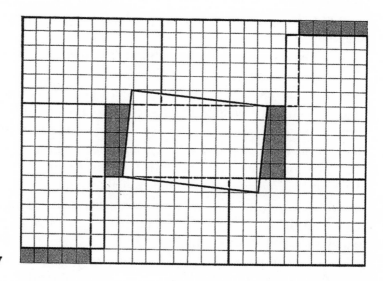

Figure 7.27

ANSWER 7.6—(February 1960)

A number of readers sent "proofs" that an obtuse triangle cannot be dissected into acute triangles, but of course it can. Figure 7.28 shows a seven-piece pattern that applies to any obtuse triangle.

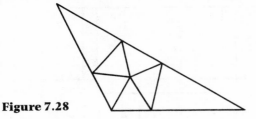

Figure 7.28

It is easy to see that seven is minimal. The obtuse angle must be divided by a line. This line cannot go all the way to the other side, for then it would form another obtuse triangle, which in turn would have to be dissected, consequently the pattern for the large triangle would not be minimal. The line dividing the obtuse angle must, therefore, terminate at a point *inside* the triangle. At this vertex, at least five lines must meet, otherwise the angles at this vertex would not all be acute. This creates the inner pentagon of five triangles, making a total of seven triangles in all. Wallace Manheimer, a Brooklyn high school teacher at the time, gave this proof as his solution to problem E1406 in *American Mathematical Monthly*, November 1960, page 923. He also showed how to construct the pattern for any obtuse triangle.

The question arises: Can any obtuse triangle be dissected into seven acute *isosceles* triangles? The answer is no. Verner E. Hoggatt, Jr., and Russ Denman (*American Mathematical Monthly*, November 1961, pages 912–13) proved that eight such triangles are sufficient for all obtuse triangles, and Free Jamison (ibid., June-July 1962, pages 550–52) proved that eight are also necessary. These articles can be consulted for details as to conditions under which less than eight-piece patterns are possible. A right triangle and an acute nonisosceles triangle can each be cut into nine acute isosceles triangles, and an acute isosceles triangle can be cut into four congruent acute isosceles triangles similar to the original.

A square can be cut into eight acute triangles as shown in Figure 7.29. If the dissection has bilateral symmetry, points P and P' must lie

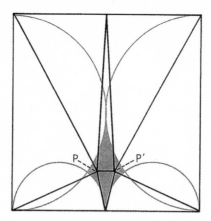

Figure 7.29

within the shaded area determined by the four semicircles. Donald L. Vanderpool pointed out in a letter that asymmetric distortions of the pattern are possible with point P anywhere outside the shaded area provided it remains outside the two large semicircles.

About 25 readers sent proofs, with varying degrees of formality, that the eight-piece dissection is minimal. One, by Harry Lindgren, appeared in *Australian Mathematics Teacher*, vol. 18, 1962, pages 14–15. His proof also shows that the pattern, aside from the shifting of points P and P' as noted above, is unique.

H.S.M. Coxeter pointed out the surprising fact that for any rectangle, even though its sides differ in length by an arbitrarily small amount, the line segment PP' can be shifted to the center to give the pattern both horizontal and vertical symmetry.

Free Jamison found in 1968 that a square can be divided into 10 acute isosceles triangles. See *The Fibonacci Quarterly* (December 1968) for a proof that a square can be dissected into any number of acute isosceles triangles equal to or greater than 10.

Figure 7.30 shows how the pentagram (regular five-pointed star) and the Greek cross can each be dissected into the smallest possible number of acute triangles.

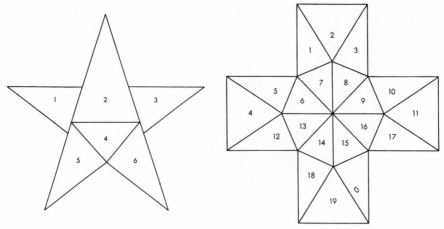

Figure 7.30

ANSWER 7.7—(April 1974)

Figure 7.31 shows one way of placing 27 dominoes on an 8 × 10 field to form 26 holes. Found by Capt. John C. Huval, it was published in *Mathematics Magazine* for November 1972. Many trivial variations can be produced by sliding one domino, by switching to adjacent dominoes or by rotating a 3 × 3 pattern of three dominoes.

Kenneth M. Brown and Jon Petersen each proved that Murray Pearce's formulas for the gunport problem cannot be exceeded. It remains an open question whether there are rectangles for which Pearce's upper bounds cannot be achieved. Petersen and Douglas W. Oman independently found a procedure showing that all rectangles with areas smaller than 224 could meet the upper bounds. The general case is undecided, with the 14 × 16 rectangle being a likely candidate for the smallest counterexample. According to Pearce's conjecture, it should be possible to cover it with 75 dominoes that leave 74 holes.

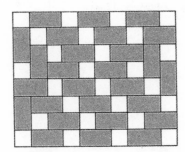

Figure 7.31

GEOMETRIC PUZZLES

Figure 7.32 shows how two congruent polycubes (one shaded and one transparent) can be fitted together. By extension of the ends any finite number of such pieces can be nested in this manner so that every pair "touches" in the sense that they share a common surface and there are no interior holes. Extended to infinity, an infinite number of congruent polycubes, of order infinity, can mutually "touch."

If we drop the requirement of congruency but add the requirement of convexity, it has been known since about 1900 that an infinite number of noncongruent convex solids can mutually "touch." It is not known whether an infinite number of congruent convex solids can mutually touch, but Scott Kim has recently shown (although not published) how this arrangement can be achieved with an arbitrarily large finite number of such solids.

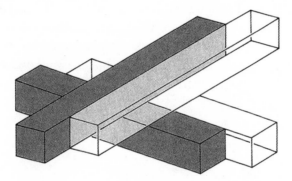

Figure 7.32

Robert Ammann, Greg Frederickson, and Jean L. Loyer each found an 18-sided polygonal tetrad with bilateral symmetry [see Figure 7.33], thus improving on the 22-sided solution I had published. (See problem 8.15 for related work.)

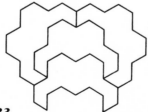

Figure 7.33

An 11-sided convex polygon can be formed with unit-sided squares and equilateral triangles, as shown in Figure 7.34, top left. The angles possible for a convex polygon formed with the pieces are 60, 90, 120, and 150 degrees. For a polygon with the maximum number of sides, all angles must be 150 degrees. The number of sides will then be 12. Figure 7.34, top right, shows the smallest example.

Several readers "proved" that an 11-sided polygon could not be formed with squares and equilateral triangles of unit sides. The flaw, of course, was failing to realize that a side could be more than one unit long.

Wade Philpott pointed out that *any* convex pentagon formed with unit equilateral triangles can be used as the core of an 11-sided polygon. Simply place unit squares next to each triangle and complete the perimeter with six triangles. The solution I gave leads to an infinite family of 11-sided polygons, shown in the middle of Figure 7.34. At the bottom are two other examples with different inner pentagons. The problem derives from one posed by Joseph Malkewitch in *Mathematics Magazine* and answered by Michael Goldberg in the May 1969 issue, page 158.

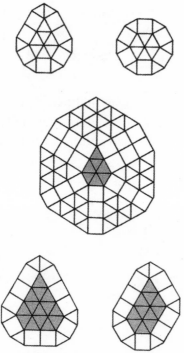

Figure 7.34

ANSWER 7.10—(June 1961)

Figure 7.35 shows how the 6 × 10 rectangle, formed with the 12 pentominoes, can be cut into two parts and the parts refitted to make the 7 × 9 rectangle with three interior holes. Figure 7.36 shows the only two possible patterns for the 6 × 10 rectangle in which all 12 pieces touch the border. The second of these patterns is also remarkable in that it can be divided into two congruent halves.

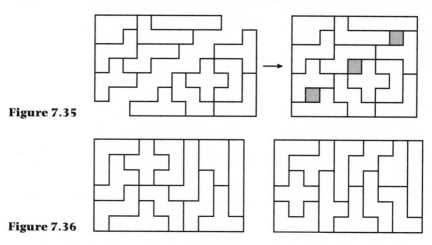

Figure 7.35

Figure 7.36

ANSWER 7.11—(May 1973)

A solution to the farm problem, enclosing 128 square units, is shown in Figure 7.37.

I learned later that this problem had been proposed by R. J. French in *The Fairy Chess Review*, vol. 4, 1939, page 43. French said the area was more than 120. I have not been able to determine whether the problem was answered in subsequent issues.

After I published Knuth's 128 solution, Yoichi Kotani sent a proof, along with 1,440 solutions, that 128 is the maximum. Robert Reid Dalmau of Lima, Peru, sent the same set of solutions. In 1978 Takakazu Shimauchi published in Japanese a proof that 128 is the maximum (*Sugaku Seminar*, March 1978, pages 11–16).

For references on pentomino farm problems in the *Journal of Recreational Mathematics*, see the issues for January 1968, pages 55–61; October 1968, pages 234–35; July 1969, pages 187–88; and vol. 17, no. 1, 1984–85, pages 75–77. If the 12 pieces are allowed to touch only at corners and all edges are required to be horizontal and vertical, a farm

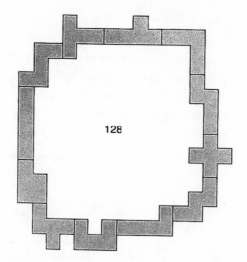

Figure 7.37

of 160 square units is the largest known. If the pieces are allowed any orientation and corner touching, the area can be raised to slightly more than 161.

ANSWER 7.12—(April 1972)

The maximum number of nonoverlapping triangles that can be produced by seven, eight, and nine lines are 11, 15, and 21 respectively [see Figure 7.38]. These are thought, although not yet proved, to be maximal solutions.

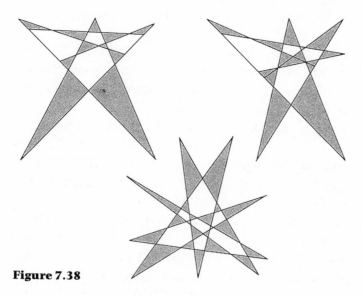

Figure 7.38

ANSWER 7.13—(April 1981)

Figure 7.39 shows how to cut a square into nine acute triangles. The solution is unique. If triangulation is taken in the topological sense, so that a vertex is not allowed to be on the side of a triangle, then there is no solution with nine triangles, although there is a solution for 8 triangles, for 10, and for all higher numbers. This curious result has been proved in an unpublished paper by Charles Cassidy and Graham Lord of Laval University in Quebec.

Figure 7.40 shows four ways to cut an equilateral triangle into five isosceles triangles. The first pattern has no equilateral triangle among the five, the second and third patterns both have one equilateral triangle, and the fourth pattern has two equilateral triangles. The four patterns, devised by Robert S. Johnson, appear in *Crux Mathematicorum* (vol. 4, no. 2, February 1978, page 53). A proof by Harry L. Nelson that there cannot be more than two equilateral triangles is in the same volume of the journal (no. 4, April 1978, pages 102–4).

The first three patterns of Figure 7.40 are not unique. Many readers sent alternate solutions. The largest number, 13, came from Roberto Teodoro Garrido, a civil engineer in Buenos Aires.

The task of minimizing the number of acute triangles into which a square can be cut suggests companion problems for right triangles, obtuse triangles, and scalene triangles. For right triangles the trivial answer is two; for scalene triangles the easy answer is three. Figure 7.41 shows how a square can be cut into six obtuse triangles. I believe this to be minimal.

Into how many equilateral triangles can an equilateral triangle be

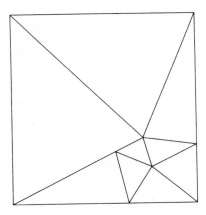

Figure 7.39

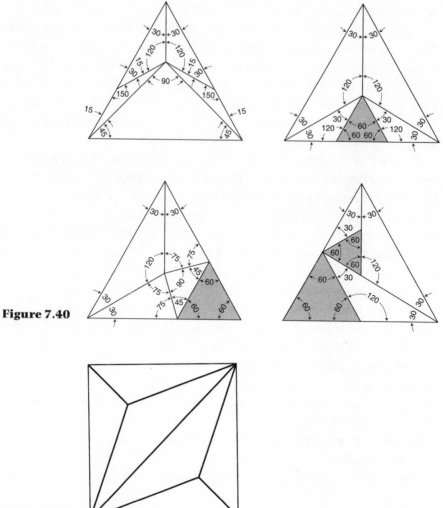

Figure 7.40

Figure 7.41

cut? Two and three are impossible, four is obvious, five is impossible, and any number above five is possible.

ANSWER 7.14—(February 1979)

I originally said, in *Scientific American* that Singleton's problem had only one solution, and the one I gave was incorrectly drawn. In 1979 I received a letter from M. den Hertog, containing much interesting material on rectangling theory. He showed that for every rectangled rectangle one could obtain another pattern by using the same side lengths. The two solutions show in Figure 7.42 appeared in the *Journal of Recreational Mathematics* (vol. 12, 1979–80, pp. 147–48) as a solution by Jean Meeus to Problem 731.

I confined rectangling to nonsquares. If the outer "rectangle" is allowed to be square, it can be cut into five rectangles for which the 10 different sides have the lengths 1 through 10, and the outside square has a side of 11. This pretty pattern, discovered by M. den Hertog, of Belgium, is shown in Figure 7.43.

I said I knew of no work on incomparable tiling of cuboids (rectangular parallelepipeds) by smaller cuboids—that is, a tiling such that given any pair of the boxes, neither will fit inside the other. Just such work was discussed by Richard Guy in his note on "How Few Incom-

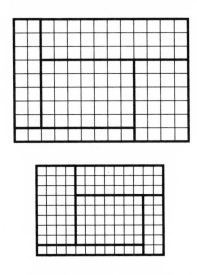

Figure 7.42

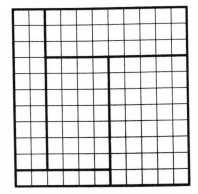

Figure 7.43

Shapes & Dissections

195

parable Cuboids Will Tile a Cube?" (*American Mathematical Monthly*, vol. 91, 1984, pages 624–29). The smallest number of incomparable cuboids that will tile a cube is known to be at least seven, and the smallest number that will tile a noncubical cuboid is six.

In a later note on "Unsolved Problems" (ibid., vol. 92, 1985, pages 725–32), Guy reports on several remarkable discoveries by Charles H. Jepsen, of Grinell College. Jepsen covers them at greater length in his paper "Tiling with Incomparable Cuboids," in *Journal of Recreational Mathematics*, vol. 59, 1986, pages 283–92. He proves that no incomparable tiling of a cuboid can have fewer than six pieces, and gives an example of such a tiling of a $3 \times 5 \times 9$ cuboid, along with a proof that its volume of 135 is minimal. He also proves that no tiling is possible of a $3 \times 3 \times C$ cuboid.

Jepsen also gives an incomparable tiling of a $10 \times 10 \times 10$ cube with seven pieces, and conjectures that its 10^3 volume is minimal. The paper closes with four unanswered questions.

chapter 8

Topology

Topology was the first area of mathematics that attracted Gardner. In the 1930s he became interested in Möbius strips, known to magicians as "Afghan bands." This lead him to read the popular math literature and to write "Topology and Magic" in Scripta Mathematica *(1951). This chapter highlights several magic tricks.*

Sometimes called "rubber-sheet geometry," topology is the study of what properties remain invariant when objects are deformed by stretching and contracting. For example, a solid with a hole in it will continue to have a hole; perhaps you are familiar with the famous example of a doughnut deformed into a coffee cup, with the doughnut's hole becoming the handle.

Connectivity is an invariant property. Map coloring, where each pair of adjoining countries must not be colored the same color, is a classic connectivity problem. You will be able to try your hand at several coloring problems here. Finally, we see examples of knot theory, a surprisingly subtle branch of topology.

Problems

PROBLEM 8.1—Crossing Theorem

With a black pencil, draw a closed curve of any shape you please. With a red pencil draw a second curve of the same kind on top of the first one, never passing through a previously created intersection. Circle all points where one curve crosses the other [see Figure 8.1]. Prove that the number of such points is even.

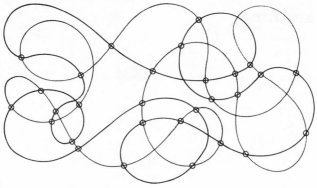

Figure 8.1

PROBLEM 8.2—Matchstick Map

The six matches in Figure 8.2 form a map requiring three colors, assuming that no two regions sharing part of a border can have the same color. Rearrange the six to form a planar map requiring four colors. Confining the map to the plane rules out the three-dimensional solution of forming a tetrahedral skeleton.

Figure 8.2

PROBLEM 8.3—Puzzling Bands

Each of the two paper structures shown in Figure 8.3 consists of a horizontal band attached to a vertical band of the same length and width. The structures are identical except that the second has a half-twist in its vertical band. If the first is cut along the broken lines, the surprising result is the large square band shown as a border of the illustration.

What results if the second structure is similarly cut along the broken lines?

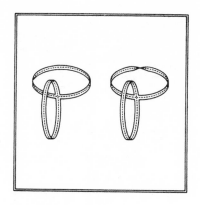

Figure 8.3

PROBLEM 8.4—Knotted String?

A string, lying on the floor in the pattern shown in Figure 8.4, is too far away for you to see how it crosses itself at points A, B, and C. What is the probability that the string is knotted? (From L. H. Longley-Cook, *Fun with Brain Puzzlers*.)

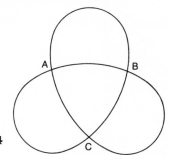

Figure 8.4

PROBLEM 8.5—The Key and the Keyhole

This frustrating topological puzzle calls for a door key and a piece of heavy cord at least a few yards long. Double the cord, push the loop through the keyhole of a door as shown in the top drawing in Figure 8.5, then put both ends of the cord through the projecting loop as in the middle drawing. Now separate the ends, one to the left, the other to the right (bottom drawing). Thread the key on the left cord and slide it near the door, and secure the cord's ends by tying them to something—the backs of two chairs, for example. Allow plenty of slack.

The problem is to manipulate the key and cord so that the key is moved from spot P on the left to spot Q on the right. After the transfer the cord must be looped through the door in exactly the same way as before.

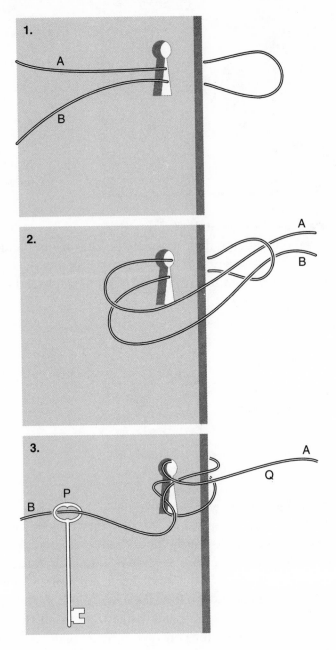

Figure 8.5

PROBLEM 8.6—Problem of the Six Matches

Professor Lucius S. Wilsun is a brilliant, though somewhat eccentric, topologist. His name had formerly been Wilson. As a graduate student he had noted that when his full name, Lucius Sims Wilson, was printed in capital letters, all the letters were topologically equivalent except for the *O*. This so annoyed him that he had his name legally changed.

When I met him for lunch recently, I found him forming patterns on the tablecloth with six paper matches. "A new topological puzzle?" I asked hopefully.

"In a way," he replied. "I'm trying to find out how many topologically distinct patterns I can make with six matches by placing them flat on the table, without crossing one match over another, and joining them only at the ends."

"That shouldn't be difficult," I said.

"Well, it's trickier than you might think. I've just worked out all the patterns for smaller numbers of matches." He handed me an envelope on the back of which he had jotted down a rough version of the chart in Figure 8.6.

"Didn't you overlook a five-match pattern?" I said. "Consider that

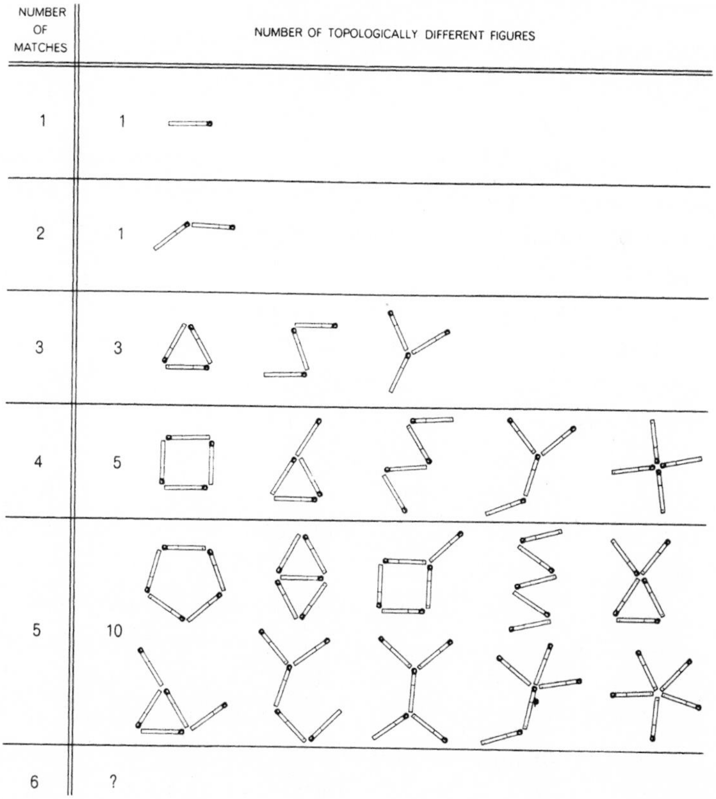

Figure 8.6

NUMBER OF MATCHES	NUMBER OF TOPOLOGICALLY DIFFERENT FIGURES	
1	1	
2	1	
3	3	
4	5	
5	10	
6	?	

Topology

third figure—the square with the tail. Suppose you put the tail *inside* the square. If the matches are confined to the plane, obviously one pattern can't be deformed into the other."

Wilsun shook his head. "That's a common misconception about topological equivalence. It's true that if one figure can be changed to another by pulling and stretching, without breaking or tearing, the two must be topologically identical—as we topologists like to say, homeomorphic. But not the other way around. If two figures are homeomorphic, it is *not* always possible to deform one into the other."

"I beg your pardon," I said.

"Don't topologize. Two figures are homeomorphic if, as you move continuously from point to point along one figure, you can make a corresponding movement from point to point—the points of the two figures must be in one-to-one correspondence, of course—along the other figure. For example, a piece of rope joined at the ends is homeomorphic with a piece of rope that is knotted before the ends are joined, although you obviously cannot deform one to the other. Two spheres that touch externally are homeomorphic with two spheres of different size, the smaller inside the other and touching at one point."

I must have looked puzzled, because he quickly added: "Look, here's a simple way to make it clear to your readers. Those match figures are on the plane, but think of them as elastic bands. You can pick them up, manipulate them any way you wish, turn them over if you please, put them back down again. If one figure can be changed to another this way, they are topologically the same."

"I see," I said. "If you think of a figure as embedded in a higher space, then it *is* possible to deform one figure into any other figure that is topologically equivalent to it."

"Precisely. Imagine the endless rope or the two spheres in a four-dimensional space. The knot can be tied or untied while the ends remain joined. The small sphere can be moved in or out of the larger one."

With this understanding of topological equivalence, determine the exact number of topologically different figures that can be formed on the plane with six matches. Remember, the matches themselves are rigid and all the same size. They must not be bent or stretched, they must not overlap, and they may touch only at their ends. But once a figure is formed it must be thought of as an elastic structure that can be picked up, deformed in three-space, then returned to the plane. The figures are not graphs in which vertices, where two matches join, keep their identity. Thus a triangle is equivalent to a square or a pentagon; a chain of two matches is equivalent to a chain of any length; the capital letters *E, F, T,* and Y are all equivalent; *R* is the same as its mirror image; and so on.

PROBLEM 8.7—Cross the Network

One of the oldest of topological puzzles, familiar to many schoolchildren, consists of drawing a continuous line across the closed network shown in Figure 8.7 so that the line crosses each of the 16 segments of the network only once. The curved line shown here does not solve the puzzle because it leaves one segment uncrossed. No "trick" solutions are allowed, such as passing the line through a vertex or along one of the segments, folding the paper, and so on.

It is not difficult to prove that the puzzle cannot be solved on a plane surface. Two questions: Can it be solved on the surface of a sphere? On the surface of a torus (doughnut)?

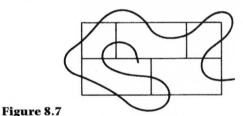

Figure 8.7

PROBLEM 8.8—Barr's Belt

Stephen Barr of Woodstock, New York, says that his dressing gown has a long cloth belt, the ends of which are cut at 45-degree angles as shown in Figure 8.8. When he packs the belt for a trip, he likes to roll it up as neatly as possible, beginning at one end, but the slanting ends disturb his sense of symmetry. On the other hand, if he folds over an end to make it square off, then the uneven thicknesses of cloth put lumps in the roll. He experimented with more complicated folds, but try as he would, he could not achieve a rectangle of uniform thickness. For example, the fold shown in the illustration produces a rectangle with three thicknesses in section A and two in section B.

"Nothing is perfect," says one of the Philosophers in James Stephens's *The Crock of Gold*. "There are lumps in it." Nonetheless,

Barr finally managed to fold his belt so that each end was straight across and part of a rectangle of uniform thickness throughout. The belt could then be folded into a neat roll, free of lumps. How did Barr fold his belt? A long strip of paper, properly cut at the ends, can be used for working on the problem.

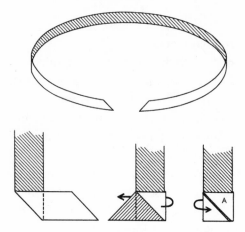

Figure 8.8

PROBLEM 8.9—Folding a Möbius Strip

Consider Stephen Barr's method of folding a Möbius strip from a square sheet of paper. The square [at left in Figure 8.9] is simply folded in half twice along the dotted lines, then edge *B* is taped to *B'*. The result is a band with a half-twist, one-sided and one-edged; it is a legitimate model of a Möbius surface even though it cannot be opened out for easy inspection.

Suppose instead of a square we use a paper rectangle twice as long as it is high [at right in Figure 8.9]. Is it possible to fold *this* into a Möbius surface that joins *B* to *B'*? One can fold or twist the paper in any way, but of course it must not be torn. Assume that the paper can be made as thin as desired. The surface must be given a half-twist that allows the entire length of edge *B* to be joined to the entire length of edge *B'*.

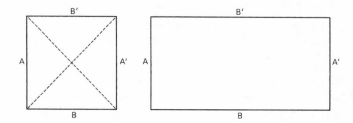

Figure 8.9

It would not be difficult to make the strip by joining A to A'; the problem is to find a way to do it by connecting the pair of longer edges.

Once the reader has either found a way to do it or concluded that it is impossible, a more interesting question arises: What is the smallest value for A/B that will allow a Möbius strip to be folded by the joining of B to B'?

Problem 8.10—Knot the Rope

Obtain a piece of clothesline rope about five-and-a-half feet long. Knot both ends to form a loop as shown in Figure 8.10. Each loop should be just large enough to allow you to squeeze a hand through it. With the loops around each wrist and the rope stretched between them, is it possible to tie a single overhand knot in the center of the rope? You may manipulate the rope any way you like, but of course you must not slide a loop off your wrist, cut the rope, or tamper with either of the existing knots. The trick is not well known except to magicians.

Figure 8.10

Problem 8.11—Reversed Trousers

Each end of a 10-foot length of rope is tied securely to a man's ankles. Without cutting or untying the rope, is it possible to remove his trousers, turn them inside out on the rope, and put them back on correctly? Party guests should try to answer this confusing topological question before initiating any empirical tests.

Problem 8.12—Ten-Penny Patterns

It turns out that 10 pennies can be arranged to form five straight lines, with four in each line. (It is assumed, of course, that each line must pass through the centers of the four coins on it.) Figure 8.11 shows five ways it can be done. Each pattern can be distorted in an infinity of ways without changing its topological structure; they are shown here in forms given by the English puzzlist Henry Ernest Dudeney to display bilateral symmetry for all of them. There is a sixth solution, topologically distinct from the other five. Can you discover it?

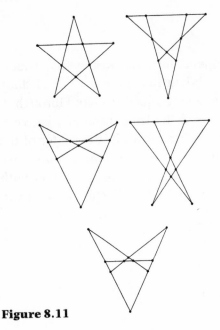

Figure 8.11

PROBLEM 8.13—Flexible Band

Gustavus J. Simmons, in charge of research and development at Rolamite Inc., Albuquerque, New Mexico, sent this curious topological problem. Work at Rolamite involves complex banded rolling systems. One of the Rolamite engineers, Virgil Erbert, was confronted in the course of his work with the problem shown in Figure 8.12. End *A* of a flexible band was fastened to an object that was too large to pass through the slot at end *B*. It was essential that the band be formed into

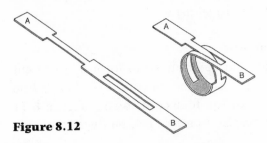

Figure 8.12

the looped configuration shown in the illustration without detaching end *A* from the object to which it was fastened. Can it be done?

It looks impossible, but the answer is yes. Draw a rough facsimile of the band on a sheet of paper, cut it out, and tape end *A* to a tabletop. The puzzle, which is not difficult, is to manipulate the strip into the looped configuration.

PROBLEM 8.14—Colored Poker Chips

What is the smallest number of poker chips that can be placed on a flat surface so that chips of three different colors are required to meet the condition that no two touching chips are the same color? As Figure 8.13 shows, the answer is obviously three.

Our problem is to determine the smallest number of chips, all of the same size, that can be placed flat on the plane so that chips of not three but four colors are necessary for meeting the same proviso.

Figure 8.13

PROBLEM 8.15—Polyomino Four-Color Problem

Polyominoes are shapes formed by joining unit squares. A single square is a monomino, two squares are a domino, three can be combined to make two types of trominoes, four make five different tetrominoes, and so on. I recently asked myself: What is the lowest order of polyomino four replicas of which can be placed so that every pair shares a common border segment? I believe, but cannot prove, that the octomino is the answer. Five solutions (there are more) were found by John W. Harris of Santa Barbara, California [see Figure 8.14]. If each piece is regarded as a region on a map, each pattern clearly requires four colors to prevent two bordering regions from having the same color.

Let us now remove the restriction to four replicas and ask: What is the lowest order of polyomino any number of replicas of which will

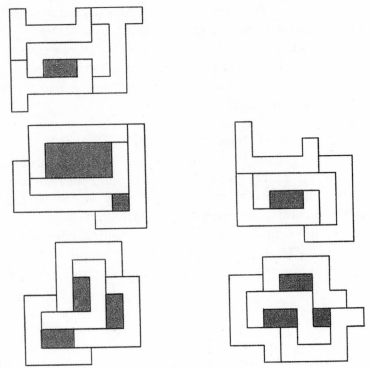

Figure 8.14

form a pattern that requires four colors? It is not necessary for any set of four to be mutually contiguous. It is only necessary that the replicas be placed so that, if each is given a color, four colors will be required to prevent two pieces of the same color from sharing a common border segment. Regions formed between the replicas are not considered part of the "map." They remain uncolored. The answer is a polyomino that is of much lower order than eight.

PROBLEM 8.16—Knotted Molecule

Enormously long chainlike molecules (long in relation to their breadth) have been discovered in living organisms. The question has arisen: Can such molecules have knotted forms? Max Delbrück of the California Institute of Technology, who received a Nobel prize in 1960, proposed the following idealized problem:

Assume that a chain of atoms, its ends joined to form a closed space curve, consists of rigid, straight-line segments each one unit long. At every node where two such "links" meet, a 90-degree angle is formed. At each end of each link, therefore, the next link may have one of four

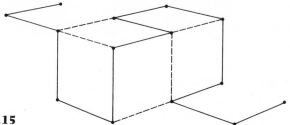

Figure 8.15

different orientations. The entire closed chain could be traced along the edges of a cubical lattice, with the proviso that at each node the joined links form a right angle [see Figure 8.15]. At no point is the chain allowed to touch or intersect itself; that is, two and only two links meet at every node.

What is the shortest chain of this type that is tied in a single overhand (trefoil) knot? In the answer I shall reproduce the shortest chain Delbrück has found. It has not been proved minimal; perhaps you will discover a shorter one. (I wish to thank John McKay for calling this problem to my attention.)

8.17—Toroidal Cannibalism

is a surface shaped like a doughnut. Imagine a torus made ...er. It is well known that if there is a hole in such a torus, ...e turned inside out through the hole.

...well, a mathematician at Monash University in Australia, ... the following problem. Two toruses, *A* and *B*, are linked as is ...own in Figure 8.16. There is a "mouth" (a hole) in *B*. We can stretch, compress, and deform either torus as radically as we please,

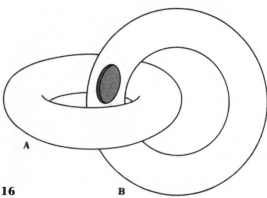

Figure 8.16   **B**

but of course no tearing is allowed. Can *B* swallow *A*? At the finish *B* must have its original shape, although it will be larger, and *A* must be entirely inside it.

PROBLEM 8.18—Toroidal Paradox

Two topologists were discussing at lunch the two linked surfaces shown at the left in Figure 8.17 which one of them had drawn on a paper napkin. You must not think of these objects as solids, like ropes or solid rubber rings. They are the surfaces of toruses, one surface of genus 1 (one hole), the other of genus 2 (two holes).

Thinking in the mode of "rubber-sheet geometry," assume that the surfaces in the illustration can be stretched or shrunk in any desired way provided there is no tearing or sticking together of separate parts. Can the two-hole torus be deformed so that one hole becomes unlinked as is shown at the right in the illustration?

Topologist *X* offers the following impossibility proof. Paint a ring on each torus as is shown by the black lines. At the left the rings are linked. At the right they are unlinked.

"You will agree," says *X*, "that it is impossible by continuous deformation to unlink two linked rings embedded in three-dimensional space. It therefore follows that the transformation is impossible."

"But it doesn't follow at all," says *Y*.

Who is right? I am indebted to Herbert Taylor for discovering and sending this mystifying problem.

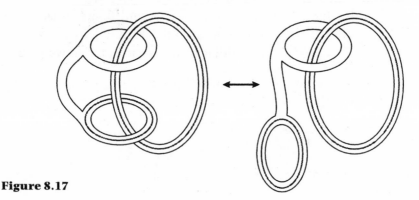

Figure 8.17

Answers

ANSWER 8.1—(July 1971)

The black curve divides the plane into a number of regions. Trace a round trip along the red curve and it is obvious that every region you enter you must also leave or you will never get back to where you started. Since each entrance and exit is a pair of crossing points, the total number of such spots must be even.

ANSWER 8.2—(July 1969)

Arrange the six matchstick as shown in Figure 8.18 to form a planar map requiring four colors.

Figure 8.18

ANSWER 8.3—(August 1968)

Cutting the second structure has the same result as cutting the first. Indeed, the result is the same large square regardless of the number of twists in the vertical band! For an additional surprise, see what happens when the untwisted band of the second structure is bisected and the twisted band is *trisected*.

ANSWER 8.4—(July 1971)

Only two of the eight possible combinations of crossings create a knot, making the probability of a knot 2/8 = 1/4.

James A. Ulrich was the first to argue that the probability of the string's being knotted is 1 because there is no way a closed loop of string can exist without its ends being tied.

ANSWER 8.5—(November 1965)

To transfer the key from one side of the door to the other, first pass the key through the loop so that it hangs as shown at the left of Figure 8.19. Seize the double cord at points *A* and *B* and pull the loop back through and out of the keyhole. This will pull two new loops out of the hole, as shown in the middle. Move the key up along the cord,

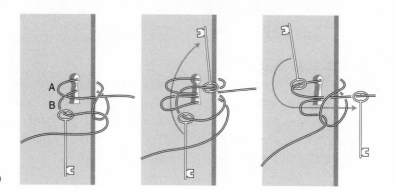

Figure 8.19

through both projecting loops. Grasp the two cords on the opposite side of the door and pull the two loops back through the hole, restoring the cord to its original state (right). Slide the key to the right, through the loop, and the job is done.

One reader, Allan Kiron, pointed out that if the cord is long enough, and one is allowed to remove the door from its hinges, the puzzle can be solved by passing a loop of cord over the entire door as if the door were a ring.

ANSWER 8.6—(February 1962)

Nineteen topologically distinct networks can be made with six matches, placing them on a plane so that no matches overlap and the matches touch only at their ends. The 19 networks are shown in Figure 8.20. If the restriction to a plane is dropped and three-space networks permitted, only one additional figure is possible: the skeleton of a tetrahedron.

Readers William G. Hoover and Victoria N. Hoover of Durham, North Carolina, Ronald C. Read of University of London, and Henry Eckhardt of Fair Oaks, California, extended the problem to seven matches and found 39 topologically distinct patterns.

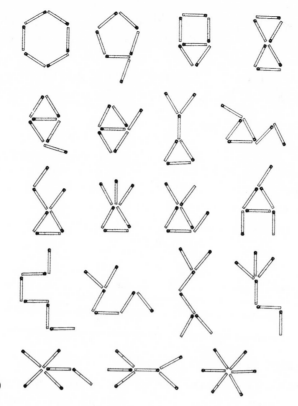

Figure 8.20

ANSWER 8.7—(November 1957)

A continuous line that enters and leaves one of the rectangular spaces must of necessity cross two line segments. Since the spaces labeled A, B, and C in Figure 8.21 are each surrounded by an odd number of segments, it follows that an end of a line must be inside each if all segments of the network are crossed. But a continuous line has only

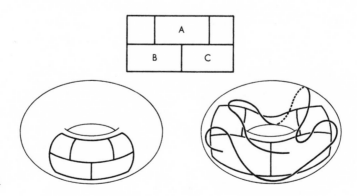

Figure 8.21

Topology

two ends, so the puzzle is insoluble on a plane surface. This same reasoning applies if the network is on a sphere or on the side of a torus [drawing at lower left]. However, the network can be drawn on the torus [drawing at lower right] so that the hole of the torus is *inside* one of the three spaces, A, B, and C. When this is done, the puzzle is easily solved.

ANSWER 8.8—(February 1960)

The simplest way to fold Barr's belt so that each end is straight across and part of a rectangle of uniform thickness is shown in Figure 8.22. This permits the neatest roll (the seams balance the long fold) and works regardless of the belt's length or the angles at which the ends are cut.

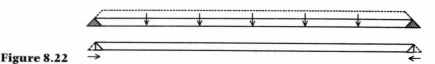

Figure 8.22

ANSWER 8.9—(November 1963)

What is the smallest value of *A/B* that allows one to join the *B* edges of the paper rectangle into a Möbius band? The surprising answer is that there *is* no minimum. The fraction *A/B* can be made as small as one pleases.

Proof is supplied by a folding technique explained in Stephen Barr's *Experiments in Topology* (New York: T. Y. Crowell, 1964). The strip is pleated as shown in Figure 8.23 to form a narrow strip with ends that show two-fold symmetry. After this narrow strip is given a half-twist the ends are joined. *Voilà!*

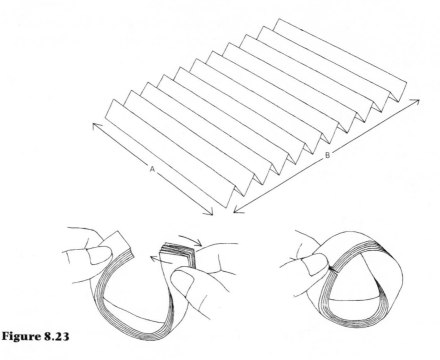

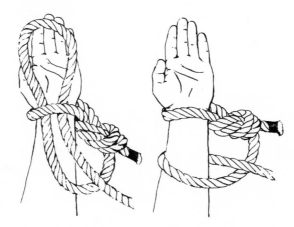

Figure 8.23

ANSWER 8.10—(November 1967)

To tie an overhand knot in the rope stretched from wrist to wrist, first push the center of the rope under the rope that circles the left wrist, as shown in Figure 8.24. Pass the loop over the left hand, then pull it back from under the rope circling the wrist. The loop will then be on the left arm, as shown at the right in the illustration. When the loop is taken off the arm, carrying it over the left hand, it will form an overhand knot in the rope.

Figure 8.24

If the loop, after it is first pushed under the rope around the wrist, is given a half-twist to the right before it is passed over the hand, the resulting knot will be a figure eight. And if the end of the loop is pushed through a ring before the loop goes over the hand, the ring will be firmly knotted to the rope after either type of knot is formed.

Van Cunningham and B. L. Schwartz each wrote to point out that the statement of the problem did not prohibit a second solution. Fold your arms, slip one hand through one loop, the other hand through the other loop, then unfold your arms.

ANSWER 8.11—(March 1965)

To reverse a man's trousers while his ankles are joined by rope, first slide the trousers off onto the rope, then push one leg through the other. The outside leg is reversed twice in this process, leaving the trousers on the rope right side out but with the legs exchanged and pointing toward the man's feet. Reach into the trousers from the waist and turn both legs inside out. The trousers are now reversed on the rope and in position to be slipped back on the man, zipper in front as originally but with the legs interchanged.

ANSWER 8.12—(February 1966)

The sixth configuration of 10 pennies having five straight lines of four coins each is shown in Figure 8.25.

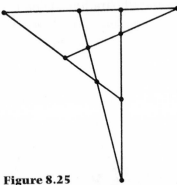

Figure 8.25

Readers Jonathan T. Y. Yeh and Eric Drew independently noticed that all six solutions can be generated from one basic pattern and a moving line. Figure 8.26 shows how a moving horizontal line gener-

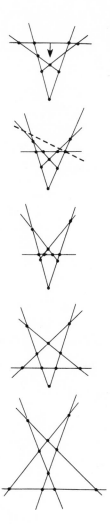

Figure 8.26

ates five patterns. The sixth is obtained by tilting the line as shown by the dotted line in the second pattern.

ANSWER 8.13—(April 1972)

Figure 8.27 shows how to form a loop with the Rolamite band while end *A* is taped to a tabletop.

Robert Neale suggested applying this to a playing card, say the joker. Use a razor blade to cut along the lines shown on the card at the left of Figure 8.28. Discard the shaded cut-out region. By carefully executing the trick bend with the little square loop, taking care not to

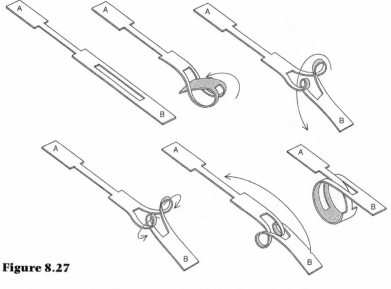

Figure 8.27

Figure 8.28

crease or tear its sides, you can produce the structure shown on the right. It is an amusing curiosity to carry in a wallet and show to friends. How the devil was it made? It looks, of course, as if the entire card had to be somehow pushed through the tiny window!

ANSWER 8.14—(May 1975)

It is easy to prove that the pattern of 11 circles [see Figure 8.29] requires at least four colors to ensure that no pair of touching circles are the same color. Assume that it can be done with three colors. Label circles 1, 2, and 3 with colors A, B, and C as shown. This determines the colors of 4, 5, and 6. We have a choice of two ways to color 7, but either way forces 11 to have the same color as 1, which it touches. Three colors are therefore not enough.

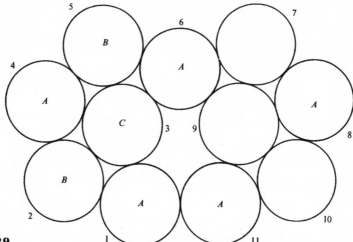

Figure 8.29

A number of readers sent proofs that 11 is the minimal number of circles. A proof by Allen J. Schwenk was published in *The American Mathematical Monthly* (vol. 83, June 1976, pp. 485–86) as a solution to part of question E 2527.

In three dimensions the least number of colors needed for any arrangement of identical spheres is not known, although it has been narrowed to 5, 6, 7, 8, or 9. For this and many other unsolved coloring problems, see "Coloring of Circles," by Brad Jackson and Gerhard Ringel (*The American Mathematical Monthly* 91, January 1984, pp. 42–49).

ANSWER 8.15—(February 1967)

Figure 8.30 shows my solution as to how as few as six dominoes can be placed so that if each is given a color, four colors are necessary to prevent two dominoes of the same color from touching along a border.

To my astonishment, two readers (Bent Schmidt-Nielsen and E. S. Ainley) found a way of doing it with as few as 11 monominoes (unit

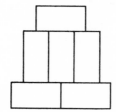

Figure 8.30

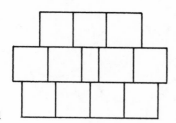

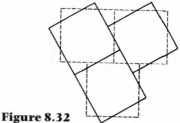

Figure 8.31

Figure 8.32

squares). Their solution is shown in Figure 8.31. Informal proofs that 11 is minimal were supplied by R. Vincent Kron and W. H. Grindley.

If we ask for the maximum number of monominoes that can be placed so each pair shares a common edge segment, the answer obviously is three. The question is not so easily answered for cubes; that is, what is the maximum number of cubes that can be placed so each pair shares a common surface? A common surface need not be an entire face, but it must be a surface, not a line or point. The answer is six. See Figure 8.32 for the pretty solution. The three cubes shown by the solid lines rest on three shown by dotted lines.

ANSWER 8.16—(November 1970)

The shortest knotted chain known that meets all the conditions specified has 36 links [see Figure 8.33]. It is reproduced from Max Delbrück's paper, "Knotting Problems in Biology," in *Mathematical Problems in the Biological Sciences: Proceedings of Symposia in Applied Mathematics*, vol. 14, 1962, pages 55–68.

If links are not required to be at right angles to their adjacent links, knots of 24 links are possible.

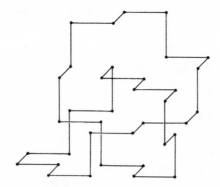

Figure 8.33

One torus can be inside another in two topologically distinct ways: the inside torus may surround the hole of the outside torus or it may not.

If two toruses are linked and one has a "mouth," it cannot swallow the other so that the eaten torus is inside in the second sense. This result can be proved by drawing a closed curve on each torus in such a way that the two curves are linked in a simple manner. No amount of deformation can unlink the two curves. If one torus could swallow the other in the manner described, however, it could disgorge the eaten torus through its mouth and the two toruses would be unlinked.

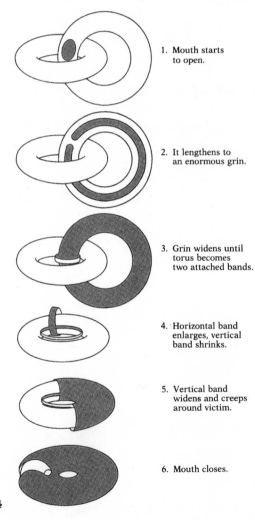

1. Mouth starts to open.

2. It lengthens to an enormous grin.

3. Grin widens until torus becomes two attached bands.

4. Horizontal band enlarges, vertical band shrinks.

5. Vertical band widens and creeps around victim.

6. Mouth closes.

Figure 8.34

This result would also unlink the two closed curves. Because unlinking is impossible, cannibalism of this kind is also impossible.

The torus with the mouth can, however, swallow the other one so that the eaten torus is inside in the first sense explained above. Figure 8.34 shows how it is done. In the process it is necessary for the cannibal torus to turn inside out.

A good way to understand what happens is to imagine that torus *A* is shrunk until it becomes a stripe of paint that circles *B*. Turn *A* inside out through its mouth. The painted stripe goes inside, but in doing so it ends up circling B's hole. Expand the stripe back to a torus and you have the final picture of the sequence.

Answer 8.18—(December 1979)

Figure 8.35 shows how a continuous deformation of the two-hole torus will unlink one of its holes from the single-hole torus. The argument for the impossibility of this task fails because if a ring is painted

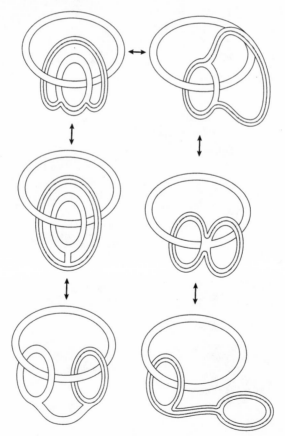

Figure 8.35

around one hole [shown by the black line], the ring becomes distorted in such a way that after the hole is unlinked the painted ring remains linked through the one-hole torus.

For a mind-boggling selection of similar problems involving linked toruses, see Herbert Taylor's article "Bicycle Tubes Inside Out," in *The Mathematical Gardner* (1981), edited by David Klarner.

part three

Algorithmic Puzzles & Games

chapter 9

Procedures

It can be argued that all good puzzles are ones that ask the reader to devise a method of solution. An algebraic problem that was posed before algebra was invented clearly required the solver to invent a plan of attack. However, after algebra was invented (in response to the need to solve such problems) solving these puzzles has become really more of an "exercise." There is little joy in solving math "puzzles" for which the method is well known.

An algorithm is a procedure for solving a problem, for transforming an input into an output. A large number of Gardner's problems ask only how to solve a problem, so the puzzle is to devise an algorithm, not to use an algorithm.

There are two classic types of algorithmic puzzles. First is the weighing problem, where you are given a scale and a set of objects/weights. You are then asked to solve a task in the fewest weighings. Second is the fair division problem. You are typically asked some variant of: How can you slice a cake into k pieces so that each recipient is satisfied that it is a fair procedure? The puzzles in this chapter emphasize these types of questions.

Problems

PROBLEM 9.1—Reaction Time

A chemist discovered that a certain chemical reaction took 80 minutes when he wore a jacket. When he was not wearing a jacket, the same reaction always took an hour and 20 minutes. Can you explain?

Problem 9.2—How Did He Know?

A customer in a restaurant found a dead fly in his coffee. He sent the waiter back for a fresh cup. After taking one sip he shouted, "This is the *same* cup of coffee I had before!" How did he know?

Problem 9.3—How Tall Is the Building?

Give at least three ways a barometer can be used to determine the height of a tall building.

Problem 9.4—Five Tires

A man traveled 5,000 miles in a car with one spare tire. He rotated tires at intervals so that when the trip ended each tire had been used for the same number of miles. For how many miles was each tire used?

Problem 9.5—Disappearing Dollar

This oldtimer still confuses almost everyone who hears it for the first time. Smith gave a hotel clerk $15 for his room for the night. When the clerk discovered that he had overcharged by $5, he sent a bellboy to Smith's room with five $1 bills. The dishonest bellboy gave only three to Smith, keeping the other two for himself. Smith has now paid $12 for his room. The bellboy has acquired $2. This accounts for $14. Where is the missing dollar?

Problem 9.6—Circular Lake

Figure 9.1 is a diagram of a deep circular lake, 300 yards in diameter, with a small island at the center. The two black spots are trees. A man who cannot swim has a rope a few yards longer than 300 yards. How does he use it as a means of getting to the island?

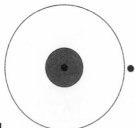

Figure 9.1

PROBLEM 9.7—Cocktail Cherry

This is one of those rare, delightful puzzles that can be solved at once if you approach it right, but that is subtly designed to misdirect your thoughts toward the wrong experimental patterns. Intelligent people have been known to struggle with it for 20 minutes before finally deciding that there is no solution.

Place four paper matches on top of the four matches that form the cocktail glass in Figure 9.2. The problem is to move two matches, and only two, to new positions so that the glass is re-formed in a different position and the cherry is *outside* the glass. The orientation of the glass may be altered but the empty glass must be congruent with the one illustrated. The drawing at A shows how two matches can be moved to turn the glass upside down. This fails to solve the problem, however, because the cherry remains inside. The drawing at B shows a way to empty the glass, but this does not solve the puzzle either because three, rather than two, matches have been moved.

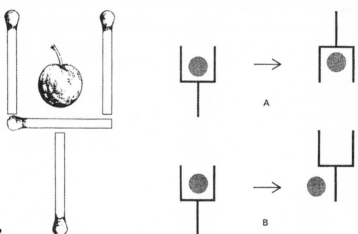

Figure 9.2

PROBLEM 9.8—Interrupted Bridge Game

A telephone call interrupts a man after he has dealt about half of the cards in a bridge game. When he returns to the table, no one can remember where he had dealt the last card. Without learning the number of cards in any of the four partly dealt hands, or the number of cards yet to be dealt, how can he continue to deal accurately, everyone getting exactly the same cards he would have had if the deal had not been interrupted?

PROBLEM 9.9—Million Points

An infinity of nontouching points lie inside the closed curve shown in Figure 9.3. Assume that a million of those points are selected at random. Will it always be possible to place a straight line on the plane so that it cuts across the curve, misses every point in the set of a million, and divides the set exactly in half so that 500,000 points lie on each side of the line? The answer is yes; prove it.

Figure 9.3

PROBLEM 9.10—Lavinia Seeks a Room

The line in Figure 9.4 represents University Avenue in a small college town where Lavinia is a student. The spots labeled *A* through *K* are buildings along the avenue in which Lavinia's 11 best friends are living.

Lavinia has been living with her parents in a nearby town, but now she wants to move to University Avenue. She would like a room or an apartment at a location *L* on the street that minimizes the sum of all its distances from her 11 friends. Assuming that a place is available at the right location, specify where Lavinia should live and prove it does make the sum of all its distances to the other locations as small as possible.

Figure 9.4

PROBLEM 9.11—Share and Share Alike

There is a simple procedure by which two people can divide a cake so that each is satisfied he has at least half: One cuts and the other chooses. Devise a general procedure so that *n* persons can cut a cake into *n* portions in such a way that everyone is satisfied he has at least 1/*n* of the cake.

ALGORITHMIC PUZZLES & GAMES

PROBLEM 9.12—Scrambled Box Tops

Imagine that you have three boxes, one containing two black marbles, one containing two white marbles, and the third, one black marble and one white marble. The boxes were labeled for their contents—BB, WW, and BW—but someone has switched the labels so that every box is now incorrectly labeled. You are allowed to take one marble at a time out of any box, without looking inside, and by this process of sampling you are to determine the contents of all three boxes. What is the smallest number of drawings needed to do this?

PROBLEM 9.13—How Did Kant Set His Clock?

It is said that Immanuel Kant was a bachelor of such regular habits that the good people of Königsberg would adjust their clocks when they saw him stroll past certain landmarks.

One evening Kant was dismayed to discover that his clock had run down. Evidently his manservant, who had taken the day off, had forgotten to wind it. The great philosopher did not reset the hands because his watch was being repaired and he had no way of knowing the correct time. He walked to the home of his friend Schmidt, a merchant who lived a mile or so away, glancing at the clock in Schmidt's hallway as he entered the house.

After visiting Schmidt for several hours Kant left and walked home along the route by which he came. As always, he walked with a slow, steady gait that had not varied in 20 years. He had no notion of how long this return trip took. (Schmidt had recently moved into the area and Kant had not yet timed himself on this walk.) Nevertheless, when Kant entered his house, he immediately set his clock correctly.

How did Kant know the correct time?

PROBLEM 9.14—Measuring with Yen

This problem was originated by Mitsunobu Matsuyama, a reader in Tokyo. He sent me a supply of Japanese one-yen coins and told me of the following remarkable facts about them that are not well known even in Japan. The one-yen coin is made of pure aluminum, has a radius of exactly one centimeter, and weighs just one gram. Thus a supply of yens can be used with a balance scale for determining the weight of small objects in grams. It also can be used on a plane surface for measuring distances in centimeters.

It is easy to see how one-yen coins can be placed on a line to measure distances in even centimeters (two centimeters, four, six, and so on), but can they also be used to measure odd distances (one, three, five, and so on)? Show how a supply of one-yen coins can be used for measuring all integral distances in centimeters along a line.

PROBLEM 9.15—Counterfeit Coins

In recent years a number of clever coin-weighing or ball-weighing problems have aroused widespread interest. Here is a new and charmingly simple variation. You have 10 stacks of coins, each consisting of 10 half-dollars [see Figure 9.5]. One entire stack is counterfeit, but you do not know which one. You do know the weight of a genuine half-dollar and you are also told that each counterfeit coin weighs one gram more than it should. You may weigh the coins on a pointer scale. What is the smallest number of weighings necessary to determine which stack is counterfeit?

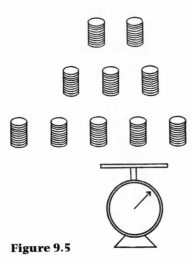

Figure 9.5

PROBLEM 9.16—Fair Division

Two brothers inherited a herd of sheep. They sold all of them, receiving for each sheep the same number of dollars as there were sheep in the herd. The money was given to them in $10 bills except for an excess amount, less than $10, that was in silver dollars. They divided the bills between them by placing them on a table and alternately taking a bill until there were none left.

"It isn't fair," complained the younger brother. "You drew first and you also took the last bill, so you got $10 more than I did."

To even things up partially the older brother gave the younger one all the silver dollars, but the younger brother was still not satisfied. "You gave me less than $10," he argued. "You still owe me some money."

"True," said the older brother. "Suppose I write you a check that will make the total amounts we each end up with exactly the same."

This he did. What was the value of the check? The information seems inadequate, but nevertheless the question can be answered.

Ronald A. Wohl, a chemist at Rutgers University, called my attention recently to this beautiful problem, which he had found in a French book. Later I discovered in my files a letter from Carl J. Coe, a retired mathematician at the University of Michigan, discussing essentially the same problem, which he said had been making the rounds among his colleagues in the 1950s. I suspect it is still not widely known.

PROBLEM 9.17—Ranking Weights

Five objects, no two the same weight, are to be ranked in order of increasing weight. You have available a balance scale but no weights. How can you rank the objects correctly in no more than seven separate weighings?

For two objects, of course, only one weighing is required. Three objects call for three weighings. The first determines that A is heavier than B. We then weigh B against C. If B is heavier, we have solved the problem in two weighings, but if C is heavier, a third weighing is required to compare C with A. Four objects can be ranked easily with no more than five weighings.

With five objects the problem ceases to be trivial. As far as I know, no general method for ranking n objects with a minimum number of weighings has yet been established.

PROBLEM 9.18—Gold Links

Lenox R. Lohr, president of the Museum of Science and Industry in Chicago, was kind enough to pass along the following deceptively simple version of a type of combinatorial problem that turns up in many fields of applied mathematics. A traveler finds himself in a strange town without funds; he expects a large check to arrive in a few weeks. His most valuable possession is a gold watch chain of 23 links.

To pay for a room he arranges with a landlady to give her as collateral one link a day for 23 days.

Naturally the traveler wants to damage his watch chain as little as possible. Instead of giving the landlady a separate link each day he can give her one link the first day, then on the second day take back the link and give her a chain of two links. On the third day he can give her the single link again and on the fourth take back all she has and give her a chain of four links. All that matters is that each day she must be in possession of a number of links that corresponds to the number of days.

The traveler soon realizes that this can be accomplished by cutting the chain in many different ways. The problem is: What is the smallest number of links the traveler needs to cut to carry out his agreement for the full 23 days? More advanced mathematicians may wish to obtain a general formula for the longest chain that can be used in this manner after n cuts are made at the optimum places.

PROBLEM 9.19—Efficient Electrician

An electrician is faced with this annoying dilemma. In the basement of a three-story house he finds bunched together in a hole in the wall the exposed ends of 11 wires, all alike. In a hole in the wall on the top floor he finds the other ends of the same 11 wires, but he has no way of knowing which end above belongs to which end below. His problem: to match the ends.

To accomplish his task he can do two things: (1) short-circuit the wires at either spot by twisting ends together in any manner he wishes; (2) test for a closed circuit by means of a "continuity tester" consisting of a battery and a bell. The bell rings when the instrument is applied to two ends of a continuous, unbroken circuit.

Not wishing to exhaust himself by needless stair-climbing, and having a passionate interest in operations research, the electrician sat down on the top floor with pencil and paper and soon devised the most efficient possible method of labeling the wires.

What was his method?

PROBLEM 9.20—Cutting a Cake

A cake has been baked in the form of a rectangular parallelepiped with a square base. Assume that the square cake is frosted on the top and four sides and that the frosting's thickness is negligible (zero). We want to cut the cake into n pieces so that each piece has the same volume and the same area of frosting. The slicing is conventional. Seen from above, the cuts are like spokes radiating from the square's center, and each cutting plane is perpendicular to the cake's base.

If n is 2, 4, or 8, the problem is easily solved by slicing the cake into two, four, or eight congruent solids. Suppose, however, that n is 7. How can we locate the required seven points on the perimeter of the cake's top? If you solve it for 7, you will be able to generalize to any n.

This pretty problem is given by H.S.M. Coxeter on page 38 of his classic *Introduction to Geometry* (Wiley, 1967). Coxeter does not answer it, but it should give you little difficulty. The general solution is surprisingly simple.

PROBLEM 9.21—Water and Wine

A familiar chestnut concerns two beakers, one containing water, the other wine. A certain amount of water is transferred to the wine, then the same amount of the mixture is transferred back to the water. Is there now more water in the wine than there is wine in the water? The answer is that the two quantities are the same.

Raymond Smullyan writes to raise the further question: Assume that at the outset one beaker holds 10 ounces of water and the other holds 10 ounces of wine. By transferring three ounces back and forth any number of times, stirring after each transfer, is it possible to reach a point at which the percentage of wine in each mixture is the same?

PROBLEM 9.22—Red, White, and Blue Weights

Here is an unusual problem involving weights and balance scales. It was invented by Paul Curry, who is well known in conjuring circles as an amateur magician.

You have six weights. One pair is red, one pair white, one pair blue. In each pair one weight is a trifle heavier than the other but otherwise appears to be exactly like its mate. The three heavier weights (one of each color) all weigh the same. This is also true of the three lighter weights.

In two separate weighings on a balance scale, how can you identify which is the heavier weight of each pair?

PROBLEM 9.23—Poisoned Glass

"Mathematicians are curious birds," the police commissioner said to his wife. "You see, we had all those partly filled glasses lined up in rows on a table in the hotel kitchen. Only one contained poison, and we wanted to know which one before searching that glass for fingerprints. Our laboratory could test the liquid in each glass, but the tests take time and money, so we wanted to make as few of them as possible. We phoned the university and they sent over a mathematics professor to help us. He counted the glasses, smiled and said:

"'Pick any glass you want, Commissioner. We'll test it first.'

"'But won't that waste a test?' I asked.

"'No,' he said, 'it's part of the best procedure. We can test one glass first. It doesn't matter which one.'"

"How many glasses were there to start with?" the commissioner's wife asked.

"I don't remember. Somewhere between 100 and 200."

What was the exact number of glasses? (It is assumed that any group of glasses can be tested simultaneously by taking a small sample of liquid from each, mixing the samples and making a single test of the mixture.)

PROBLEM 9.24—Redistribution in Oilaria

"Redistributive justice" is a phrase much heard in arguments among political philosophers. Should the ideal modern industrial state tax the rich for the purpose of redistributing wealth to the poor? Yes, says Harvard philosopher John Rawls in his influential book *A Theory of Justice*. No, says his colleague (they have adjoining offices) Robert Nozick in his controversial defense of extreme libertarianism, *Anarchism, State and Utopia*. It is hard to imagine how two respected political theorists, both believing in democracy and free enterprise, could hold such opposing views on the desirable powers of government.

The Sheik of Oilaria, in this problem devised by Walter Penney of Greenbelt, Maryland, has never heard of Rawls or Nozick, but he has proposed the following share-the-wealth program for his sheikdom. The population is divided into five economic classes. Class 1 is the poorest, class 2 is the next-poorest, and so on to class 5, which is the richest. The plan is to average the wealth by pairs, starting with classes 1 and 2, then 2 and 3, then 3 and 4, and finally 4 and 5. Averaging means that the total wealth of the two classes is redistributed evenly to everyone in the two classes.

The Sheik's Grand Vizier approves the plan but suggests that averaging begin with the two richest classes, then proceed down the scale instead of up.

Which plan would the poorest class prefer? Which would the richest class prefer?

PROBLEM 9.25—Uneven Floor

A kitchen has an uneven floor. There are no "steps," but the continuous random waviness of the linoleum is such that when one tries to place on it a small square table with four legs, one leg is usually off the floor, causing the table to wobble. If one does not mind the table-top being on a slant, is it always possible to find a place where all four legs are firmly on the floor? Or can a floor wave in such a way that no such spot is available? The problem can be answered by a simple, elegant proof.

PROBLEM 9.26—Lady on the Lake

A young lady was vacationing on Circle Lake, a large artificial body of water named for its precisely circular shape. To escape from a man who was pursuing her, she got into a rowboat and rowed to the center of the lake, where a raft was anchored. The man decided to wait it out on shore. He knew she would have to come ashore eventually. Since he could run four times as fast as she could row, he assumed that it would be a simple matter to catch her as soon as her boat touched the lake's edge.

But the girl—a mathematics major at Radcliffe—gave some thought to her predicament. She knew that once she was on solid ground she could outrun the man; it was only necessary to devise a rowing strategy that would get her to a point on shore before *he* could get there. She soon hit on a simple plan, and her applied mathematics applied successfully.

What was the girl's strategy? (For puzzle purposes it is assumed that she knows at all times her exact position on the lake.)

PROBLEM 9.27—Rotating Table

Imagine a square table that rotates about its center. At each corner is a deep well, and at the bottom of each well is a drinking glass that is either upright or inverted. You cannot see into the wells, but you can reach into them and feel whether a glass is turned up or down.

A move is defined as follows: Spin the table, and when it stops, put each hand into a different well. You may adjust the orientation of the glasses any way you like, that is, you may leave them as they are or turn one glass or both.

Now, spin the table again and repeat the same procedure for your second move. When the table stops spinning, there is no way to distinguish its corners, and so you have only two choices: you may reach into any diagonal pair of wells or into any adjacent pair. The object is to get all four glasses turned in the same way, either all up or all down. When this task is accomplished, a bell rings.

At the start, the glasses in the four wells are turned up or down at random. If they all happen to be turned in the same direction at this point, the bell will ring at once and the task will have been accomplished before any moves were made. Therefore it should be assumed that at the start the glasses are not all turned the same way. It is also assumed that you are not allowed to keep your hands in the wells and make experimental turnings to see if the bell rings. Furthermore, you must announce in advance the two wells to be probed at each step. You cannot probe one well, then decide which other well to probe.

Is there a procedure guaranteed to make the bell ring in a finite number of moves? Many people, after thinking briefly about this problem, conclude that there is no such procedure. It is a question of probability, they reason. With bad luck one might continue to make moves indefinitely. That is not the case, however. After no more than n correct moves one can be certain of ringing the bell. What is the minimum value of n, and what procedure is sure to make the bell ring in n or fewer moves?

Consider a table with only two corners and hence only two wells. In this case one move obviously suffices to make the bell ring. If there are three wells (at the corners of a triangular table), the following two moves suffice.

1. Reach into any pair of wells. If both glasses are turned the same way, invert both of them, and the bell will ring. If they are turned in different directions, invert the glass that is facing down. If the bell does not ring:

2. Spin the table and reach into any pair of wells. If both glasses are

turned up, invert both, and the bell will ring. If they are turned in different directions, invert the glass turned down, and the bell will ring.

Although the problem can be solved in a finite number of moves when there are four wells and four glasses, it turns out that if there are five or more glasses (at the corners of tables with five or more sides), there are no procedures guaranteed to complete the task in n moves.

PROBLEM 9.28—Dragon Curve

A weird cover design decorated a booklet that William G. Harter, then a candidate for a doctorate in physics at the University of California at Irvine, prepared for a National Aeronautics and Space Administration seminar on group theory that he had taught the previous summer at NASA's Lewis Research Center in Cleveland [see Figure 9.6]. The "dragon curve," as he calls it, was discovered by a NASA colleague, physicist John E. Heighway, and later analyzed by Harter, Heighway, and Bruce A. Banks, another NASA physicist. The curve is not connected with group theory, but it was used by Harter to symbolize what he calls "the proliferation of cryptic structure that one finds in this discipline." It is drawn here as a fantastic path along the lattice lines of graph paper, with each right-angle turn rounded off to make it

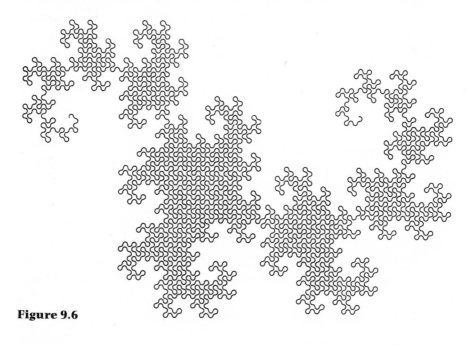

Figure 9.6

Procedures

clear that the path never crosses itself. You will see that the curve vaguely resembles a sea dragon paddling to the left with clawed feet, his curved snout and coiled tail just above an imaginary waterline.

Can you find a simple method of generating the dragon curve? In the answer I shall explain three: one based on a sequence of binary digits, one on a way of folding paper, and one on a geometric construction. It was the second procedure that led to the discovery of the curve. I shall also explain the connection with the logarithmic spiral, although this was not noticed until later and it plays no role in the construction of the curve.

PROBLEM 9.29—Playing Twenty Questions

In the well-known game Twenty Questions one person thinks of an object, such as the Liberty Bell in Philadelphia or Lawrence Welk's left little toe, and another person tries to guess the object by asking no more than 20 questions, each answerable by yes or no. The best questions are usually those that divide the set of possible objects into two subsets as nearly equal in number as possible. Thus if a person has chosen as his "object" a number from 1 through 9, it can be guessed by this procedure in no more than four questions—possibly less. In 20 questions one can guess any number from 1 through 2^{20} (or 1,048,576).

Suppose that each of the possible objects can be given a different value to represent the probability that it has been chosen. For example, assume that a deck of cards consists of one ace of spades, two deuces of spades, three threes, and on up to nine nines, making 45 spade cards in all. The deck is shuffled; someone draws a card. You are to guess it by asking yes-no questions. How can you minimize the number of questions that you will probably have to ask?

PROBLEM 9.30—Triangular Duel

Smith, Brown, and Jones agree to fight a pistol duel under the following unusual conditions. After drawing lots to determine who fires first, second, and third, they take their places at the corners of an equilateral triangle. It is agreed that they will fire single shots in turn and continue in the same cyclic order until two of them are dead. At each turn the man who is firing may aim wherever he pleases. All three duelists know that Smith always hits his target, Brown is 80 percent accurate and Jones is 50 percent accurate.

Assuming that all three adopt the best strategy, and that no one is killed by a wild shot not intended for him, who has the best chance to survive? A more difficult question: What are the exact survival probabilities of the three men?

Answers

ANSWER 9.1—(August 1968)
Eighty minutes are the same as one hour and 20 minutes.

ANSWER 9.2—(August 1968)
The customer had sugared his coffee before he found the dead fly.

ANSWER 9.3—(August 1968)
Here are five ways to use the barometer to determine the building's height:

1. Lower the barometer by a string from the roof to the street, pull it up, and measure the string.
2. Same as above except instead of pulling the barometer up, let it swing like a pendulum and calculate the length of the string by the pendulum's frequency. (Thanks to Dick Akers for this one.)
3. Drop the barometer off the roof, note the time it takes to fall, and compute the distance from the formula for falling bodies.
4. On a sunny day, find the ratio of the barometer's height to the length of its shadow and apply this ratio to the length of the building's shadow.
5. Find the superintendent and offer to give him the barometer if he will tell you the height of the building.

Solution (1) is quite old (I heard it from my father when I was a boy), but the most complete discussion of the problem, with all solutions except (2), is in Alexander Calandra's book *The Teaching of Elementary Science and Mathematics* (Ballwin, MO: ACCE Reporter, 1969). Calandra's earlier discussion of the problem, in the teacher's edition of *Current Science*, was the basis of a *New York Times* story, March 8, 1964, page 56.

ANSWER 9.4—(April 1969)
Each tire is used 4/5 of the total time. Therefore each tire has been used for 4/5 of 5,000 miles, or 4,000 miles.

ANSWER 9.5—(August 1968)

Adding the bellboy's $2 to the $12 Smith paid for his room produces a meaningless sum. Smith is out $12, of which the clerk has $10 and the bellboy $2. Smith got back $3, which, added to the $12 held by the clerk and the bellboy, accounts for the full amount of $15.

ANSWER 9.6—(July 1971)

He ties one end of the rope to the tree at the edge of the lake, walks around the lake holding the other end of the rope and ties that end to the same tree. The doubled rope is now firmly stretched between the two trees, making it easy for the man to pull himself through the water, by means of the rope, to the island.

William B. Friedman proposed placing the rope so that half of it is high above the water and the other half still higher. The man could then walk along the lower rope, holding on to the upper one, and not even get wet. If the rope is hemp, wrote P. H. Lyons, the man could smoke it and fly to the island.

ANSWER 9.7—(November 1967)

Figure 9.7 shows how two matches are moved to re-form the cocktail glass with the cherry outside.

Figure 9.7

ANSWER 9.8—(February 1967)

He deals the bottom card to himself, then continues dealing from the bottom counterclockwise.

ANSWER 9.9—(November 1965)

It is easy to show that for any finite set of points on the plane there must be an infinity of straight lines that divide the set exactly in half. The following proof for the six points in Figure 9.8 applies to any finite number of points.

Consider every line determined by every pair of points. Pick a new point, A, that lies outside a closed curve surrounding all the other points and that does *not* lie on any of the lines. Draw a line through point A. As this line is rotated about point A, in the direction shown, it must pass over one point at a time. (It cannot pass two points simultaneously; this would mean that point A lay on a line determined by those two points.) After it has passed half of the points inside the curve it will divide the set of points in half. Since A can be given an infinity of positions, there is an infinity of such lines.

This problem is based on a quickie problem contributed by Herbert Wills to *The Mathematical Gazette*, May 1964.

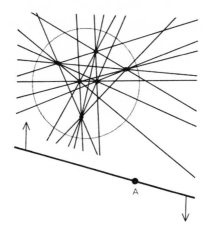

Figure 9.8

ANSWER 9.10—(April 1981)

Consider the two outermost spots, A and K. A spot L on the line anywhere between A and K (inclusive) will have the same sum of its distances to A and K. Clearly this is smaller than the sum if L is not between A and K. Now consider B and J, the next pair of spots as you move inward. As before, to minimize the sum of L's distances from B and J, L must be between B and J. Since L is then also between A and K, its position will minimize the sum of its distances to A, K, B, and J.

Continue in this way, taking the spots by pairs as you move inward to new nested intervals along the line. The last pair of spots is E and G. Between them is the single spot F. Any spot between E and G will minimize the distances to all spots except F. Obviously if you want also to minimize the distance to F, the spot must be exactly at F. In brief, Lavinia should move into the same building where her friend Frank lives.

To generalize, for any even number of spots on a straight line, any spot between the two middle spots will have a minimal sum of distances to all spots. For any odd number of spots the center spot is the desired location. The problem appeared in "No Calculus, Please," by J. H. Butchart and Leo Moser (*Scripta Mathematica*, vol. 18, nos. 3–4, September–December 1952, pages 221–36).

Thomas Szirtes made the following comment on the problem about Lavinia's room: "The solution is interesting because it is contrary to intuition. According to the rules, the location of the minimum-sum distance is independent of the relative or even the absolute distances among the points. One would feel, for example, that if the rightmost point K receded 100 miles, it would somehow 'pull' the minimum sum distance point to the right. But this is not the case. In fact, all points to the right of F could recede to infinity and all points to the left of F could be infinitely close to F, and the minimum-sum distance point still would be at F!"

Answer 9.11—(May 1959)

Several procedures have been devised by which n persons can divide a cake in n pieces so that each is satisfied that he has at least $1/n$ of the cake. The following system has the merit of leaving no excess bits of cake.

Suppose there are five persons: A, B, C, D, E. A cuts off what he regards as 1/5 of the cake and what he is content to keep as his share. B now has the privilege, if he thinks A's slice is more than 1/5, of reducing it to what he thinks is 1/5 by cutting off a portion. Of course if he thinks it is 1/5 or less, he does not touch it. C, D, and E in turn now have the same privilege. The last person to touch the slice keeps it as his share. Anyone who thinks that this person got less than 1/5 is naturally pleased because it means, in his eyes, that more than 4/5 remains. The remainder of the cake, including any cut-off pieces, is now divided among the remaining four persons in the same manner, then among three. The final division is made by one person cutting and the other choosing. The procedure is clearly applicable to any number of persons.

For a discussion of this and other solutions, see the section "Games of Fair Division," pages 363–68, in *Games and Decisions*, by R. Duncan Luce and Howard Raiffa (John Wiley and Sons, Inc., 1957).

ANSWER 9.12—(February 1957)

You can learn the contents of all three boxes by drawing just one marble. The key to the solution is your knowledge that the labels on all three of the boxes are incorrect. You must draw a marble from the box labeled "black-white." Assume that the marble drawn is black. You know then that the other marble in this box must be black also, otherwise the label would be correct. Since you have now identified the box containing two black marbles, you can at once tell the contents of the box marked "white-white": you know it cannot contain two white marbles, because its label has to be wrong; it cannot contain two black marbles, for you have identified that box; therefore it must contain one black and one white marble. The third box, of course, must then be the one holding two white marbles. You can solve the puzzle by the same reasoning if the marble you draw from the "black-white" box happens to be white instead of black.

ANSWER 9.13—(June 1961)

Immanuel Kant calculated the exact time of his arrival home as follows. He had wound his clock before leaving, so a glance at its face told him the amount of time that had elapsed during his absence. From this he subtracted the length of time spent with Schmidt (having checked Schmidt's hallway clock when he arrived and again when he left). This gave him the total time spent in walking. Since he returned along the same route, at the same speed, he halved the total walking time to obtain the length of time it took him to walk home. This added to the time of his departure from Schmidt's house gave him the time of his arrival home.

Winston Jones, of Johannesburg, South Africa, wrote to suggest another solution. Mr. Schmidt, in addition to being Kant's friend, was also his watchmaker. So while Kant sat and chatted with him, he repaired Kant's watch.

ANSWER 9.14—(April 1981)

Integral distances in centimeters along a line can be measured by one-yen Japanese coins, each with a radius of one centimeter, in the manner shown in Figure 9.9.

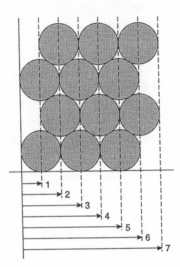

Figure 9.9

ANSWER 9.15—(February 1957)

The counterfeit stack can be identified by a single weighing of coins. You take one coin from the first stack, two from the second, three from the third, and so on to the entire 10 coins of the tenth stack. You then weigh the whole sample collection on the pointer scale. The excess weight of this collection, in number of grams, corresponds to the number of the counterfeit stack. For example, if the group of coins weighs seven grams more than it should, then the counterfeit stack must be the seventh one, from which you took seven coins (each weighing one gram more than a genuine half-dollar). Even if there had been an eleventh stack of 10 coins, the procedure just described would still work, for no excess weight would indicate that the one remaining stack was counterfeit.

ANSWER 9.16—(November 1967)

We were told that the two brothers who inherited a herd of sheep sold each sheep for the same number of dollars as there were sheep. If the number of sheep is n, the total number of dollars received is n^2. This was paid in $10 bills plus an excess, less than 10, in silver dollars.

By alternately taking bills, the older brother drew both first and last, and so the total amount must contain an odd number of 10's. Since the square of any multiple of 10 contains an even number of 10's, we conclude that n (the number of sheep) must end in a digit the square of which contains an odd number of 10's. Only two digits, 4 and 6, have such squares: 16 and 36. Both squares end in 6, and so n^2 (the total amount received for the sheep) is a number ending in 6. The excess amount consisted of six silver dollars.

After the younger brother took the $6 he still had $4 less than his brother, so to even things up the older brother wrote a check for $2. It is surprising how many good mathematicians will work the problem correctly up to this last step, then forget that the check must be $2 instead of $4.

ANSWER 9.17—(June 1964)

Five objects can be ranked according to weight with no more than seven weighings on a balance scale:

1. Weigh A against B. Assume that B is heavier.
2. Weigh C against D. Assume that D is heavier.
3. Weigh B against D. Assume that D is heavier. We now have ranked three objects: $D > B > C$.
4. Weigh E against B.
5. If E is heavier than B, we now weigh it against D. If E is lighter than B, we weigh it against A. In either case E is brought into the series so that we obtain a rank order of four objects. Assume that the order is $D > B > E > A$. We already know (from step 2) how the remaining object C compares with D. Therefore we have only to find C's place with respect to the rank order of the other three. This can always be done in two weighings. In this case:
6. Weigh C against E.
7. If C is heavier than E, weigh it against B. If C is lighter than E, weigh it against A.

The general problem of ranking n weights with a minimum number of weighings (or n tournament players with a minimum number of no-draw two-person contests) was first proposed by Hugo Steinhaus. He discusses it briefly in the 1950 edition of *Mathematical Snapshots* and includes it as Problem 52 (with $n = 5$) in *One Hundred Problems in Elementary Mathematics* (New York: Basic Books, 1964). In *Mathe-*

matical Snapshots (New York: Oxford University Press, 1968), Steinhaus gives a formula that provides correct answers through $n = 11$. (For 1 through 11 the minimum number of weighings are 0, 1, 3, 5, 7, 10, 13, 16, 19, 22, 26.) The formula predicts 29 weighings for 12 objects, but it has been proved that the minimum number is 30.

The general problem is discussed by Lester R. Ford and Selmer M. Johnson, both of the Rand Corporation, in "A Tournament Problem," in *The American Mathematical Monthly*, May 1969. For a more recent discussion of this and closely related problems, see Section 5-3-1 of Donald Knuth's *Sorting and Searching* (Reading, MA: Addison-Wesley, 1971).

ANSWER 9.18—(November 1963)

The traveler with a 23-link gold chain can give his landlady one link a day for 23 days if he cuts as few as 2 links of the chain. By cutting the fourth and eleventh links he obtains two segments containing one link and segments of 3, 6, and 12 links. Combining these segments in various ways will make a set of any number of lengths from 1 to 23.

The formula for the maximum length of chain that can be handled in this way with n cuts is

$$(n + 1)2^{n+1} - 1$$

Thus one cut (link 3) is sufficient for a chain of 7 links, three cuts (links 5, 14, 31) for a chain of 63 links, and so on.

ANSWER 9.19—(November 1957)

On the top floor the electrician shorted five pairs of wires (the shorted pairs are connected by broken lines in Figure 9.10), leaving one free wire. Then he walked to the basement and identified the lower ends of the shorted pairs by means of his "continuity tester." He labeled the ends as shown, then shorted them in the manner indicated by the dotted lines.

Back on the top floor, he removed all the shorts but left the wires twisted at insulated portions so that the pairs were still identifiable. He then checked for continuity between the free wire (which he knew to be the upper end of F) and some other wire. When he found the other wire, he was able at once to label it E_2 and to identify its mate as E_1. He next tested for continuity between E_1 and another end which,

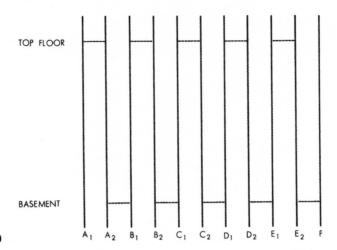

TOP FLOOR

BASEMENT

Figure 9.10

A_1 A_2 B_1 B_2 C_1 C_2 D_1 D_2 E_1 E_2 F

when found, could be marked D_2 and its mate D_1. Continuing in this fashion, the remaining ends were easily identified. The procedure obviously works for any odd number of wires.

J. G. Fletcher of Princeton, New Jersey, was the first to send a method of applying the above procedure, with a slight modification, to any even number of wires except two. Assume there is a twelfth wire on the far right in Figure 9.10. The same five pairs are shorted on the top floor, leaving two free wires. In the basement, the wires are shorted as before, and the twelfth wire is labeled G. Back on the top floor, G is easily identified as the only one of the two free wires in which no continuity is found. The remaining 11 wires are than labeled as previously explained.

In some ways a more efficient procedure, which takes care of all cases except two wires (two wires have no solution), was sent in by D. N. Buell, Detroit; R. Elsdon-Dew, Durban, South Africa; Louis Katz and Fremont Reizman, physics students at the University of Wisconsin; and Danforth K. Gannett, Denville, New Jersey. Mr. Gannett explained it clearly with the diagram for 15 wires shown in Figure 9.11. The method of labeling is as follows:

1. Top floor: short wires in groups of 1, 2, 3, 4, . . . Label the groups A, B, C, D,. . . . The last group need not be complete.
2. Basement: identify the groups by continuity tests. Number the wires and short them in groups Z, Y, X, W, V,. . . .
3. Top floor: remove the shorts. Continuity tests will now uniquely identify all wires. Wire 1 is of course A. Wire 3 is the only wire in

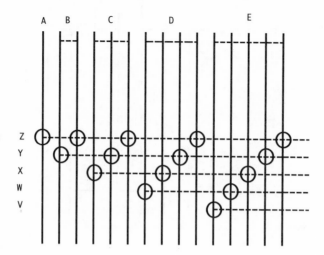

Figure 9.11

group B that has continuity with 1. Its mate will be 2. In group C, only wire 6 connects with 1. Only 5 connects with 2. The remaining wire in C will be 4. And so on for the other groups.

The chart can be extended to the right as far as desired. To determine the procedure for *n* wires, simply cover the chart beyond the *n*th wire.

ANSWER 9.20—(November 1975)

To cut a square cake, frosted on the top and four sides, into *n* pieces of equal volume and equal frosting area, we need only mark the perimeter into *n* equal parts and cut the cake in the usual manner [see Figure 9.12]. To understand why this is so, we adopt a stratagem given by Norman N. Nelson and Forest N. Fisch in their article "The Classical Cake Problem" (*The Mathematics Teacher* 66, November 1973, pp. 659–61). Imagine the cake cut along its diagonals into four congruent triangular prisms, and the prisms arranged in a row as shown at the right in the illustration. The baseline is the cake's perimeter. Divide the cake into seven equal portions, *A, B, C, D, E, F, G*, as shown by the seven gray lines. Those lines correspond to the seven cutting planes in the solution.

It is easy to see that the area of icing on each portion is the same. The sums of the one or two rectangles that form the sides of each portion clearly are equal. The tops of each portion (composed of one or two triangles) are also equal in area because the sums of the bases of

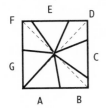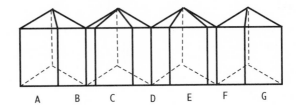

Figure 9.12

these triangles are equal, and all triangles have the same altitude. Finally, the volumes of the seven portions are the same because their heights are equal and their tops are equal in area.

The solution obviously generalizes to *n* portions. We cannot, however, apply the procedure to a cake on which the frosting has a finite thickness, because the four corners of such a cake introduce complexities.

Stephen I. Warshaw of the Lawrence Livermore Laboratory of the University of California generalized the cake-cutting problem. The procedure given for the square cake, he showed, also applies to any cake with a polygonal base each side of which is tangent to the same circle. This includes all triangles and rhombuses, all regular polygons (as well as the limiting case of the circle), and all polygons of unequal sides that meet the proviso. Even more surprising, Warshaw found that this generalization led to the solving of a sticky problem in his work: "How to find a simple way to divide a discrete mesh cell into parts of given proportions in computer simulation studies involving hydrodynamic motions. The solution, once seen, is 'obvious.'"

ANSWER 9.21—(May 1959)

Regardless of how much wine is in one beaker and how much water is in the other, and regardless of how much liquid is transferred back and forth at each step (provided it is not all of the liquid in one beaker), it is impossible to reach a point at which the percentage of wine in each mixture is the same. This can be shown by a simple inductive argument. If beaker A contains a higher concentration of wine than beaker B, then a transfer from A to B will leave A with the higher concentration. Similarly a transfer from B to A—from a weaker to a stronger mixture—is sure to leave B weaker. Since every transfer is one of these two cases, it follows that beaker A must always contain a mixture with a higher percentage of wine than B. The only way to equalize the concentrations is by pouring all of one beaker into the other.

There is a fallacy in the above solution. It assumes that liquids are infinitely divisible, whereas they are composed of discrete molecules. P. E. Argyle of Royal Oak, British Columbia, set me straight with the following letter:

Sirs:

Your solution to the problem of mixing wine and water seems to ignore the physical nature of the objects involved. When a sample of fluid is taken from a mixture of two fluids, the proportion of one fluid present in the sample will be different from its proportion in the mixture. The departure from the "correct" amount will be of the order $\pm \sqrt{n}$, where n is the number of molecules expected to be present.

Consequently it is possible to have equal amounts of wine in the two glasses. The probability of this occurring becomes significant after the expected lack of equality in the mixture has been reduced to the order of $\sqrt{n}$. This requires only 47 double interchanges for the problem as it was stated. . . .

ANSWER 9.22—(February 1970)

One way to solve the problem of the six weights—two red, two white, and two blue—is first to balance a red and a white weight against a blue and a white weight.

If the scales balance, you know there are a heavy and a light weight on each pan. Remove both colored weights, leaving the white weights, one on each side. This establishes which white weight is the heavier. At the same time it tells you which of the other two weights used before (one red, one blue) is heavy and which is light. This in turn tells you which is heavy and which is light in the red-blue pair not yet used.

If the scales do not balance on the first weighing, you know that the white weight on the side that went down must be the heavier of the two whites, but you are still in the dark about the red and blue. Weigh the original red against the *mate* of the original blue (or the original blue against the *mate* of the original red). As C. B. Chandler (who sent this simple solution) put it, the result of the second weighing, plus the memory of which side was heavier in the first weighing, is now sufficient to identify the six weights.

For readers who liked working on this problem, Ben Braude, a New York City dentist and amateur magician, devised the following variation. The six weights are alike in all respects (including color) except that three are heavy and three light. The heavy weights weigh the same and the light weights weigh the same. Identify each in three separate weighings on a balance scale.

As Thomas O'Beirne pointed out, Braude's problem has two essentially different types of solutions: one in which pairs are weighed against pairs, and one in which each weighing involves a single weight on each side. John Hamilton gave this concise chart for the four possibilities of the simpler method. (It appeared in the March 1970 issue of the magic periodical, *The Pallbearers Review*.)

1	2	3	4
a\B	a\B	a—b	a—b
c\D	c—d	b—c	b\C
e\F	d\E	c\D	D—E

A capital letter indicates a heavy weight, a small letter a light weight. A horizontal line means balance, a slanted line shows how the scale tips.

ANSWER 9.23—(November 1965)

This is how I originally answered the problem:

The most efficient procedure for testing any number of glasses of liquid to identify a single glass containing poison is a binary procedure. The glasses are divided as nearly in half as possible. One set is tested (by mixing samples from all the glasses and testing the mixture). The set known to include the poison glass is then again divided as nearly in half as possible, and the procedure is repeated until the poison glass is identified. If the number of glasses is from 100 to 128 inclusive, as many as seven tests might be required. From 129 to 200 glasses might take eight tests. The number 128 is the turning point, because it is the only number between 100 and 200 that is in the doubling series: 1, 2, 4, 8, 16, 32, 64, 128, 256,. . . . There must have been 129 glasses in the hotel kitchen, because only in that case (we were told that the number was between 100 and 200) will the initial testing of one glass make no difference in applying the most efficient testing procedure. Testing 129 glasses, by halving, could demand eight tests. But if a single glass is tested first, the remaining 128 glasses require no more than seven tests, so that the total number of tests remains the same.

When the above answer appeared, many readers pointed out that the police commissioner was right, and the mathematician wrong. Regardless of the number of glasses, the most efficient testing procedure is to divide them as nearly in half as possible at each step and test the glasses in either set. When the probabilities are worked out, the expected number of tests of 129 glasses, if the halving procedure is followed, is 7.0155+. But if a single glass is tested first, the expected number is 7.9457+. This is a rise of .930+ test, so the commissioner was almost right in considering the mathematician's procedure a waste of one test. Only if there had been 129 glasses, however, do we have a plausible excuse for the error, so, in a way, the problem was correctly answered even by those readers who proved that the mathematician's test procedure was inefficient.

The problem appears in Barr's *Second Miscellany of Puzzles*.

ANSWER 9.24—(December 1979)

Surprising as it may first seem, both the richest and the poorest classes in Oilaria would prefer pair averaging from the top down. Those in the richest class would prefer to be averaged with the next-richest class before the latter is reduced in wealth by averaging. Those in the poorest class would prefer being averaged with the next-poorest class after the latter has been increased in wealth by averaging.

An example will make this clear. Assume that the wealth of the five classes is in the proportions 1 : 3 : 4 : 7 : 13. Averaging from the bottom up changes the proportions to 2 : 3 : 5 : 9 : 9. Averaging from the top down changes the proportions to 3 : 3 : 5 : 7 : 10.

Robert Summers, professor of economics at the University of Pennsylvania, sent the following comments on the Oilaria problem:

Of course, it is trivial to show that both the low and high income people would prefer the Grand Vizier's variant to the original income redistribution plan proposed by the Sheik. It should be enough simply to observe that both the poor and the rich would like their incomes averaged with others having as large an income as possible. The Grand Vizier's variant—averaging from above—adjusts upward the income of the next-to-the-bottom class before it is combined with the bottom class; similarly, the income of the next-to-the-top class is combined with the top class before it is lowered through the averaging process. Whether or not another income class, the second or next to last or any one in between, would like the redistribution from either or both averaging methods depends upon the particular income distribution. The fact that the so-called Lorenz curves cross reflects this indeterminacy.

There are several interesting questions that go beyond the exercise as you gave it: (1) If the two redistribution schemes were repeated over and over again, how would the limiting distributions obtained by averaging from above and below compare? Is one better than the other from the standpoint of any particular class? Repeating the arithmetic over and over again with any starting income distribution will suggest that if either averaging process is repeated indefinitely, a completely equal—egalitarian—distribution will eventually result. Any one with an initial income below the average will be pleased with the iterated redistribution done either way; anyone above the average will dislike either redistribution. This can be shown in general by drawing upon some theorems about doubly stochastic matrices. (2) Suppose a random pair of persons is selected from Oilaria's population, and each person in the pair is given half of the total income of the two. If this process were repeated to cover every possible pairing, what would the resulting distribution be like? Suppose the random pairings, independent from trial to trial, occurred an infinite number of times. What would the limiting distribution be? Strong conjecture: Again an egalitarian distribution would result.

ANSWER 9.25—(May 1973)

A square table can always be placed somewhere on a wavy floor with all four legs touching the floor. To prove this, put the table anywhere. Assume that only three legs, A, B, C, are on the floor and D is off [see Figure 9.13]. It is always possible for three legs to touch the floor because three points, anywhere in space, mark the corners of a triangle. Rotate the table 90 degrees around its center, keeping legs A

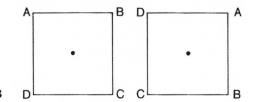

Figure 9.13

and B always on the floor. This brings the table to a position where C is now the only leg that does not touch the floor.

During the rotation D has moved to the floor and C has left it. But D must have touched the floor before C left, otherwise there would be a position at which only A and B would touch the floor, and we know that it is always possible for three legs to touch. At some point in the rotation, therefore, all four legs must have been in contact with the floor. A similar argument can be applied to wobbly rectangular tables by giving them 180-degree rotations.

Many readers called attention to two tacit assumptions that are necessary to make this proof valid:

1. The table, like all normal tables, has four legs of equal length, their lower ends at the corners of a square.
2. The legs of the table are sufficiently long and the unevenness of the floor is sufficiently mild, so while the table is rotated, there is never a moment at which three legs cannot be made to touch the floor.

The theorem is actually useful. Suppose you have a circular table with four legs that wobbles a bit when you move it to a porch. If you don't mind the table's surface being on a slight slant, you don't have to search for something to slip under a leg: Just rotate the table to a stable position. If you have to stand on a four-legged stool or chair to replace a light bulb and the floor is uneven, you can always rotate the stool or chair to make it steady.

ANSWER 9.26—(November 1965)

If the girl's objective is to escape by reaching the shore as quickly as possible, her best strategy is as follows. First she rows so that the lake's center, marked by the raft, is always between her and the man on shore, the three points maintaining a straight line. At the same time, she moves shoreward. Assuming that the man follows his optimum strategy of always running in the same direction around the lake, with a speed four times as fast as the girl can row, the girl's optimum path is a semicircle with a radius of $r/8$, where r is the lake's radius. At the end of this semicircle, she will have reached a distance of $r/4$ from the lake's center. That is the point at which the angular velocity she must maintain to keep the man opposite her just equals his angular velocity, leaving her no reserve energy for moving outward. (If during this period the man should change direction, she can do as well or better by mirror reflecting her path.)

As soon as the girl reaches the end of the semicircle, she heads straight for the nearest spot on the shore. She has a distance of $3r/4$ to go. He has to travel a distance of πr to catch her when she lands. She escapes, because when she reaches the shore he has gone a distance of only $3r$.

Suppose, however, the girl prefers to reach the shore not as *soon* as possible but at a spot as *far away* as possible from the man. In this case her best strategy, after she reaches a point $r/4$ from the lake's center, is to row in a straight line that is tangent to the circle of radius $r/4$, moving in a direction opposite to the way the man is running. This was first explained by Richard K. Guy in his article, "The Jewel Thief," in *NABLA*, vol. 8, September 1961, pages 149–50. (This periodical, a bulletin of the Malayan Mathematical Society, had published the minimum-time solution of the problem in its July 1961, issue, page 112.)

Using elementary calculus, Guy shows that the girl can always escape even when the man runs 4.6+ as fast as the girl rows. The same results are given by Thomas H. O'Beirne in *The New Scientist*, no. 266, December 21, 1961, page 753, and by W. Schurman and J. Lodder in "The Beauty, the Beast, and the Pond," *Mathematics Magazine*, vol. 47, March 1974, pages 93–95.

ANSWER 9.27—(February 1979)

The following five moves will suffice to get all four glasses in the wells of the rotating table turned either all up or all down.

1. Reach into any diagonal pair of wells. If the glasses are not both turned up, adjust them so that they are. If the bell does not ring:
2. Spin the table and reach into any adjacent pair of wells. If both glasses are turned up, leave them that way; otherwise invert the glass that is turned down. If the bell fails to ring, you know that now three glasses are turned up and one is turned down.
3. Spin the table and reach into any diagonal pair of wells. If one of the glasses is turned down, invert it and the bell will ring. If both are turned up, invert one so that the glasses are arranged in the following pattern:

Up	Down
Up	Down

4. Spin the table and reach into any adjacent pair of wells. Invert both glasses. If they were both turned in the same direction, the bell will ring; otherwise the glasses are now arranged in the following pattern:

Up	Down
Down	Up

5. Spin the table, reach into any diagonal pair of wells and invert both glasses. The bell will ring.

Ronald L. Graham of Bell Laboratories and Persi Diaconis of Harvard University have together considered two ways of generalizing the problem. One is to assume that the player has more than two hands. If the number of glasses n is greater than 4, the problem is solvable with $n - 2$ hands if and only if n is not a prime number. Therefore for five glasses (5 is a prime) the problem is not solvable with $5 - 2$, or 3, hands. It may be that if n is a composite number, then for some values of n the problem can be solved with fewer than $n - 2$ hands. The problem can also be generalized by replacing glasses with objects that have more than two positions.

Ted Lewis and Stephen Willard, writing on "The Rotating Table" in *Mathematics Magazine* (vol. 53, 1980, pp. 174–75), were the first to publish a solution for the general case involving a rotating polygonal table with n wells, and a player with k hands whom they called a "bell ringing octopus." They showed that the player can always force the bell to ring in a finite number of steps if and only if k is equal to or greater than $(1 - 1/p)n$, where p is the largest prime factor of n.

This result was also reached independently by a number of readers. Unless I have misplaced some letters they were Lyle Ramshaw, Richard Litherland, Eugene Gover and Nishan Krikorian, Richard Ahrens and John Mason, and Sven Eldring. The most detailed proof, along with related results and conjectures, can be found in "Probing the Rotating Table," by Ramshaw and William T. Laaser, in *The Mathematical Gardner*, edited by David A. Klarner (Wadsworth, 1981, pp. 288–307).

Two amusing robot variations were proposed by Albert G. Stanger as Problem 1598, "Variations on the Rotating Table Problem," in the *Journal of Recreational Mathematics* (vol. 19, 1987, pp. 307–8).

1. Instead of reaching into the wells, indicate to a robot the two you want probed. The robot does the reaching, then says either "Same" (if both glasses have the same orientation) or "Different." You then command the robot either to flip no glass, flip both, or flip only one which he selects at random.
2. This is the same as the previous variant except that the robot may tell a lie at any time.

In his solution (ibid., vol. 20, 1988, pp. 312–14), Stanger showed that in both variants the bell can be made to ring in a finite number of steps. He gives a five-step solution for the first variation and a seven-step solution for the second. In the second strategy, it makes no difference if the robot always tells the truth, always lies, or mixes truths and lies, because the strategy pays no attention to what the robot says! This surprising strategy, in which you pay no attention to anything, provides a seven-step solution to the original problem. It had been sent to me earlier in a March 1979 letter from Miner S. Keeler, president of the Keeler Brass Company in Grand Rapids, Michigan, who expressed surprise that I had not mentioned it when I answered the original problem. Here is how it works:

1. Invert any diagonal pair.
2. Invert any adjacent pair.
3. Invert any diagonal pair.
4. Invert any single glass.
5. Invert any diagonal pair.
6. Invert any adjacent pair.
7. Invert any diagonal pair.

If you don't believe these steps are certain to ring the bell after at most seven steps, try it out by placing four playing cards on the table in any pattern of up and down. Regardless of the initial pattern, you'll find that the strategy will produce either four up cards or four down cards in at most seven moves.

In a note added to Stanger's answer to his problem, Douglas J. Lanska generalized the robot variation to p wells and q hands. He pointed out that if the robot always lies or always tells the truth it is not necessary to know which is the case. Simply follow the n-step strategy on the assumption that the robot is a truthteller. If the bell fails to ring after n steps, you know the robot is a liar. Then repeat the same n

steps, but now reverse the robot's answers. The bell will ring in another n steps, or $2n$ steps in all.

Gover and Krikorian, in a later letter, reported some results on rotating cubes and other regular polyhedra.

I first heard of the original problem from Robert Tappay, a Toronto friend, who was told of it by a mathematician at the University of Quebec. He in turn had found it in a test given to Russian mathematics students. Someone in Poland, he told Tappay, had written his doctor's thesis on how people go about trying to solve the problem.

Answer 9.28—(February 1967)

Each dragon curve can be described by a sequence of binary digits, with 1's standing for left turns and 0's for right turns as the curve is traced on graph paper from tail to snout. The formula for each order is obtained from the formula for the next-lowest order by the following recursive technique: add 1, then copy all the digits preceding that 1 but change the center digit of the set. The order-1 dragon has the formula 1. In this case, after adding a 1 there is only one digit on the left, and since it is also the "center" digit we change it to 0 to obtain 110 as the order-2 formula. To get the order-3 formula add 1, followed by 110 with the center digit changed: 1101100. Higher-order formulas are obtained in the same way. It is easy to see that each dragon consists of two replicas of dragons of the next-lowest order, but joined head to head so that the second is drawn from snout to tail.

Figure 9.14 shows dragon curves of orders 0 to 6. All dragons are drawn from tail to snout and are here turned so that each is swimming to the right, the tips of his snout and tail touching the waterline. If each 1 is taken as a symbol of a right turn instead of a left and each 0 as a left turn, the formula produces dragons that face the opposite way. The spots on each curve correspond to the central 1's in the formulas for the successive orders from 1 to the order of the curve. These spots, on a dragon of any order, lie on a logarithmic spiral.

The dragon curve was discovered by physicist John E. Heighway as the result of an entirely different procedure. Fold a sheet of paper in half, then open it so that the halves are at right angles and view the sheet from the edge. You will see an order-1 dragon. Fold the same sheet twice, always folding in the same direction, and open it so that every fold is a right angle. The sheet's opposite edges will have the shapes of order-2 dragons, each a mirror image of the other. Folding

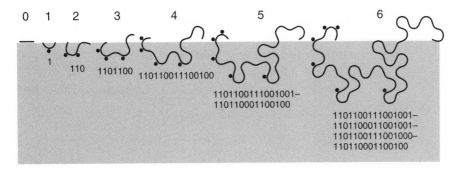

0   1   2   3   4   5   6

1

110   1101100   110110011100100

1101100111001001–
110110001100100

1101100111001001–
1101100011001001–
1101100111001000–
110110001100100

Figure 9.14

the paper in half three times generates an order-3 dragon, as illustrated in Figure 9.15. In general *n* folds produce an order-*n* dragon.

]The binary formula can be applied, of course, to the folding of a strip of paper (adding-machine tape works nicely) into models of higher-order dragons. Let each 1 stand for a "mountain fold," each 0 for a "valley fold." Start at one end of the strip, making the folds according to the formula. When the strip is opened until each fold is a right angle, it will have the shape of the dragon corresponding to the formula you used.

Physicist Bruce A. Banks discovered the geometric construction shown in Figure 9.16. It begins with a large right angle. Then at each step each line segment is replaced by a right angle of smaller segments in the manner illustrated. This is analogous to the construction of the "snowflake curve," as explained in my *Sixth Book of Mathematical Games from Scientific American* (1971, Chapter 22). The reader should be able to see why this gives the same result as paper folding.

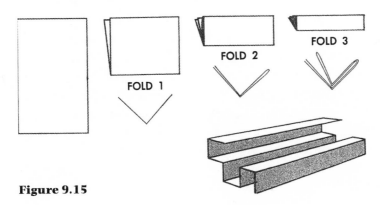

FOLD 1

FOLD 2

FOLD 3

Figure 9.15

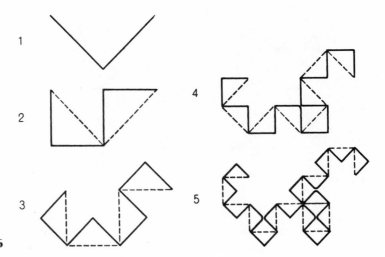

Figure 9.16

William G. Harter, the third of the three physicists who first analyzed the dragon curve, has found a variety of fantastic ways in which dragons can be fitted together snugly, like pieces of a jigsaw puzzle, to cover the plane or to form symmetrical patterns. They can be joined snout to snout, tail to tail, snout to tail, back to back, back to abdomen, and so on. Figure 9.17 shows a tail-to-tail-to-tail-to-tail arrangement of four right-facing order-6 dragons. If you wish to produce an eye-dazzling pattern, fit together in this way four order-12 dragons like the one shown earlier in Figure 9.6. If the four curves are each infinite in length, they completely fill the plane in the sense that every unit edge of the lattice is traversed exactly once. For dragon-joining experiments it is best to draw your dragons on transparent paper that can be overlapped in various ways.

Donald E. Knuth, a Stanford University computer scientist, and Chandler Davis, a University of Toronto mathematician, have made the most extensive study of dragon curves. Their two-part article "Number Representation and Dragon Curves" (*Journal of Recreational Mathematics*, vol. 3, April 1970, pages 66–81, and vol. 3, July 1970, pages 133–49) is filled with material on ways of obtaining the dragon-curve sequence, variations and generalizations of the curve, and its properties. See also the article by Knuth and his wife, Jill, in the same journal, vol. 6, Summer 1971, pages 165–67, which describes how they used three types of ceramic tiles to cover a wall of their house with an order-9 dragon.

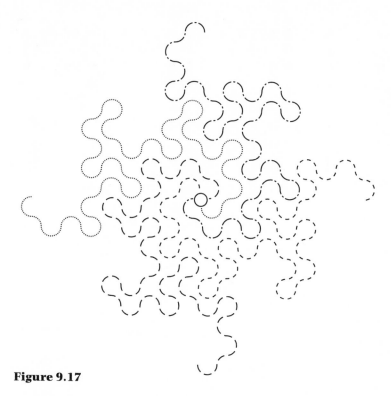

Figure 9.17

Since I introduced the dragon curve, the term "fractal," coined by Benoit Mandelbrot, has become standard, and of course the dragon curve is a fractal. Books on fractals, including Mandelbrot's classic *Fractal Geometry of Nature* (Freeman, 1982), are appearing so rapidly that I will make no effort to list any here. A selective bibliography can be found at the end of Chapter 3, "Mandelbrot's Fractals," in my *Penrose Tiles to Trapdoor Ciphers* (1988). For fascinating generalizations of the dragon curve to three dimensions, see "Wire Bending," by Michel Mendes France and J. O. Shallit, in the *Journal of Combinatorial Theory*, Series A, vol. 50, January 1989, pages 1–23.

ANSWER 9.29—(June 1961)

The first step is to list in order the probability values for the nine cards: 1/45, 2/45, 3/45, . . . The two lowest values are combined to form a new element: 1/45 plus 2/45 equals 3/45. In other words, the probability that the chosen card is either an ace or deuce is 3/45. There are now eight elements: the ace-deuce set, the three, the four, and so on up to nine. Again the two lowest probabilities are com-

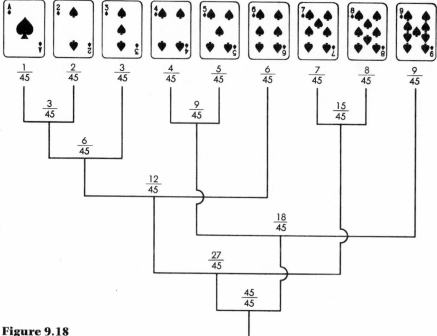

Figure 9.18

bined: the ace-deuce value of 3/45 and the 3/45 probability that the card is a three. This new element, consisting of aces, deuces, and threes, has a probability value of 6/45. This is greater than the values for either the fours or fives, so when the two lowest values are combined again, we must pair the fours and fives to obtain an element with the value of 9/45. This procedure of pairing the lowest elements is continued until only one element remains. It will have the probability value of 45/45, or 1. The chart in Figure 9.18 shows how the elements are combined. The strategy for minimizing the number of questions is to take these pairings in reverse order. Thus the first question could be: Is the card in the set of fours, fives, and nines? If not, you know it is in the other set so you ask next: Is it a seven or eight? And so on until the card is guessed.

Note that if the card should be an ace or deuce it will take five questions to pinpoint it. A binary strategy, of simply dividing the elements as nearly as possible into halves for each question, will ensure that no more than four questions need be asked, and you might even guess the card in three. Nevertheless, the previously described procedure will give a slightly lower expected minimum number of questions in the long run; in fact, the lowest possible. In this case, the minimum number is three.

The minimum is computed as follows: Five questions are needed if the card is an ace. Five are also needed if the card is a deuce, but there are two deuces, making 10 questions in all. Similarly, the three threes call for three times four, or 12, questions. The total number of questions for all 45 cards is 135, or an average of three questions per card.

This strategy was first discovered by David A. Huffman, an electrical engineer at MIT, while he was a graduate student there. It is explained in his paper "A Method for the Construction of Minimum-Redundancy Codes," *Proceedings of the Institute of Radio Engineers*, vol. 40, pages 1098–1101, September 1952. It was later rediscovered by Seth Zimmerman, who described it in his article on "An Optimal Search Procedure," *American Mathematical Monthly*, vol. 66, pages 690–93, October 1959. A good nontechnical exposition of the procedure will be found in John R. Pierce, *Symbols, Signals and Noise* (Harper & Brothers, 1961), beginning on page 94.

The eminent mathematician Stanislaw Ulam, in his biography *Adventures of a Mathematician* (Scribner's, 1976, page 281), suggested adding the following rule to the 20-question game. The person who answers is permitted one lie. What is the minimum number of questions required to determine a number between 1 and 1 million? What if he lies *twice*?

The general case is far from solved. If there are no lies, the answer is of course 20. If just one lie, 25 questions suffice. This was proved by Andrzej Pelc, in "Solution of Ulam's Problem on Searching with a Lie," in *Journal of Combinatorial Theory*, Series A, vol. 44, January 1987, pages 129–40. The author also gives an algorithm for finding the minimum number of needed questions for identifying any number between 1 and *n*. A different proof of the 25 minimum is given by Ivan Niven in "Coding Theory Applied to a Problem of Ulam," *Mathematics Magazine*, vol. 61, December 1988, pages 275–81.

When two lies are allowed, the answer of 29 questions was established by Jurek Czyzowicz, Andrzec Pelc, and Daniel Mundici in the *Journal of Combinatorial Theory*, Series A, vol. 49, November 1988, pages 384–88. In the same journal (vol. 52, September 1989, pages 62–76), the same authors solved the more general case of two lies and any number between 1 and 2^n. Wojciech Guziki, ibid., vol. 54, 1990, pages 1–19, completely disposed of the two-lie case for any number between 1 and n.

How about *three* lies? This has been answered only for numbers between 1 and 1 million. The solution is given by Alberto Negro and Matteo Sereno, in the same journal, vol. 59, 1992. It is 33 questions, and that's no lie.

The four-lie case remains unsolved even for numbers in the 1 to 1 million range. Of course if one is allowed to lie every time, there is no way to guess the number. Ulam's problem is closely related to error-correcting coding theory. Ian Stewart summarized the latest results in "How to Play Twenty Questions with a Liar," in *New Scientist* (October 17, 1992), and Barry Cipra did the same in "All Theorems Great and Small," *SIAM News* (July 1992, page 28).

ANSWER 9.30—(August 1958)

In the triangular pistol duel the poorest shot, Jones, has the best chance to survive. Smith, who never misses, has the second-best chance. Because Jones's two opponents will aim at each other when their turns come, Jones's best strategy is to fire into the air until one opponent is dead. He will then get the first shot at the survivor, which gives him a strong advantage.

Smith's survival chances are the easiest to determine. There is a chance of 1/2 that he will get the first shot in his duel with Brown, in which case he kills him. There is a chance of 1/2 that Brown will shoot first, and since Brown is 4/5 accurate, Smith has a 1/5 chance of surviving. So Smith's chance of surviving Brown is 1/2 added to 1/2 × 1/5 = 3/5. Jones, who is accurate half the time, now gets a crack at Smith. If he misses, Smith kills him, so Smith has a survival chance of 1/2 against Jones. Smith's overall chance of surviving is therefore 3/5 × 1/2 = 3/10.

Brown's case is more complicated because we run into an infinite series of possibilities. His chance of surviving against Smith is 2/5 (we saw above that Smith's survival chance against Brown was 3/5, and since one of the two men must survive, we subtract 3/5 from 1 to obtain Brown's chance of surviving against Smith). Brown now faces fire from Jones. There is a chance of 1/2 that Jones will miss, in which case Brown has a 4/5 chance of killing Jones. Up to this point, then, his chance of killing Jones is 1/2 × 4/5 = 4/10. But there is a 1/5 chance that Brown may miss, giving Jones another shot. Brown's chance of surviving is 1/2; then he again has a 4/5 chance of killing Jones, so his chance of surviving on the second round is 1/2 × 1/5 × 1/2 × 4/5 = 4/100.

If Brown misses again, his chance of killing Jones on the third round will be 4/1,000; if he misses once more, his chance on the fourth round will be 4/10,000, and so on. Brown's total survival chance against Jones is thus the sum of the infinite series:

$$\frac{4}{10} + \frac{4}{100} + \frac{4}{1,000} + \frac{4}{10,000} \ldots$$

This can be written as the repeating decimal .444444 . . . , which is the decimal expansion of 4/9.

As we saw earlier, Brown had a 2/5 chance of surviving Smith; now we see that he has a 4/9 chance to survive Jones. His overall survival chance is therefore 2/5 × 4/9 = 8/45.

Jones's survival chance can be determined in similar fashion, but of course we can get it immediately by subtracting Smith's chance, 3/10, and Brown's chance, 8/45, from 1. This gives Jones an overall survival chance of 47/90.

The entire duel can be conveniently graphed by using the tree diagram shown in Figure 9.19. It begins with only two branches because Jones passes if he has the first shot, leaving only two equal possibilities: Smith shooting first or Brown shooting first, with intent to kill. One branch of the tree goes on endlessly. The overall survival chance of an individual is computed as follows:

1. Mark all the ends of branches at which the person is sole survivor.
2. Trace each end back to the base of the tree, multiplying the probabilities of each segment as you go along. The product will be the probability of the event at the end of the branch.
3. Add together the probabilities of all the marked end point events. The sum will be the overall survival probability for that person.

In computing the survival chances of Brown and Jones, an infinite number of endpoints are involved, but it is not difficult to see from the diagram how to formulate the infinite series that is involved in each case.

When I published the answer to this problem I added that perhaps

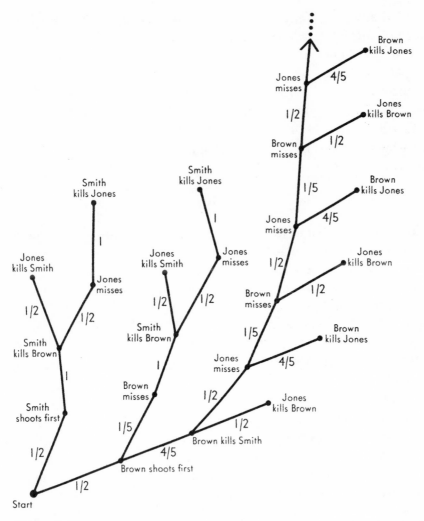

Figure 9.19

there is a moral of international politics in this somewhere. This prompted the following comment from Lee Kean of Dayton, Ohio:

Sirs:

We must not expect that in international politics nations will behave as sensibly as individuals. Fifty-fifty Jones, against his own best interests, will blaze away when able at the opponent he imagines to be most dangerous. Even so, he still has the best chance of survival, 44.722 per cent. Brown and Smith find their chances reversed. Eighty-twenty

Brown's chances are 31.111 per cent and sure-shot Smith comes in last with 24.167 per cent. Maybe the moral for international politics is even better here.

The problem, in variant forms, appears in several puzzle books. The earliest reference known to me is Hubert Phillips's *Question Time*, 1938, Problem 223. A different version of the problem can be found in Clark Kinnaird's *Encyclopedia of Puzzles and Pastimes*, 1946, but the answer is incorrect. Correct probability figures for Kinnaird's version are given in *The American Mathematical Monthly*, December 1948, page 640.

chapter 10

Scheduling & Planning

Operations Research (OR) is the branch of applied mathematics that solves problems of scheduling tasks to optimize the completion of an overall goal. Optimization might mean quickest completion or minimum cost. Sometimes the question of the existence of any solution can be very difficult.

In this chapter, there are classical OR problems such as refueling trucks and motion planning. There are also many examples from the recreational mathematics literature. Here, we find counter jumping, train shunting, coin arranging, and sliding block puzzles, each with a new twist.

The problems of OR are often—but not always—very difficult, and the best-known algorithms are often no better than exhaustive analysis. (These problems are known to computer scientists as "NP-complete" problems.) So in this chapter you may find that only by exploring many false starts can the solution be found. Of course you should be alert to special features, such as parity checks, which can reduce the case analysis.

Problems

PROBLEM 10.1—Two Spheres in One Pipe

A 10-foot piece of cylindrical iron pipe has an interior diameter of four inches. If a steel sphere three inches in diameter is inserted into the pipe at end *A* and a steel sphere two inches in diameter is inserted at end *B*, is it possible, with the help of a rod, to push each sphere through the entire length of pipe so that it emerges at the other end?

PROBLEM 10.2—Egg-Timers

With a 7-minute hourglass and an 11-minute hourglass, what is the quickest way to time the boiling of an egg for 15 minutes? (From Karl Fulves.)

PROBLEM 10.3—Jumping Pennies

One of the oldest and best coin puzzles calls for placing eight pennies in a row on the table [see Figure 10.1] and trying to transform them, in four moves, into four stacks of two coins each. There is one proviso: on each move a single penny must "jump" exactly two pennies (in either direction) and land on the next single penny. The two jumped pennies may be two single coins side by side or a stacked pair. Eight is the smallest number of pennies that can be paired in this manner.

You will find it a pleasant and easy task to solve the problem, but now comes the amusing part. Suppose two more coins are added to make a row of 10? Can the 10 be doubled in five moves? Many people, after finding a solution for eight, will give up in despair when presented with 10 and refuse to work on the problem. Yet it can be solved instantly if one has the right insight. Indeed, as will be explained in the answer section, a solution for eight makes trivial the generalization to a row of $2n$ pennies ($n > 3$), to be doubled in n moves.

Figure 10.1

PROBLEM 10.4—Crowning the Checkers

Numerous variants on the preceding problem have been proposed by Dudeney and other puzzle inventors. The following variation on the theme, which I believe is new, was suggested and solved by W. Lloyd Milligan of Columbia, South Carolina.

An even number of checkers, n, are placed in a row. First move a checker over one checker to make a king, then move a checker over two checkers, then a checker over three checkers, and so on, each time increasing by one the number of checkers to be passed over. The objective is to form $n/2$ kings in $n/2$ moves.

Can you prove that the problem cannot be solved unless n is a multiple of 4 and give a simple algorithm (procedure) for obtaining a solu-

Scheduling & Planning 271

tion in all cases where *n* is a multiple of 4? A solution is easily found by trial and error when *n* is 4 or 8, but for *n* = 16 or higher it is not so easy without a systematic method.

PROBLEM 10.5—Inverting a Triangle

Pack 10 pennies to form a triangle [at left in Figure 10.2]. This is the famous "tetractys" of the ancient Pythagoreans and today's familiar pattern for the 10 bowling pins. The problem is to turn this triangle upside down [at right in Figure 10.2] by sliding one penny at a time to a new position in which it touches two other pennies. What is the minimum number of required moves? Most people solve the problem quickly in four moves, but it can be done in three. The problem has an interesting generalization. A triangle of three pennies obviously can be inverted by moving one coin and a triangle of six pennies by moving two. Since the 10-coin triangle calls for a shift of three, can the next largest equilateral triangle, 15 pennies arranged like the 15 balls at the start of a billiard game, be inverted by moving four pennies? No, it requires five. Nevertheless, there is a remarkably simple way to calculate the minimum number of coins that must be shifted, given the number of pennies in the triangle. Can you discover it?

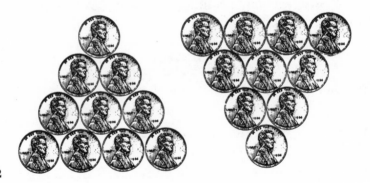

Figure 10.2

PROBLEM 10.6—Collating the Coins

Arrange three pennies and two dimes in a row, alternating the coins as shown in Figure 10.3. The problem is to change their positions to those shown at the bottom of the illustration in the shortest possible number of moves.

A move consists of placing the tips of the first and second fingers on any two touching coins, *one of which must be a penny and the other a dime*, then sliding the pair to another spot along the imaginary line shown in the illustration. The two coins in the pair must touch at all times. The coin at left in the pair must remain at left; the coin at right must remain at right. Gaps in the chain are allowed at the end of any move except the final one. After the last move the coins need not be at the same spot on the imaginary line that they occupied at the start.

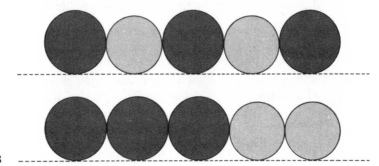

Figure 10.3

PROBLEM 10.7—Sliding Pennies

When pennies are closely packed on the plane, their centers mark the points of a triangular lattice, a fact that underlies scores of coin puzzles of many different types. For example, start with six pennies closely packed in a rhomboid formation [at left in Figure 10.4]. In three moves try to form a circular pattern [at right] so that if a seventh coin were placed in the center, the six pennies would be closely packed around it. On each move a single coin must be slid to a new position so as to touch two other coins that rigidly determine its new position.

This has a tricky aspect when you show it to someone. If he fails to solve it, demonstrate the solution slowly and challenge him to repeat it. But when you put the coins back in their starting position, set them in the mirror-image form of the original rhomboid. There is a good chance that he will not notice the difference, with the result that when he tries to duplicate your three moves, he will soon be in serious trouble.

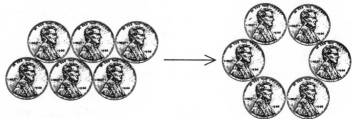

Figure 10.4

PROBLEM 10.8—Two Bookcases

Robert Abes, of the Courant Institute of Mathematical Sciences at New York University, originated this problem. A room 9 × 12 feet contains two bookcases that hold a collection of rare erotica. Bookcase *AB* is 8½ feet long, and bookcase *CD* is 4½ feet long. The bookcases are positioned so that each is centered along its wall and one inch from the wall.

The owner's young nephews are coming for a visit. He wishes to protect them and the books from each other by turning both bookcases around to face the wall. Each bookcase must end up in its starting position but with its ends reversed [see Figure 10.5]. The bookcases are so heavy that the only way to move them is to keep one end on the floor as a pivot while the other end is swung in a circular arc. The bookcases are narrow from front to back, and for purposes of the problem we idealize them to straight line segments. The ends of the bookcases cannot pass through walls midswing, or through each other. What is the minimum number of swings required to reverse the two bookcases?

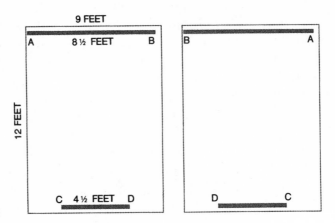

Figure 10.5

PROBLEM 10.9—Flight around the World

A group of airplanes is based on a small island. The tank of each plane holds just enough fuel to take it halfway around the world. Any desired amount of fuel can be transferred from the tank of one plane to the tank of another while the planes are in flight. The only source of fuel is on the island, and for the purposes of the problem it is assumed that there is no time lost in refueling either in the air or on the ground.

What is the smallest number of planes that will ensure the flight of one plane around the world on a great circle, assuming that the planes have the same constant ground speed and rate of fuel consumption and that all planes return safely to their island base?

PROBLEM 10.10—Crossing the Desert

An unlimited supply of gasoline is available at one edge of a desert 800 miles wide, but there is no source on the desert itself. A truck can carry enough gasoline to go 500 miles (this will be called one "load"), and it can build up its own refueling stations at any spot along the way. These caches may be any size, and it is assumed that there is no evaporation loss.

What is the minimum amount (in loads) of gasoline the truck will require in to cross the desert? Is there a limit to the width of a desert the truck can cross?

PROBLEM 10.11—Trip around the Moon

The year is 1984. A moon base has been established and an astronaut is to make an exploratory trip around the moon. Starting at the base, he is to follow a great circle and return to the base from the other side. The trip is to be made in a car built to travel over the satellite's surface and having a fuel tank that holds just enough fuel to take the car a fifth of the way around the moon. In addition the car can carry one sealed container that holds the same amount of fuel as the tank. This may be opened and used to fill the tank or it may be deposited, unopened, on the moon's surface. No fraction of the container's contents may be so deposited.

The problem is to devise a way of making the round trip with a minimum consumption of fuel. As many preliminary trips as desired may be made, in either direction, to leave containers at strategic spots where they can be picked up and used later, but eventually a complete circuit must be made all the way around in one direction. Assume that there is an unlimited supply of containers at the base. The car can always be refueled at the base from a large tank. For example, if it arrives at the base with a partly empty tank, it can refill its tank without wasting the fuel remaining in its tank.

To work on the problem, it is convenient to draw a circle and divide it into twentieths as shown in Figure 10.6. Fuel used in preliminary trips must of course be counted as part of the total amount consumed.

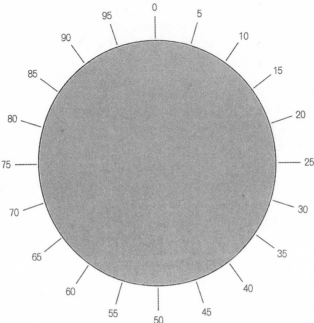

Figure 10.6

For example, if the car carried a container to point 90, left it there and returned to base, the operation would consume one tank of fuel.

This problem is similar in some respects to the earlier problem of crossing a desert in a truck, but it demands a quite different analysis.

PROBLEM 10.12—Counter-Jump "Aha!"

Draw a 5 × 6 array of spots on a sheet of paper, then rule a line as is shown in Figure 10.7 to divide the array into two triangular halves of 15 spots each. On the spots above the line [shown black] place 15 pennies or any other kind of small object.

The task is to move all the pennies from above the line to the spots below the line. Each move is a jump of one counter over an adjacent counter to an unoccupied spot immediately beyond it on the other side. Jumps may be to the left or the right and up or down but not diagonal. For example, as a first move the penny at the fourth spot on the top row may jump to the top white spot, or it may jump down to the third spot from the top of its column. All the jumps are like the jumps in checkers except that they are confined to horizontal and vertical directions and the jumped pieces are not removed.

We are not concerned with transferring the pennies to the white

ALGORITHMIC PUZZLES & GAMES

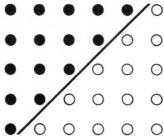

Figure 10.7

spots in a minimum number of moves, only with whether the transfer can be made at all. There are three questions:

1. Can the task be accomplished?
2. If a penny is removed from a black spot, can the 14 pennies that remain be jumped to white spots?
3. If two pennies are removed from black spots, can the remaining 13 be jumped to white spots?

This problem was devised by Mark Wegman of the Thomas J. Watson Research Center of the International Business Machines Corporation. It is of special interest because all three questions can be answered quickly by an "Aha!" insight well within the grasp of a 10 year old.

PROBLEM 10.13—Switching Cars

The efficient switching of railroad cars often poses frustrating problems in the field of operations research. The switching puzzle depicted in Figure 10.8 is one that has the merit of combining simplicity with surprising difficulty.

The tunnel is wide enough to accommodate the locomotive but not wide enough for either car. The problem is to use the locomotive for switching the positions of cars A and B, then return the locomotive to its original spot. Each end of the locomotive can be used for pushing or pulling, and the two cars may, if desired, be coupled to each other.

The best solution is the one requiring the fewest operations. An "operation" is here defined as any movement of the locomotive between stops, assuming that it stops when it reverses direction, meets a car to push it, or unhooks from a car it has been pulling. Movements of the two switches are not counted as operations.

A convenient way to work on the puzzle is to place a penny, a dime, and a nickel on the illustration and slide them along the tracks,

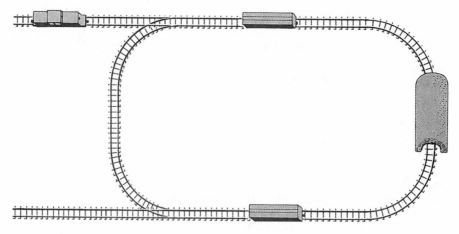

Figure 10.8

remembering that only the coin representing the locomotive can pass
through the tunnel. In the illustration, the cars were drawn in positions too close to the switches. While working on the problem,
assume that both cars are far enough east along the track so that there
is ample space between each car and switch to accommodate both the
locomotive and the other car.

No "flying switch" maneuvers are permitted. For example, you are
not permitted to turn the switch quickly just after the engine has
pushed an unattached car past it, so that the car goes one way and the
engine, without stopping, goes another way.

PROBLEM 10.14—Traffic Flow in Floyd's Knob

Robert Abbott author of *Abbott's New Card Games* (New York:
Stein and Day, 1963), provided the curious street map reproduced in
Figure 10.9, accompanied by the following story:

"Because the town of Floyd's Knob, Indiana, had only thirty-seven
registered automobiles, the mayor thought it would be safe to appoint
his cousin, Henry Stables, who was the town cutup, as its traffic commissioner. But he soon regretted his decision. When the town awoke
one morning, it found that a profusion of signs had been erected estab-
lishing numerous one-way streets and confusing restrictions on turns.

"The citizens were all for tearing down these signs until the police
chief, another cousin of the mayor, made a surprising discovery.
Motorists passing through town became so exasperated that sooner or
later they made a prohibited turn. The police chief found that the

ALGORITHMIC PUZZLES & GAMES

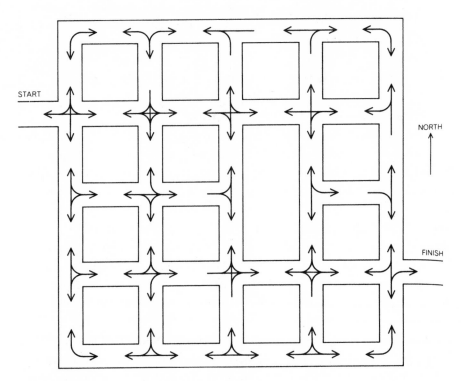

Figure 10.9

town was making even more money from these violations than from its speed trap on an outlying country road.

"Of course everyone was overjoyed, particularly because the next day was Saturday and Moses MacAdam, the county's richest farmer, was due to pass through town on his way to the county seat. They expected to extract a large fine from Moses, believing it to be impossible to drive through town without at least one traffic violation. But Moses had been secretly studying the signs. When Saturday morning came, he astonished the entire town by driving from his farm through town to the county seat without a single violation!

"Can you discover the route Moses took? At each intersection you must follow one of the arrows. That is, you may turn in a given direction only when there is a curved line in that direction, and you may go straight only when there is a straight line to follow. No turns may be made by backing the car around a corner. No U-turns are permitted. You may leave an intersection only at the head of an arrow. For instance, at the first intersection after leaving the farm, you have only

Scheduling & Planning

two choices: to go north or to go straight. If you go straight, at the next intersection you must either go straight or turn south. True, there is a curved line to the north, but there is no arrow pointing north, so you are forbidden to leave that intersection in a northerly direction."

PROBLEM 10.15—Three-Dimensional Maze

Three-dimensional mazes are something of a rarity. Psychologists occasionally use them for testing animal learning, and from time to time toy manufacturers market them as puzzles. A two-level space maze through which one tried to roll a marble was sold in London in the 1890s; it is depicted in *Puzzles Old and New* by "Professor Hoffmann" (London, 1893). Currently on sale in this country is a cube-shaped, four-level maze of a similar type. Essentially it is a cube of transparent plastic divided by transparent partitions into 64 smaller cubes. By eliminating various sides of the small cubical cells one can create a labyrinth through which a marble can roll. It is a simple maze, easily solved.

Robert Abbott, author of the book *Abbott's New Card Games* (New York: Stein and Day, 1963), recently asked himself: How difficult can a 4 × 4 × 4 cubical space maze, constructed along such lines, be made? The trickiest design he could achieve is shown in Figure 10.10.

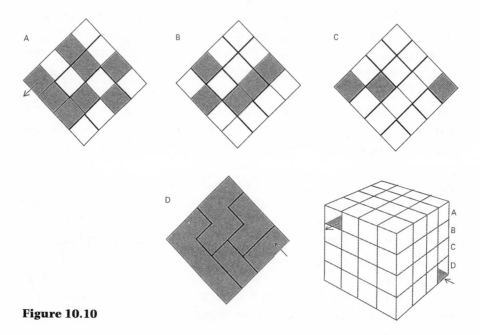

Figure 10.10

ALGORITHMIC PUZZLES & GAMES

You are asked not to make a model but to see how quickly you can run the maze without one.

On each of the four levels shown at the left in the illustration, solid black lines represent side walls. Shading indicates a floor; no shading, no floor. Hence a small square cell surrounded on all sides by black lines and unshaded is a cubical compartment closed on four sides but open at the bottom. To determine if it is open or closed at the top it is necessary to check the corresponding cell on the next level above. The top level (A) is of course completely covered by a ceiling.

Think of diagrams A through D as floor plans of the four-level cubical structure shown at the right in the illustration. First see if you can find a path that leads from the entrance on the first level to the exit on the top level. Then see if you can determine the shortest path from the entrance to the exit.

PROBLEM 10.16—Tour of Arrows

Sketch a large 4 × 4 checkerboard on a sheet of paper, obtain 16 paper matches, and you are set to work on this new solitaire puzzle. The matches represent arrows that point in the direction of the match head. Put a single spot on both sides of one match, two spots on both sides of eight matches, and three spots on both sides of seven matches. When a match is placed on a square of the board pointing north, south, east, or west, the single spot means it points to the immediately adjacent cell, two spots mean it points to the second cell, and three spots mean it points to the third cell.

Seven matches can be placed to map a closed tour [see Figure 10.11]. Start at any match of the seven and place your finger on the cell to which it points. The arrow on that cell gives the next "move." Follow the arrows until you return (in seven moves) to where you started. The problem is to place all 16 matches, one to a cell so that they map a closed tour that visits every cell. There are just two solutions, not counting rotations and reflections.

The tour will have a length of $1 + (2 \times 8) + (3 \times 7) = 38$. It is not hard to prove that this is the longest closed tour that can be made on the board by using any combination of the three types of arrows. Brian R. Barwell, a British engineer who introduced the problem in the *Journal of Recreational Mathematics* (October 1969), found that only one other maximum-length tour is possible. It requires six 3-arrows, ten 2-arrows, and no 1-arrow. You are invited to search for all three patterns.

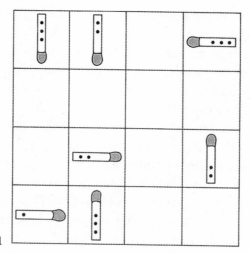

Figure 10.11

The arrows are, of course, merely a convenient way to map a maximum-length, closed tour by a chess rook, which lands on each cell exactly once. (Queen tours of this type are less interesting because there are so many of them; bishop tours cannot close and cannot visit all cells; and knight tours cannot vary in length.) The 2 × 2 field is trivial, and the 3 × 3 is easily analyzed. (Its maximum tour has a length of 14.) As far as I know, the 5 × 5 and all higher squares have yet to be investigated.

PROBLEM 10.17—Eight-Block Puzzle

For all sliding-block puzzles in which unit squares are shifted about inside a rectangle by virtue of a "hole" that is also a unit square, there is a quick parity check for determining if one pattern can be obtained from another. For example, on the simplest nontrivial square field shown in Figure 10.12 can the pattern at the left (with the blocks in descending order) be changed to the pattern at the right? To answer this we switch pairs of numbers (by removing and replacing blocks) until the desired pattern is achieved, counting the switches as we go along. This can be done in helter-skelter fashion, with no attempt at efficiency. If the number of switches is even (as it always will be in this case), the change by sliding is possible. Otherwise it is not.

But what is the *smallest* number of sliding moves sufficient to make this change? Surprisingly little work has been done on methods for minimizing such solutions. The problem given here—reversing the order of the digits—can be shown to require at least 26 moves. If

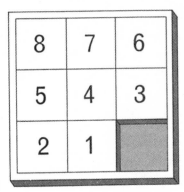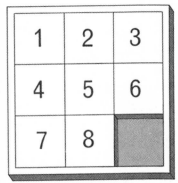

Figure 10.12

each square takes the shortest path to its destination, 16 moves are used. But 4 and 5 are adjacent and cannot be exchanged in fewer than four moves, and the same reasoning applies to 3 and 6. This lifts the lower limit to 20. Two moves are lost by the opening move of 1 or 3 and two more by the last move of 8 or 6, since in each case a square must occupy a cell outside its shortest path. This raises the lower limit to 24. Finally, if one constructs a tree graph for opening lines of play, it is apparent that two more moves must be lost by the ninth move. The puzzle therefore cannot be solved in fewer than 26 moves. Because the hole returns to its original position, it can be shown that every solution will have an even number of moves.

The best solution on record (it is the solution to problem 253 in Henry Ernest Dudeney's posthumous collection, *Puzzles and Curious Problems*) requires 36 moves. Recorded as a chain of digits to show the order in which the pieces are moved, it is as follows: 12543 12376 12376 12375 48123 65765 84785 6. I have good reason to believe, however, that it can be done in fewer moves.

To work on the problem one can move small cardboard squares, numbered 1 through 8, on a square field sketched on paper, or work with nine playing cards on a rectangular field.

PROBLEM 10.18—Red-Faced Cube

Recreational mathematicians have devoted much attention in the past to chessboard "tours" in which a chess piece is moved over the board to visit each square once and only once, in compliance with various constraints. John Harris, of Santa Barbara, has devised a fascinating new kind of tour—the "cube-rolling tour"—that opens up a wealth of possibilities.

Scheduling & Planning

To work on two of Harris's best problems, obtain a small wooden cube from a set of children's blocks or make one of cardboard. Its sides should be about the same size as the squares of your chessboard or checkerboard. Paint one side red. The cube is moved from one square to an adjacent one by being tipped over an edge, the edge resting on the line dividing the two cells. During each move, therefore, the cube makes one quarter-turn in a north, south, east, or west direction.

Problem 1. Place the cube on the northwest corner of the board, red side up. Tour the board, resting once only on every cell and ending with the cube red side up in the northeast corner. At no time *during* the tour, however, is the cube allowed to rest with the red side up. (Note: It is not possible to make such a tour from corner to diagonally opposite corner.)

Problem 2. Place the cube on any cell, an uncolored side up. Make a "reentrant tour" of the board (one that visits every cell once and returns the cube to its starting square) in such a way that at no time during the tour, including at the finish, will the cube's red side be up.

Both problems have unique solutions, not counting rotations and reflections of the path.

PROBLEM 10.19—Rolling Cubes

For this beautiful combinatorial puzzle, invented by John Harris of Santa Barbara, California, you must obtain eight unit cubes. On each cube, color one face and make the opposite face black. (Of course, you may distinguish the two faces in any other way you like.) Place the cubes in a shallow 3 × 3 box (or on a 3 × 3 matrix) with the middle cell vacant and all cubes black on top [see Figure 10.13].

A move consists of rolling a cube to an empty cell by tipping it over

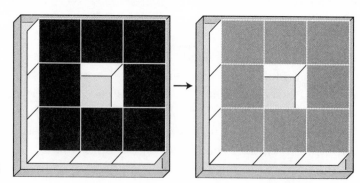

Figure 10.13

one of its four bottom edges so that it makes a quarter-turn. The problem is to invert all eight cubes so their colored sides are up and the center cell is vacant as before. This is to be done in a minimum number of moves. For recording solutions, you can use *U, D, L, R* for roll up, down, left, and right, and start all solutions with *URD*. (Any other way of starting is symmetrically the same.)

PROBLEM 10.20—Escott's Sliding Blocks

This remarkable sliding-block puzzle [see Figure 10.14] was invented by Edward Brind Escott, an American mathematician who died in 1946. It appeared in the August 1938 issue of a short-lived magazine called *Games Digest*. No solution was published. The problem is to slide the blocks one at a time, keeping them flat on the plane and inside the rectangular border, until block 1 and block 2 have exchanged places with block 7 and block 10, so that at the finish the two pairs of blocks are in the position shown at the right, with the other pieces anywhere on the board. No block is allowed to rotate, even if there is space for it to do so; each must keep its original orientation as it moves up, down, left, or right.

It is the most difficult sliding-block puzzle I have seen in print. The solution will be in 48 moves, counting each movement of one block as a single move even if it goes around a corner.

Escott was an expert on number theory who contributed frequently to mathematical journals. He taught at several schools and colleges in the Midwest and in his later years was an insurance company actuary in Oak Park, Illinois.

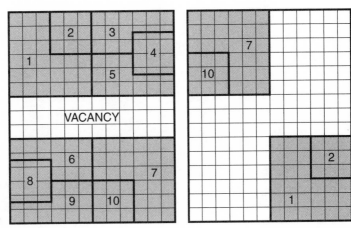

Figure 10.14

PROBLEM 10.21—Triangular Peg Solitaire

Peg solitaire, traditionally played on a square lattice, is an old recreation that is the topic of a considerable literature. As far as I know, similar problems based on the triangular lattice have received only the most superficial attention. The simplest starting position that is not trivial is the Pythagorean 10-spot triangle. To make it easy to record solutions, draw 10 spots on a sheet of paper, spacing them so that when pennies are placed on the spots, there will be some space between them, and number the positions {see Figure 10.15].

The problem: Remove one penny to make a "hole," then reduce the coins to a single penny by jumping. No sliding moves are allowed. The jumps are as in checkers, over an adjacent coin to an empty spot immediately beyond. The jumped coin is removed. Note that jumps can be made in six directions: in either direction parallel with the base, and in either direction parallel with each of the triangle's other two sides. As in checkers, a continuous chain of jumps counts as one "move." After some trial-and-error tests you'll discover that the puzzle is indeed solvable, but of course the recreational mathematician is not happy until he solves it in a minimum number of moves. Here, for example, is a solution in six moves after removal of the coin on spot 2:

1. 7–2
2. 9–7
3. 1–4

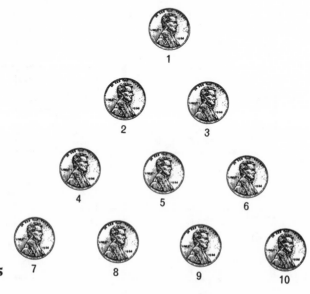

Figure 10.15

4. 7–2
5. 6–4, 4–1, 1–6
6. 10–3

There is a better solution in five moves. Can you find it? If so, go on to the 15-spot triangle. The novelty company S. S. Adams has been selling a peg version of this for years under the trade name Ke Puzzle Game, but no solutions are provided with the puzzle.

PROBLEM 10.22—Time the Toast

Even the simplest of household tasks can present complicated problems in operational research. Consider the preparation of three slices of hot buttered toast. The toaster is the old-fashioned type, with hinged doors on its two sides. It holds two pieces of bread at once but toasts each of them on one side only. To toast both sides it is necessary to open the doors and reverse the slices.

It takes three seconds to put a slice of bread into the toaster, three seconds to take it out, and three seconds to reverse a slice without removing it. Both hands are required for each of these operations, which means that it is not possible to put in, take out, or turn two slices simultaneously. Nor is it possible to butter a slice while another slice is being put into the toaster, turned, or taken out. The toasting time for one side of a piece of bread is 30 seconds. It takes 12 seconds to butter a slice.

Each slice is buttered on one side only. No side may be buttered until it has been toasted. A slice toasted and buttered on one side may be returned to the toaster for toasting on its other side. The toaster is warmed up at the start. In how short a time can three slices of bread be toasted on both sides and buttered?

Answers

ANSWER 10.1—(August 1968)

Yes, if the two spheres are pushed through the tube at different times.

ANSWER 10.2—(April 1969)

Start the 7- and 11-minute hourglasses when the egg is dropped into the boiling water. When the sand stops running in the 7-glass, turn it over. When the sand stops in the 11-glass, turn the 7-glass again. When the sand stops again in the 7-glass, 15 minutes will have elapsed.

The above solution is the quickest one, but requires two turns of hourglasses. When the problem first appeared. I thoughtlessly asked for the "simplest" solution, having in mind the shortest one. Several dozen readers called attention to the following solution which is longer (22 minutes) but "simpler" in the sense of requiring only one turn. Start the hourglasses together. When the 7-minute one runs out, drop the egg in the boiling water. When the 11-minute hourglass runs out, turn it over. When it runs out a second time, the egg has boiled for 15 minutes.

If you enjoyed that problem, here is a slightly harder one of the same type, from Howard P. Dinesman's *Superior Mathematical Puzzles* (London: Allen and Unwin, 1968). What is the quickest way to measure 9 minutes with a 4-minute hourglass and a 7-minute hourglass?

ANSWER 10.3—(February 1966)

To double the eight pennies in a row into four stacks of two coins each, number them from one to eight and move 4 to 7, 6 to 2, 1 to 3, 5 to 8. For 10 pennies, simply double the pennies at one end (for example, move 7 to 10), to leave a row of eight that can be solved as before. Clearly, a row of $2n$ pennies can be solved in n moves by doubling the coins at one end until eight remain, then solving for the eight.

ANSWER 10.4—(April 1970)

If n is odd, it is obvious there is no solution. If n is even but not a multiple of 4, an odd number of checkers must be jumped on the final move. This would necessarily leave a single checker in the row, therefore the assumption that there is a solution when n is not a multiple of 4 must be false.

If there are $4n$ checkers, the problem can be solved by working it backward according to the following algorithm. Start with $n/2$ kings in a row. Take the top checker from either of the two middle kings, jump over the largest group of kings and put down the checker as a single man. On the next backward move take the top checker from the other middle king and jump in the same direction as before, jumping one fewer checker. Follow this procedure until all kings in the direction of the first jump are eliminated. Take the top checker from the inside king and jump in the same direction as the previous jumps, moving it over the proper number of checkers. Continue this procedure, always in the same direction, until all the kings are reduced to single men. When these moves are taken in reverse order, they provide one solution (there are many others) to the original problem.

ANSWER 10.5—(February 1966)

A triangle of 10 pennies is inverted by moving three coins as shown in Figure 10.16. In working on the general problem, for equilateral formations of any size, you may have realized that the problem is one of drawing a bounding triangle (like the frame used to group the 15 balls for a game of billiards), inverting it, and placing it over the figure so that it encloses a maximum number of coins. In every case the smallest number of coins that must be moved to invert the pattern is obtained by dividing the number of coins by 3 and ignoring the remainder.

Figure 10.16

ANSWER 10.6—(June 1961)

The dime and penny puzzle can be solved in four moves as follows. Coins are numbered from left to right.

1. Move 3, 4 to the right of 5 but separated from 5 by a gap equal to the width of two coins.
2. Move 1, 2 to the right of 3, 4, with coins 4 and 1 touching.
3. Move 4, 1 to the gap between 5 and 3.
4. Move 5, 4 to the gap between 3 and 2.

If it were permissible to shift two coins of the same kind, the puzzle could be solved easily in three moves: slide 1, 2 to left, fill the gap with 4, 5, then move 5, 3 from right to left end. But with the proviso that each shifted pair must include a dime and penny it is a baffling and pretty problem. H. S. Percival, of Garden City, New York, was the first to call it to my attention.

Puzzles involving alternating coins that are to be collated by shifting them in pairs have many variations and generalizations. Here are some references:

- Jan M. Gombert, "Coin Strings," *Mathematics Magazine* (November–December 1969), pp. 244–47.
- Yeong-Wen Hwang, "An Interlacing Transformation Problem," *The American Mathematical Monthly*, vol. 67 (December 1960), pp. 974–76.
- James Achugbue and Francis Shin "Some New Results on a Shuffling Problem," *The Journal of Recreational Mathematics*, vol. 12, no. 2 (1979–80), pp. 126–29.

ANSWER 10.7—(February 1966)

Six pennies in rhomboid formation [Figure 10.17] can be formed into a circle, in accordance with the rules given, by numbering them as shown and moving 6 to touch 4 and 5, 5 to touch 2 and 3 from below, 3 to touch 5 and 6.

(Essentially the same problem appeared in the August 1958 column, except the starting position was a triangle, adding one additional move to the solution.)

Figure 10.17

ANSWER 10.8—(April 1974)

Eight swings are enough to reverse those two bookcases. One solution: (1) Swing end *B* clockwise 90 degrees; (2) swing *A* clockwise 30 degrees; (3) swing *B* counterclockwise 60 degrees; (4) swing *A* clockwise 30 degrees; (5) swing *B* clockwise 90 degrees; (6) swing *C* clockwise 60 degrees; (7) swing *D* counterclockwise 300 degrees; (8) swing *C* clockwise 60 degrees.

"If you moved bookcase *AB* in fewer than five swings," writes Robert Abes, who originated this problem, "then you put an end through a wall mid-swing, or (more likely) wound up with its front side still facing out. If you moved bookcase *CD* without a 300-degree second swing, you either wasted a swing or scooped a hollow out of a wall. Thanks to Jim Lewis for helping me move the large bookcase."

Wayne E. Russell noted that the bookcase problem did not rule out the possibility that the room was much higher than the cases. He showed that by raising one end of the large case high enough it could be reversed in three swings. Johannes Sack discovered the surprising fact that the minimum-move solution does not correspond to a solution with a minimum expenditure of energy. The given three-move reversal of the small bookcase carries the case at least 33 feet. If four moves are used (*D* counterclockwise 90 degrees, *C* counterclockwise 60 degrees, *D* counterclockwise 60 degrees, *C* clockwise 30 degrees), the case is carried only 18.8 feet, an energy saving of 43 percent.

ANSWER 10.9—(August 1958)

Three airplanes are quite sufficient to ensure the flight of one plane around the world. There are many ways this can be done, but the following seems to be the most efficient. It uses only five tanks of fuel, allows the pilots of two planes sufficient time for a cup of coffee and a sandwich before refueling at the base, and there is a pleasing symmetry in the procedure.

Planes A, B, and C take off together. After going 1/8 of the distance, C transfers 1/4 tank to A and 1/4 to B. This leaves C with 1/4 tank; just enough to get back home.

Planes A and B continue another 1/8 of the way, then B transfers 1/4 tank to A. B now has 1/2 tank left, which is sufficient to get him back to the base where he arrives with an empty tank.

Plane A, with a full tank, continues until it runs out of fuel 1/4 of the way from the base. It is met by C which has been refueled at the base. C transfers 1/4 tank to A, and both planes head for home.

The two planes run out of fuel 1/8 of the way from the base, where they are met by refueled plane B. Plane B transfers 1/4 tank to each of the other two planes. The three planes now have just enough fuel to reach the base with empty tanks.

The entire procedure can be diagramed as shown in Figure 10.18, where distance is the horizontal axis and time the vertical axis. The

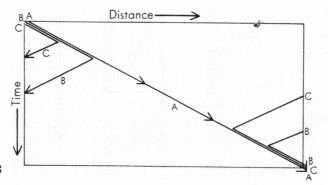

Distance ⟶

Time

Figure 10.18

right and left edges of the chart should, of course, be regarded as joined.

ANSWER 10.10 —(May 1959)

The following analysis of the desert-crossing problem appeared in an issue of *Eureka*, a publication of mathematics students at the University of Cambridge. Five hundred miles will be called a "unit"; gasoline sufficient to take the truck 500 miles will be called a "load"; and a "trip" is a journey of the truck in either direction from one stopping point to the next.

Two loads will carry the truck a maximum distance of $1\frac{1}{3}$ units. This is done in four trips by first setting up a cache at a spot $\frac{1}{3}$ unit from the start. The truck begins with a full load, goes to the cache, leaves $\frac{1}{3}$ load, returns, picks up another full load, arrives at the cache, and picks up the cache's $\frac{1}{3}$ load. It now has a full load, sufficient to take it the remaining distance to one unit.

Three loads will carry the truck $1\frac{1}{3}$ plus $\frac{1}{5}$ units in a total of nine trips. The first cache is $\frac{1}{5}$ unit from the start. Three trips put $\frac{6}{5}$ loads in the cache. The truck returns, picks up the remaining full load, and arrives at the first cache with $\frac{4}{5}$ load in its tank. This, together with the fuel in the cache, makes two full loads, sufficient to carry the truck the remaining $1\frac{1}{3}$ units, as explained in the preceding paragraph.

We are asked for the minimum amount of fuel required to take the truck 800 miles. Three loads will take it $766\frac{2}{3}$ miles ($1\frac{1}{3}$ plus $\frac{1}{5}$ units), so we need a third cache at a distance of $33\frac{1}{3}$ miles ($\frac{1}{15}$ unit) from the start. In five trips the truck can build up this cache so that when the truck reaches the cache at the end of the seventh trip, the combined fuel of truck and cache will be three loads. As we have seen, this is sufficient to take the truck the remaining distance of 766²/₃ miles. Seven trips are made between starting point and first cache, using ⁷/₁₅ load of gasoline. The three loads of fuel that remain are just sufficient for the rest of the way, so the total amount of gasoline consumed will be 3⁷/₁₅, or a little more than 3.46 loads. Sixteen trips are required.

Proceeding along similar lines, four loads will take the truck a distance of 1¹/₃ plus ¹/₅ plus ¹/₇ units, with three caches located at the boundaries of these distances. The sum of this infinite series diverges as the number of loads increases; therefore the truck can cross a desert of any width. If the desert is 1,000 miles across, seven caches, 64 trips, and 7.673 loads of gasoline are required.

Hundreds of letters were received on this problem, giving general solutions and interesting sidelights. Cecil G. Phipps, professor of mathematics at the University of Florida, summed matters up succinctly as follows:

"The general solution is given by the formula:

$$d = m \left(1 + 1/3 + 1/5 + 1/7 + \ldots \right),$$

where d is the distance to be traversed and m is the number of miles per load of gasoline. The number of depots to be established is one less than the number of terms in the series needed to exceed the value of d. One load of gasoline is used in the travel between each pair of stations. Since the series is divergent, any distance can be reached by this method although the amount of gasoline needed increases exponentially.

"If the truck is to return eventually to its home station, the formula becomes:

$$d = m \left(1/2 + 1/4 + 1/6 + 1/8 + \ldots \right)$$

This series is also divergent and the solution has properties similar to those for the one-way trip."

Many readers called attention to three previously published discussions of the problem:

- C. G. Phipps, "The Jeep Problem: A More General Solution," *American Mathematical Monthly*, vol. 54, no. 8, October 1947, pp. 458–62.
- G. G. Alway, "Crossing the Desert," *Mathematical Gazette*, vol. 41, no. 337, October 1947, p. 209.
- Olaf Helmer, *Problem in Logistics: The Jeep Problem*, Project Rand Report No. RA-15015, December 1, 1946.

(This was the first unclassified report of the Rand publications, issued when the project was still under the wing of Douglas Aircraft Company. It is the clearest analysis of the problem, including the return-trip version, that I have seen.)

ANSWER 10.11—(June 1964)

The moon can be circled with a consumption of 23 tankfuls of fuel.

1. In five trips, take five containers to point 90, return to base (consumes five tanks).
2. Take one container to point 85, return to point 90 (one tank).
3. Take one container to point 80, return to point 90 (one tank).
4. Take one container to point 80, return to point 85, pick up the container there and take it to point 80 (one tank).
5. Take one container to point 70, return to point 80 (one tank).
6. Return to base (one tank).

This completes all preliminary trips in the reverse direction. There is now one container at point 70, one at point 90. Ten tanks have been consumed.

7. Take one container to point 5, return to base (half a tank).
8. In four trips, take four containers to point 10, return to base (four tanks).
9. Take one container to point 10, return to point 5, pick up the container there and leave it at point 10 (one tank).
10. In the next two trips take two containers to point 20, return to point 10 (two tanks).
11. Take one container to point 25, return to point 20 (one tank).
12. Take one container to point 30, return to point 25, pick up the container there and carry it to point 30 (one tank).
13. Proceed to point 70 (two tanks).
14. Proceed to point 90 (one tank).
15. Proceed to base (half a tank).

The car arrives at base with its tank half-filled. The total fuel consumption is 23 tanks.

I found this problem, in the story form of circling a mountain, as Problem 50 in H. E. Dudeney's *Modern Puzzles* (1926); it is reprinted as Problem 77 in my edition of Dudeney's *536 Puzzles and Curious*

Problems (New York: Scribner's, 1967). The above solution, which was supplied in variant forms by many readers, is essentially the same as Dudeney's.

It is not, however, minimal. Wilfred H. Shepherd, Manchester, England, first reduced the fuel consumption to $22\,7/12$ containers. This was further reduced by Robert L. Elgin, Altadena, California, to $22\,9/16$ containers. His solution can be varied in trivial ways, but it is believed to be minimal.

Elgin's solution is best explained by dividing the circle into 80 equal parts. A container is picked up every time you move away from home base and left on the ground every time you turn to move toward the base. Any available container is emptied into the car's tank every time the car runs out of fuel. Assume that each container holds $80/5 = 16$ units of fuel. The solution follows:

1. Take one container to point 73, return to base (consumes 14 units).
2. Two containers to 75, return to base (20 units).
3. Two containers to 72, return to base (32 units).
4. One container to $69\frac{1}{2}$, back to 75 (16 units).
5. One container to $67\frac{1}{2}$, back to $69\frac{1}{2}$, forward to $67\frac{1}{2}$, back to 72 (16 units).
6. One container to 64, back to $67\frac{1}{2}$, forward to 66, back to $67\frac{1}{2}$, forward to 66 (16 units).
7. One container to 57, back to 64 (16 units).
8. Return to base (16 units).
9. Five containers to 8, return to base (80 units).
10. One container to 10, back to 8, forward to 10, back to 8 (16 units).
11. One container to 16, back to 8 (16 units).
12. One container to $16\frac{1}{2}$, back to 16, forward to $16\frac{1}{2}$, back to 10 (16 units).
13. One container to $21\frac{1}{4}$, back to $16\frac{1}{2}$ (16 units).
14. One container to 25, back to $21\frac{1}{4}$, forward to 25 (16 units).
15. One container to 41 (16 units).
16. Proceed to 57 (16 units).
17. Proceed to 73 (16 units).
18. Proceed to base (7 units).

Total fuel consumption is $361/16 = 22\,9/16$ units.

ANSWER 10.12—(December 1979)

The aha! insight that solves the counter-jumping puzzle is to color nine spots black as is shown in Figure 10.19. It is obvious that, no matter how many jumps are made, a penny on any black spot can go only to another black spot.

There are six black spots above the line and only three below it. Therefore, by the pigeonhole principle, there must be three pennies above the line that have nowhere to go below the line. The task of moving all the pennies to spots below the line cannot be accomplished unless at least three pennies, on three black spots, are removed from the top triangular array. Remove any three such pennies and the transfer of the remaining 12 is a simple task.

Benjamin L. Schwartz wrote to point out that the proof of impossibility holds even if the allowed moves are extended to include diagonal jumps, or bishop moves along diagonals without jumping. David J. Abineri sent a different proof of impossibility also based on the pigeonhole principle.

Number the columns 1, 2, 3, 4, 5, 6. Spots in columns 1, 3, and 5 are black. No matter how a penny jumps, orthogonally or diagonally, it must remain on a black spot. At the start nine pennies occupy black spots. But there are only six black spots above the line.

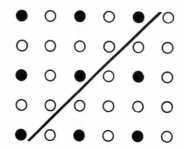

Figure 10.19

ANSWER 10.13—(October 1960)

The locomotive can switch the positions of cars A and B, and return to its former spot, in sixteen operations:

1. Locomotive moves right, hooks to car A.
2. Pulls A to bottom.
3. Pushes A to left, unhooks.
4. Moves right.
5. Makes a clockwise circle through tunnel.

6. Pushes B to left. All three are hooked.
7. Pulls A and B to right.
8. Pushes A and B to top, A is unhooked from B.
9. Pulls B to bottom.
10. Pushes B to left, unhooks.
11. Circles counterclockwise through tunnel.
12. Pushes A to bottom.
13. Moves left, hooks to B.
14. Pulls B to right.
15. Pushes B to top, unhooks.
16. Moves left to original position.

This procedure will do the job even when the locomotive is not permitted to pull with its front end, provided that at the start the locomotive is placed with its back toward the cars.

Howard Grossman, New York City, and Moises V. Gonzalez, Miami, Florida, each pointed out that if the lower siding is eliminated completely, the problem can still be solved, although two additional moves are required, making 18 in all. Can you discover how it is done?

For additional train-switching puzzles, see the puzzle books by Sam Loyd, Ernest Dudeney, and others. You may also wish to read A. K. Dewdney's "Computer Recreations" column in *Scientific American* (June 1987), and "Reversing Trains, A Turn of the Century Sorting Problem," by Nancy Amato, et al., *Journal of Algorithms*, vol. 10, 1989, pages 413–28.

ANSWER 10.14—(October 1962)

To drive through Floyd's Knob without a traffic violation, take the following directions at each successive intersection (the letters stand for North, South, East, West): E-E-S-S-E-N-N-N-E-S-W-S-E-S-S-W-W-W-W-W-N-N-E-S-W-S-E-E-E-E-N-E.

Robert Abbott has also written a marvelous book, *Mad Mazes* (Bob Adams, 1990), that contains 20 mazes unlike any you've encountered before. I had the pleasure of writing the book's introduction.

ANSWER 10.15—(November 1963)

The simplest paper-and-pencil way to solve Robert Abbott's three-dimensional maze is to place a spot in each cell and then draw lines from spot to spot to represent all open corridors. Since a maze involves

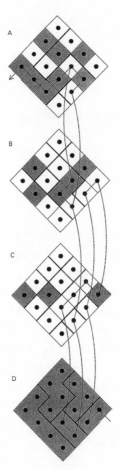

Figure 10.20

only topological properties of the pattern, it does not matter how these lines twist and turn as long as they connect the spots properly.

The next step is to erase all blind-alley lines and all loops that do no more than take one from spot to spot in a roundabout way when a shorter path is available. Eventually only the shortest route remains. This path is shown in Figure 10.20. Note that two loops near the top offer two different routes of equal length. Each of the long curved lines connecting the spots is, of course, only one unit in length in the actual maze, therefore the entire maze can be run by a path 19 units long.

An alternate method of finding the shortest path in any type of maze is to make a model of the network out of string. Each segment of string must have a length that is in the same proportion to the length of the corridor it represents, and it must be labeled in some way so that the corridor can be identified. After the model is completed, pick up the "start" of the network with one thumb and finger and the "end" of the network with the other thumb and finger. Pull the string taut. Roundabout loops and blind alleys hang loose. The taut portion of the model traces the path of minimum length!

A third method is to label the starting cell 1. Put a 2 in all cells that can be reached in one step. Put a 3 in all cells that can be reached in one step from each 2-cell. Continue in this manner, numbering every cell once. If you return to a cell already labeled, do not give it a larger number. After all cells are labeled, start at the final cell and move backward through the numbered cells, taking them in reverse order, to trace out a minimal-length path.

There is now a large literature on these and other algorithms for finding shortest routes in mazes or on graphs. Some references follow; you will find many earlier articles on the topic listed in them:

- E. F. Moore, "The Shortest Path through a Maze," *Proceedings of an International Symposium on the Theory of Switching*, Part II, April 2–5, 1957. (Reprinted in *Annals of the Computation Laboratory of Harvard University*, vol. 30, 1959, pp. 285–92.)
- C. Y. Lee, "An Algorithm for Path Connections and Its Applications," *I.R.E. Transactions on Electronic Computers*, vol. EC-10, September 1961, pp. 346–65.
- G. B. Dantzig, "All Shortest Routes in a Graph," *Operations Research Technical Report 66–3*, Stanford University, November 1966.
- T. C. Hu and W. T. Torres, "Shortcut in the Decomposition Algorithm for Shortest Paths in a Network," *IBM Journal of Research and Development*, vol. 13, no. 4, July 1969, pp. 387–90.
- Richard Bellman, Kenneth L. Cooke, and Jo Ann Lockett, *Algorithms, Graphs, and Computers* (New York: Academic Press, 1970), pp. 94–100.

ANSWER 10.16—(May 1973)

The three ways of forming maximum-length arrow tours on the 4 × 4 field are shown in Figure 10.21.

Edward N. Peters, on the faculty of the University of Rochester Medical School, discovered a general procedure for constructing maximum-length rook tours on square boards of any size. See his paper "Rooks Roaming Round Regular Rectangles," in *Journal of Recreational Mathematics*, vol. 6, 1973, pages 169–73.

Frederick Hartmann of Rolling Hills Estate, California, extended the analysis to nonsquare rectangular boards, but so far as I know, his

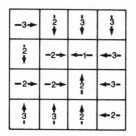 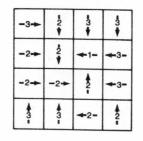

Figure 10.21

results remain unpublished. When the board is $n \times 1$, it reduces to the "worst-route" problem for a postman delivering mail to a row of n houses (see my *Sixth Book of Mathematical Games from Scientific American*, 1971, Chapter 23). Maximum-length rook tours on these linear boards are unique from $n = 1$ through 4, then increase in number steadily as n exceeds 4. For $n = 7$, for example, there are 18 such tours.

Hartmann gave an algorithm for constructing at least one maximum-length tour on any rectangular board. If m and n are the lengths of the sides, with m equal to or greater than n, and C is obtained from the table shown in Figure 10.22 the formula for the length of the tour is

$$\frac{n(3m^2 + n^2 - 10)}{6} + C$$

For square boards of side n the formula reduces to

$$\frac{2n^3 - 5n}{3} + C$$

with $C = 1$ for odd n and $C = 2$ for even n.

Figure 10.23 (from Hartmann) gives the maximum-length rook tours for values of m and n through 12. Neither Peters nor Hartmann attempted the much more difficult task of finding a formula for the number of distinct tours on a given board.

m	n	$\{n/2\}$	C
even	even	—	2
odd	odd	—	1
even	odd	even	3/2
even	odd	odd	1/2
odd	even	even	0
odd	even	odd	1

$\{n/2\}$ indicates the greatest integer contained in the $\{\ \}$.

Figure 10.22

n \ m	2	3	4	5	6	7	8	9	10	11	12
1	2	4	8	12	18	24	32	40	50	60	72
2	4	8	16	24	36	48	64	80	100	120	144
3		14	26	38	56	74	98	122	152	182	218
4			38	54	78	102	134	166	206	246	290
5				76	104	136	174	216	264	316	374
6					136	174	220	270	328	390	460
7						218	270	330	396	470	550
8							330	396	474	556	650
9								472	558	652	756
10									652	756	872
11										870	996
12											1,134

Figure 10.23

ANSWER 10.17—(March 1965)

The sliding-block puzzle can be solved in 30 moves. I had hoped I could list the names of all readers who found a 30-move solution, but the letters kept coming until there were far too many names for the available space. All together readers found 10 different 30-movers. They are shown paired in Figure 10.24, because, as many readers pointed out, each solution has its inverse, obtained by substituting for each digit its difference from 9 and taking the digits in reverse order. Note that of the four possible two-move openings, only 3,6 does not lead to a minimum-move solution. Solutions 2a and 3b proved to be the easiest to find. The most elusive solution, 5a, was discovered by only 10 readers. Only two readers, H. L. Fry and George E. Raynor, found all 10 without the help of a computer.

Donald Michie of the University of Edinburgh has been using this eight-block puzzle in his work on game-learning machines. His colleague Peter Schofield, of the university's computer unit, had written a program for determining minimum solutions for all the 20,160 pat-

1a.	34785	21743	74863	86521	47865	21478
1b.	12587	43125	87431	63152	65287	41256
2a.	34785	21785	21785	64385	64364	21458
2b.	14587	53653	41653	41287	41287	41256
3a.	34521	54354	78214	78638	62147	58658
3b.	14314	25873	16312	58712	54654	87456
4a.	34521	57643	57682	17684	35684	21456
4b.	34587	51346	51328	71324	65324	87456
5a.	12587	48528	31825	74316	31257	41258
5b.	14785	24786	38652	47186	17415	21478

Figure 10.24

terns that begin and end with the hole in the center. (Of these patterns, 60 require 30 moves, the maximum for center-hole problems.) With the aid of this program Schofield was able to find all 10 solutions, but this did not rule out the possibility of others, or even of a shorter solution. The matter was first laid to rest by William F. Dempster, a computer programmer at the Lawrence Radiation Laboratory of the University of California at Berkeley, with a program for an IBM 7094. It first ran off all solutions of 30 moves or fewer, printing out the 10 solutions in 2½ minutes. A second run, for all solutions of 34 moves or fewer, took 15 minutes. It produced 112 solutions of 32 moves and 512 solutions of 34 moves. There are therefore 634 solutions superior to the 36-mover given by Henry Ernest Dudeney, who first posed the problem. The ten 30-movers were later confirmed by about a dozen other computer programs. It is not yet known if there are starting and ending patterns, with the hole in the corner or side cell, that require more than 30 moves.

ANSWER 10.18—(November 1965)

Solutions to the cube-rolling tour problems are shown in Figure 10.25. In the first solution the red side of the cube is up only on the top corner squares. In the second, the dot marks the start of the tour, with the red side down.

Cube-rolling tours are a fascinating field that, so far as I know, only

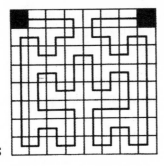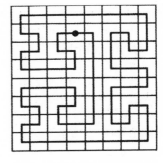

Figure 10.25

John Harris has investigated in any depth. The problems one can devise are endless. Two of Harris's best: What reentrant tour has the red face on top as often as possible? Is there a reentrant tour that starts and ends with red on top but does not have red on top throughout the tour? One can invent problems in which more than one face is colored red, or faces may have different colors, or they may be marked with an *A* so that the orientation of the face is taken into account. What about rolling a standard die over a checkerboard to meet various provisos? Or a die numbered in a nonstandard way? See "Single Vacancy Rolling Cube Problems," by Harris, in *The Journal of Recreational Mathematics*, vol. 7, Summer 1974.

A novel board game based on rolling cubes was marketed in 1971 by Whitman (a subsidiary of Western Publishing Company) under the name of Relate. The board is a 4 × 4 checkerboard. The pieces are four identically colored cubes, two for each player. If each cube were numbered like a die, faces 1 and 2 would be one color, faces 3 and 5 a second color, 4 a third color, and 6 a fourth color. One player's pair of cubes is distinguished from the other player's by having a black spot on each face.

Play begins by alternately placing a cube on a cell, in any orientation, provided that each cube shows a different color on top. Assuming the cells are numbered left to right, one player puts his cubes on cells 3 and 4, the other on cells 13 and 14. These are called the player's starting cells. Players then take turns rolling one of their cubes to an orthogonally adjacent cell. There are three rules:

1. A player's two cubes must at all times have different colors on top.
2. If a player moves so the top color of his cube matches an opponent's cube, the opponent, on his next move, must move the matching cube to a new cell and so that it shows a new color on top.

3. If a move cannot be made without violating rules 1 and 2, a player must turn one of his cubes to a different color, but without leaving the cell it occupies. This counts as a move.

The winner is the first to occupy simultaneously the starting cells of his opponent. If one cube is on an opponent's starting cell, it still must move if forced by the opponent to do so.

I am indebted to John Gough, of Victoria, Australia, for calling this game to my attention. As he points out, the game suggests that there are unexplored possibilities for board games with rolling cubes on square lattices, or rolling tetrahedrons or octahedrons on triangular lattices.

ANSWER 10.19—(May 1975)

In his article "Single Vacancy Rolling Cube Problems" *Journal of Recreational Mathematics*, vol. 7, Summer 1974, pp. 220–24), John Harris gave a 38-move solution to his rolling cube puzzle. About a dozen readers who wrote computer programs for the problem found a unique minimal-move solution in 36 moves. (Reversals, rotations, and reflections are not considered different.) The solution is:

URD LLD RRU LDL URD RUL DLU URD RUL DRD LUL DRU

Harris ends his article with a difficult problem that also involves eight cubes on an order-3 matrix. Color the cubes so that when they are on the matrix, with the center cell vacant, every exposed face is red and all hidden faces are uncolored. There will be just 24 red sides and 24 uncolored sides. The problem is to roll the cubes until they are back on the same eight cells, with the center cell vacant but with all red sides hidden and all visible sides uncolored.

Harris reported a solution in 84 moves, then later lowered it to 74. In May 1981 I received a letter from Hikoe Enomoto, Kiyoshi Ishihata, and Satoru Kawai, all in the Department of Information Science, University of Tokyo. Their computer program produced the following minimal-move solution in 70 steps:

DRUUL	*RDLUR*	*ULURD*	*DLURD*
LDLDR	*LDRUL*	*DLULD*	*DRRUL*
RULDR	*RULDL*	*ULURD*	
RURDL	*DDRUU*	*DLURU*	

For two earlier rolling-cube puzzles invented by Harris, and a description of a board game based on rolling cubes, see Chapter 9 of my *Mathematical Carnival* (Knopf, 1975). For biographical information about Harris, see Steven Levy's *Hackers: Heroes of the Computer Revolution* (Doubleday, 1984).

ANSWER 10.20—(February 1970)

When I originally presented Escott's sliding-block puzzle in my column, I gave a solution in 66 moves, but many readers succeeded in lowering the number to 48. This is now the shortest known solution.

There is no unique 48-move solution. The one given in Figure 10.26 (sent by John W. Wright) is typical. The letters *U*, *D*, *L*, and *R* stand for up, down, left, and right. In each case the numbered piece moves as far as possible in the indicated direction.

Move	Block	Direction
1	6	U, R
2	1	D
3	5	L
4	6	L
5	4	D
6	5	R
7	2	D
8	3	L
9	5	U
10	2	R
11	6	U, L
12	4	L, U
13	7	U
14	10	R
15	9	R
16	8	D
17	1	D
18	7	L
19	2	D
20	4	R, D
21	5	D
22	3	R
23	6	U
24	5	L, U

Move	Block	Direction
25	4	U
26	2	U
27	10	U
28	9	R
29	7	U
30	1	U
31	8	U, R
32	1	D
33	7	L
34	10	L
35	2	L
36	4	D
37	2	R, D
38	3	D
39	6	R
40	5	R
41	7	U
42	10	L, U, L, U
43	8	U, L, U, L
44	4	L, U
45	2	D, L
46	9	U
47	2	R
48	1	R

Figure 10.26

Scheduling & Planning

Since the initial pattern has twofold symmetry, every solution has its inverse. In this case the inverse starts with piece 5 moving down and to the left instead of piece 6 moving up and to the right and continues with symmetrically corresponding moves.

ANSWER 10.21—(February 1966)

The 10-coin triangle, with positions numbered as shown earlier in Figure 10.15, is reduced to one coin in five moves by first removing the penny at spot 3, then jumping: 10–3, 1–6, 8–10–3, 4–6–1–4, 7–2. The solution is unique except for the fact that the triple jump can be made clockwise or counterclockwise. Of course, the initial vacancy may be at any of the six spots not at a corner or in the center.

Nine moves are minimal for the 15-coin triangle. The vacancy must be in the middle of a side, and the first two moves must be 1–4, 7–2, or one of the five other pairs which are symmetrically equivalent. With the first two moves specified, a computer program by Malcolm E. Gillis, Jr., of Slidell, Louisiana found 260 solutions.

The following solution is one of many that ends with a dramatic five-jump sweep. The positions are numbered left to right, top to bottom: (1) 11–4, (2) 2–7, (3) 13–4, (4) 7–2, (5) 15–13, (6) 12–14, (7) 10–8, (8) 3–10, (9) 1–4, 4–13, 13–15, 15–6, 6–4.

I know of no computer analysis of the 21-triangle or any triangle of higher order. John Harris, of Santa Barbara, California proved that nine moves are necessary for the 21-triangle, and the following solution, from Edouard Marmier, Zurich, proves that nine moves are also sufficient: (1) 1–4, (2) 7–2, (3) 16–7, (4) 6–1, 1–4, 4–11, (5) 13–6, 6–4, 4–13, (6) 18–16, 16–7, 7–18, 18–9, (7) 15–6, 6–13, (8) 20–18, 18–9, 9–20, (9) 21–19.

Triangular solitaire obviously can be played on triangular matrices with borders other than the triangle: hexagons, rhombuses, six-pointed stars, and so on. It also can be played with varying rules such as: (1) Jumps parallel to one side of the unit cell can be prohibited. See the 15-triangle solitaire problem in Maxey Brooke, *Fun for the Money* (Scribner's, 1963, page 12). (2) Sliding moves, as well as jumps, may be allowed, as in the game of Chinese checkers.

If isometric (triangular) solitaire is played in the classic way, allowing jumps only, and in all six directions, there are elegant procedures for testing whether a given pattern can be derived from another. The procedures are extensions of those which have been developed for square solitaire. (See the chapter on peg solitaire in my *Unexpected Hanging and Other Mathematical Diversions*, 1969.)

As in square solitaire, the procedures do not provide actual solutions, nor do they prove that a solution exists, but they are capable of showing certain problems to be unsolvable. Much unpublished work on this has been done by Mannis Charosh, Harry O. Davis, John Harris, and Wade E. Philpott. The methods are all based on a commutative group which determines a pattern's parity. By coloring the lattice points with three colors, and following various rules, one can quickly determine the impossibility of certain problems. For example, a hexagonal field with a center vacancy cannot be reduced to one counter in the center unless the side of the hexagon is a number of the form $3n + 2$. A clear explanation of one such procedure is given by Irvin Roy Hentzel in "Triangular Puzzle Peg," *Journal of Recreational Mathematics*, vol. 6, Fall 1973, pages 280–83.

The only reference in print that I have been able to find on isometric solitaire that predates my February 1966 column in *Scientific American* is a 1930 pamphlet titled *Puzzle Craft*, edited by Lynn Rohrbough and published by the Cooperative Recreation Service, Delaware, Ohio. It describes a 15-hole triangular board with the initial hole at 13 and an 11-move solution. However, Jerry Slocum, coauthor of *Puzzles Old and New* (1986), a book on mechanical puzzles, has a copy of an 1891 American patent for a triangular board, and the idea is probably older than that.

A version called "Think a Jump" is sold with an hexagonal board of four holes to the side, but with jump lines removed between the two middle holes on each side. A parity check shows that the task is impossible with the hole in the center as well as at 12 other spots. Aside from these 13 impossible holes, all other spots are solvable— that is, all pegs but one can be removed. *Puzzles Old and New* gives a version (with solution) based on the 15-triangle with two extra holes adjacent to each of its corners.

The regular hexagon of side 3 is a promising field to analyze, but as Michael Merchant showed in 1976 (private communication), this pattern is unsolvable from any hole. The smallest regular hexagon on which it is possible to end with a peg at the initial hole is the 61-hole board of side 5. In fact, *n*-to-*n* solutions on this board are possible for every *n*.

The puzzle may be varied by changing the rules. Maxey Brooke, in his book on coin puzzles, considers the 15-triangle with the added proviso that no horizontal jumps are allowed except along the base line. (He gives a 12-move solution.) Sliding moves, as in checkers, may be permitted in addition to jumps, and minimum-move solutions may be sought. An article in *Crux Mathematicorum* (vol. 4, 1978, pages 212–16) analyzes the 15-triangle when jumps are allowed from any corner hole to the middle hole on the opposite side.

Much work has been done on low-order triangles, following traditional rules, to determine when solutions are possible when both starting and finishing holes are specified. As we have seen, the only possible solution on the 10-triangle starts with the hole at one of the symmetrically equivalent spots, 2, 3, 4, 6, 8, 9, and ends with the peg at an adjacent side hole.

The 15-triangle has just four spots that are symmetrically different, for example 1, 2, 4, and 5. It is possible to reduce the pegs to one, from any hole on the board, but when start and stop spots are specified, difficult tasks arise, some still unsolved. The Hentzels, in their paper "Triangular Puzzle Peg" (1985–86) show that if the final peg is on an interior spot (5, 8, or 9), the initial hole must be at the middle of a side. It is impossible, therefore, to end with the peg at the initial hole if the hole is one of the three interior ones. It *is* possible to start and end at the same spot from any outside hole, the corner being the most difficult to solve.

I gave a nine-move solution, the minimum, to the task of beginning and ending on a middle outside hole. Ten moves are required in going from a corner hole to a peg on the same spot. Here is one solution: 6–1, 4–6, 1–4, 7–2, 10–3, 13–4, 15–13, 12–14–15, 2–7, 11–4–6–1. Change the first move to 3–10 and you have a 10-move solution (the minimum) when the first and last spots are adjacent to a corner.

The fullest analysis in print of possible solutions when start and stop spots are specified is the 1983–84 article by Benjamin Schwartz and Hayo Ahlburg, "Triangular Peg Solitaire—A New Result." Many pairs of spots, such as 4-to-6, are eliminated easily by coloring checks. The authors prove that 5-to-5 is impossible. Cases 5-to-1 and 5-to-7 (and their symmetrical equivalents) remain unsolved, and are probably impossible.

In 1984 I had dinner at a restaurant where a 15-triangle board gave an infuriating problem that I had not seen before. It asked for a solution to the task of starting anywhere and leaving exactly eight pegs on the board, with no more jumps possible. It is easy to leave 10 pegs: 14–5, 2–9, 12–5, 9–2 (the shortest stalemate). If n is the number of pegs left in a stalemate, it can take all values from 1 through 10 except 9. Leaving just eight pegs is a difficult task, with a solution that I shall leave as a problem for you! Benjamin Schwartz has proved that the solution is unique.

The 21-triangle has five nonequivalent starting spots. It can be solved from any starting hole. Harry Davis showed in unpublished research that the shortest solution is nine moves. Here is one that starts with the hole at 13, and ends with a nine-jump sweep: 6–13, 7–9, 16–7, 4–11, 10–8, 21–10, 18–9, 20–18–16–7, 1–4–11–13–15–6–4–13–6–1. All solutions from other starting holes have minimum moves of 10. I know of no published work on what pairs of start and stop positions have solutions when they are not eliminated by parity checks.

Little work has been done on the 28-triangle aside from the application of parity checks. It is not possible to start and end on the central hole. Indeed, it can be shown that center-to-center solutions are impossible on a triangle of any size that has a central hole.

Harry Davis, in collaboration with Wade Philpott, has shown that the 36-triangle has a solution from any starting hole, and that 14-move solutions are minimal. Here is one of Davis's 14-movers with the initial hole at 13: 6–13, 1–6, 10–3, 21–10, 36–21, 4–1–6–15–28, 19–10, 11–4–6–15, 22–11, 31–16–7–9, 32–19–8–10–21–19, 24–11–13–24, 34–36–21–34–32, 29–31–18–20–33–31.

The largest possible sweep obtainable in actual play on the 36-triangle is 15 jumps. It can be achieved, as Davis showed with this 17-move solution starting and ending with the initial hole at 1: 4–1, 13–4, 24–13, 20–18, 34–19, 32–34, 18–20, 30–32–19–8, 7–2, 16–7, 29–16, 15–26, 35–33–20, 6–15, 21–10, 36–21, 1–4–11–22–24–11–13–4–6–13–15–26–28–15–6–1.

ANSWER 10.22—(June 1961)

Three slices of bread—A, B, C—can be toasted and buttered on the old-fashioned toaster in two minutes. Figure 10.27 shows the way to do it.

After this solution appeared, I was staggered to hear from five read-

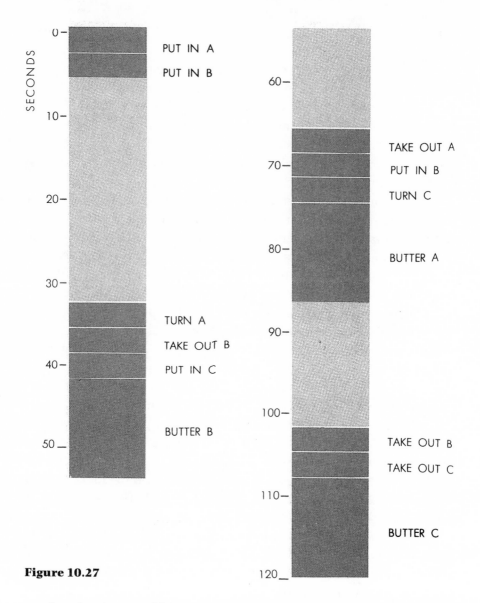

SECONDS

0 — PUT IN A

PUT IN B

10 —

20 —

30 —

TURN A

TAKE OUT B

40 — PUT IN C

BUTTER B

50 —

60 —

TAKE OUT A

70 — PUT IN B

TURN C

80 — BUTTER A

90 —

100 —

TAKE OUT B

TAKE OUT C

110 —

BUTTER C

120 —

Figure 10.27

ers that the time could be cut to 111 seconds. What I had overlooked was the possibility of partially toasting one side of a slice, removing it, then returning it later to complete the toasting. Solutions of this type arrived from Richard A. Brouse, a programing systems analyst with IBM, San Jose, California; R. J. Davis, Jr., of General Precision Inc., Little Falls, New Jersey; John F. O'Dowd, Quebec; Mitchell P. Marcus, Binghamton, New York; and Howard Robbins, Vestal, New York.

Davis's procedure is as follows:

SECONDS	OPERATION
1–3	Put in slice A.
3–6	Put in B.
6–18	A completes 15 seconds of toasting on one side.
18–21	Remove A.
21–23	Put in C.
23–36	B completes toasting on one side.
36–39	Remove B.
39–42	Put in A, turned.
42–54	Butter B.
54–57	Remove C.
57–60	Put in B.
60–72	Butter C.
72–75	Remove A.
75–78	Put in C.
78–90	Butter A.
90–93	Remove B.
93–96	Put in A, turned to complete the toasting on its partially toasted side.
96–108	A completes its toasting.
108–111	Remove C.

All slices are now toasted and buttered, but slice A is still in the toaster. Even if A must be removed to complete the entire operation, the time is only 114 seconds.

Robbins pointed out that near the end, while A is finishing its toasting, one can use the time efficiently by eating slice B.

chapter 11

Games

Taking turns, or alternating moves, readily distinguishes a game from a puzzle. Indeed there is an "opponent" of sorts in any puzzle; the poser is challenging the solver. For a game, however, the difficulty is not a single challenge. Instead the difficulty is the alternation: Is there a first move, such that for every possible second move of the opponent, there is a third move, such that for every possible fourth move . . . ?

In this chapter, games will be cast as puzzles, by asking questions about the outcomes. Can the first player win in this position? For what position can the first player win? Even so, such game puzzles are harder than most problems because of the large number of tentative statements that must be made due to the alternation. (Logicians would say games are difficult because of the alternating logical quantifiers, "there exists" and "for every.") Of course many of the problems can be solved by an insight that shortcuts exhaustive analysis.

We reserve problems related to chess for the next chapter.

Problems

PROBLEM 11.1—Modified Ticktacktoe

A single cell is added to a ticktacktoe board [see Figure 11.1]. If the game is played in the usual manner, the first player easily obtains three in a row by playing first as shown. If there were no extra cell,

Figure 11.1

the second player could stop a win only by taking the center. But now the first player can move as shown, winning obviously on his next move.

Let us modify the game by a new proviso. A player can win on the bottom row only by taking all *four* cells. Can the first player still force a win?

Problem 11.2—Knockout Geography

Knockout Open-End Geography is a word game for any number of players. The first player names any one of the 50 states. The next player must name a different state that either ends with the initial letter of the preceding state or begins with the last letter of the preceding state. For instance, if the first player gives Nevada, the next player can either affix Alaska or prefix Wisconsin. In other words, the chain of states remains open at both ends. When a player is unable to add to the chain, he is eliminated and the next player starts a new chain with a new state. No state can be named more than once in the same game. The game continues until only the winner remains.

David Silverman asks: If you are the first to name a state in a three-player game, what state can you name that will guarantee your winning? We assume that all players play rationally and without collusion to trap the first player.

Problem 11.3—Casey at the Bat

During a baseball game in Mudville, Casey was Mudville's lead-off batter. There were no substitutions or changes in the batting order of the nine Mudville men throughout the nine-inning game. It turned out that Casey came to bat in every inning. What is the least number of runs Mudville could have scored? Charles Vanden Eynden of the University of Arizona originated this amusing problem.

Problem 11.4—Draw Poker

Two men play a game of draw poker in the following curious manner. They spread a deck of 52 cards face up on the table so that they can see all the cards. The first player draws a hand by picking any five cards he chooses. The second player does the same. The first player now may keep his original hand or draw up to five cards. His discards are put aside out of the game. The second player may now draw likewise. The person with the higher hand then wins. Suits have equal value, so that two flushes tie unless one is made of higher cards. After a while the players discover that the first player can always win if he draws his first hand correctly. What hand must this be?

PROBLEM 11.5—Wild Ticktacktoe

A. K. Austin of Hull, England, has written to suggest a wild variation of ticktacktoe. It is the same as the standard game except that each player, at each turn, may mark either a naught or a cross. The first player to complete a row of three (either three naughts or three crosses) wins the game.

Standard ticktacktoe is a draw if both sides play rationally. This is not true of the unusual variant just described. Assuming that both players adopt their best strategy, who is sure to win: the first or the second player?

PROBLEM 11.6—Word Guessing Game

About 1965 Anatol W. Holt, a mathematician who likes to invent new games, proposed the following word game. Two people each think of a "target word" with the same number of letters. Beginners should start with three-letter words and then go on to longer words as their skill improves. Players take turns calling out a "probe word" of the agreed length. The opponent must respond by saying whether the number of "hits" (right letter at the right position) is odd or even. The first to guess his opponent's word is the winner. To show how logical analysis can determine the word without guesswork, Holt has supplied the following example of six probe words given by one player:

Even	Odd
DAY	SAY
MAY	DUE
BUY	TEN

If you knew the target word and compared it letter by letter with any word on the even list, you would find that an even number of letters (zero counts as even) in each probe word would match letters at the same positions in the target word; words on the odd list would match the target word in an odd number of positions. Find the target word.

PROBLEM 11.7—Coin Game

The two-person game shown in Figure 11.2 has been designed to illustrate a principle that is often of decisive importance in the end games of checkers, chess, and other mathematical board games. Place a penny on the spot numbered 2, a dime on spot 15. Players alternate turns, one moving the penny, the other the dime. Moves are made along a solid black line to an adjacent spot. The penny player always moves first. His object is to capture the dime by moving onto the spot occupied by the dime. To win he must do so before he makes his seventh move. If after six of his moves he has failed to catch the dime, he loses.

There is a simple strategy by which one player can always win. Can you discover it?

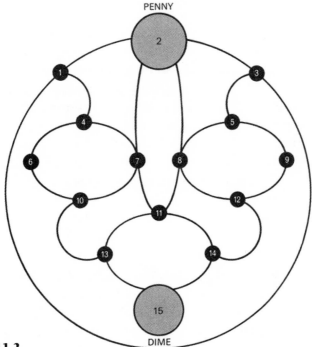

Figure 11.2

PROBLEM 11.8—Googol

In 1958 John H. Fox, Jr., of the Minneapolis-Honeywell Regulator Company, and L. Gerald Marnie, of the Massachusetts Institute of Technology, devised an unusual betting game which they call Googol. It is played as follows: Ask someone to take as many slips of paper as he pleases, and on each slip write a different positive number. The numbers may range from small fractions of 1 to a number the size of a "googol" (1 followed by a hundred 0's) or even larger. These slips are turned facedown and shuffled over the top of a table. One at a time you turn the slips face up. The aim is to stop turning when you come to the number that you guess to be the largest of the series. You cannot go back and pick a previously turned slip. If you turn over all the slips, then of course you must pick the last one turned.

Most people will suppose the odds against your finding the highest number to be at least five to one. Actually if you adopt the best strategy, your chances are a little better than one in three. Two questions arise. First, what is the best strategy? (Note that this is not the same as asking for a strategy that will maximize the *value* of the selected number.) Second, if you follow this strategy, how can you calculate your chances of winning?

When there are only two slips, your chance of winning is obviously 1/2, regardless of which slip you pick. As the slips increase in number, the probability of winning (assuming that you use the best strategy) decreases, but the curve flattens quickly, and there is very little change beyond 10 slips. The probability never drops below 1/3. Many players will suppose that they can make the task more difficult by choosing very large numbers, but a little reflection will show that the sizes of the numbers are irrelevant. It is only necessary that the slips bear numbers that can be arranged in increasing order.

The game has many interesting applications. For example, a girl decides to marry before the end of the year. She estimates that she will meet 10 men who can be persuaded to propose, but once she has rejected a proposal, the man will not try again. What strategy should she follow to maximize her chances of accepting the top man of the 10, and what is the probability that she will succeed?

The strategy consists of rejecting a certain number of slips of paper (or proposals), then picking the next number that exceeds the highest number among the rejected slips. What is needed is a formula for determining how many slips to reject, depending on the total number of slips.

PROBLEM 11.9—Whim

In his Pulitzer Prize–winning book *Gödel, Escher, Bach*, Douglas R. Hofstadter introduces the concept of self-modifying games. These are games in which on his turn a player is allowed, instead of making a legal move, to announce a new rule that modifies the game. The new rule is called a metarule. A rule that modifies a metarule is a metametarule, and so on. Hofstadter gives some chess examples. Instead of moving, a player might announce that henceforth a certain square may never be occupied, or that all knights must move in a slightly different manner, or any other metarule that is on a list of allowed alterations of the game.

The basic idea is not entirely new; before 1970 John Horton Conway proposed a whimsical self-modifying variation of nim that he called Whim. Nim is a two-person game played with counters that are arranged in an arbitrary number of piles, with an arbitrary number of counters in each pile. Players take turns removing one or more counters from any one pile. In normal nim the person taking the last counter wins; in misère, or reverse nim, the person taking the last counter loses. The strategy for perfect play has long been known. You will find it in the chapter on nim in my *Scientific American Book of Mathematical Puzzles & Diversions* (Simon and Schuster, 1959).

Whim begins without any decision on whether the game is normal or misère. At any time in the game, however, either player may, instead of making a move, announce whether the game is normal or misère. This "whim move" is made only once; from then on the game's form is frozen. It is well known that in nim the strategy is the same for both forms of the game until near the end, so that you may be tempted to suppose whim strategy is easily analyzed. Try playing a few games and you will find it is not as simple as it seems!

Suppose you are the first to play in a game with many piles and many counters in each pile, and position is a nim loss for you. You should at once make the whim move because it leaves the position unchanged and you become the winning player. Suppose, however, you are first to play and the position is a win for you. You dare not make a winning move because this allows your opponent to invoke the whim and leave you with a losing position. Hence you must make a move that would lose in ordinary nim. For the same reason your opponent must follow with a losing move. Of course, if one player fails to make a losing move, the other wins by invoking the whim.

As the game nears the end, reaching the point where the winning strategy diverges for normal or for reverse nim, it may be necessary to invoke the whim to win. How is this determined? And how can one decide at the start of a game who has the win when both sides play as well as possible? Conway's strategy is easy to remember but, as he once remarked, hard to guess even by someone well versed in nim theory.

PROBLEM 11.10—Tri-Hex

Ticktacktoe is played on a pattern that can be regarded as nine cells arranged in eight rows of three cells to a row. It is possible, however, to arrange nine cells in 9 or even 10 rows of three. Thomas H. O'Beirne of Glasgow, author of *Puzzles and Paradoxes* (Oxford, 1965), experimented with topologically distinct patterns of nine rows to see if any were suitable for ticktacktoe play. He found trivial wins for the first player on all regular configurations except the one shown in Figure 11.3.

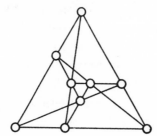

Figure 11.3

To play Tri-Hex, as O'Beirne calls this game, one player can use four pennies and the other four dimes. No fifth move is allowed the first player. Players take turns placing a coin on a spot, and the first to get three of his coins in a row wins. If both players make their best moves, is the game a win for the first player or the second, or is it a draw as in ticktacktoe?

The role played by such configurations as this in modern geometry is entertainingly discussed by Harold L. Dorwart in *The Geometry of Incidence* (Prentice-Hall, 1966) and in the instruction booklet for his puzzle kit, *Configurations*, now available from the makers of the logic game WFF'N PROOF. In addition to its topological and combinatorial properties, the pattern shown here has an unusual metric structure: every line of three is divided by its middle spot into segments with lengths in the golden ratio.

PROBLEM 11.11—Blades of Grass

According to a recent book by two Soviet mathematicians, the following method of fortune-telling was once popular in certain rural areas of the former U.S.S.R. A girl would hold in her fist six long blades of grass, the ends protruding above and below. Another girl would tie the six upper ends in pairs, choosing the pairs at random, and then tie the six lower ends in a like manner. If this produced one large ring, it indicated that the girl who did the tying would be married within a year.

A pencil-and-paper betting game (a pleasant way to decide who pays for drinks) can be based on this procedure. Draw six vertical lines on a sheet of paper. The first player joins pairs of upper ends in any manner, then folds back the top of the paper to conceal the connecting lines from his opponent. The second player now joins pairs of bottom ends as shown at the left in Figure 11.4. The sheet is unfolded to see if the second player has won by forming one large closed loop. (The illustration at the right of Figure 11.4 shows such a win.) If even money is bet, whom does the game favor and what is his probability of winning?

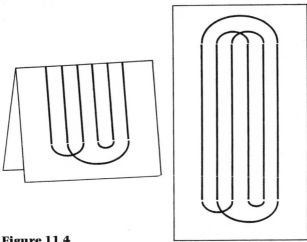

Figure 11.4

PROBLEM 11.12—Hip

The game of Hip, so named because of the hipster's reputed disdain for "squares," is played on a 6 × 6 checkerboard as follows:

One player holds 18 red counters; his opponent holds 18 black counters. They take turns placing a single counter on any vacant cell

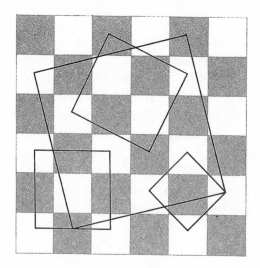

Figure 11.5

of the board. Each tries to avoid placing his counters so that four of them mark the corners of a square. The square may be any size and tipped at any angle. There are 105 possible squares, a few of which are shown in Figure 11.5.

A player wins when his opponent becomes a "square" by forming one of the 105 squares. The game can be played on a board with actual counters, or with pencil and paper. Simply draw the board, then register moves by marking X's and O's on the cells.

For months after I had devised this game I believed that it was impossible for a draw to occur in it. Then C. M. McLaury, a mathematics student at the University of Oklahoma, demonstrated that the game could end in a draw. The problem is to show how the game can be drawn by dividing the 36 cells into two sets of 18 each so that no four cells of the same set mark the corners of a square.

PROBLEM 11.13—Integer Choice

Two mathematicians are drinking beer. After each has been served a glass, they arrive at who pays for the round by simultaneously writing any positive integer on slips of paper. They compare the slips. Whoever wrote the larger integer has to pay for the round, unless the integer is larger by only 1. In that case the person who wrote the smaller integer pays for that round and the next as well. If both players have chosen the same integer, they play again.

To describe the game positively: The person who writes the smaller number scores a point, unless it is smaller by only 1. In that case the other player scores two points. For example, if the numbers were 12 and 20, 12 would win a point. If the numbers were 12 and 13, 13 would win two points.

The game is fair, but what strategy is best in the sense that no other strategy can beat it in the long run, and if any different strategy is followed, there is a counterstrategy to beat it?

The answer is surprising. I shall not have space for a proof, but I shall give the strategy and references to where the proof can be found. The game was devised by N. S. Mendelsohn and Irving Kaplansky, but it was Paul Halmos who called it to my attention.

PROBLEM 11.14—Turnablock

John Horton Conway's pathbreaking book *On Numbers and Games* (1976) brought him a flood of correspondence suggesting new games that could be analyzed by his remarkable methods. One such suggestion, from H. W. Lenstra of Amsterdam, led Conway to develop a new family of games, one of the best of which he calls Turnablock.

Turnablock is played on any $n \times n$ checkerboard, using n^2 counters with sides of different colors. The counters included in the board game Othello (a new name for the old British game of Reversi) can be used for Turnablock, or counters can be made by pairing poker chips of different colors. I shall call the two colors black and white. Here I shall describe only the simplest nontrivial version of the game, the one played on the 3×3 board.

At the start of the game the counters may be arranged in any pattern, but for the purposes of this problem assume that the game starts with the alternating-color pattern shown in Figure 11.6. Each player moves by turning over all the counters in any $a \times b$ rectangular block, where a and b are any two positive integers from 1 through 3. Thus a player's block may be a single counter, a 1×2 "domino" (oriented horizontally or vertically) or any larger configuration up to the entire 3×3 board.

There is one essential rule in Turnablock: A block may be reversed only if there is a black counter in its lower right-hand corner. It is assumed that both players are seated on the same side of the board; otherwise the player on one side may turn a block only if there is a black counter in what for him is the upper left-hand corner. The two players take turns making moves; each player must turn a block when it is his move, and the player whose move leaves all the counters with

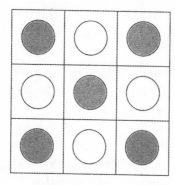

Figure 11.6

the white side up is the winner. The rule guarantees that eventually all the counters will be turned with their white side up and the game will end. With the starting pattern shown in the illustration the first player can always win if he plays correctly. What are his winning first moves, and what must his playing strategy be?

Problem 11.15—A Map-Coloring Game

This problem comes to me from its originator, Steven J. Brams, a political scientist at New York University. He is the author of *Game Theory and Politics* (1975), *Paradoxes in Politics* (1976), and *The Presidential Election Game* (1978). His *Biblical Games* (MIT Press, 1980) is a surprising application of game theory to Old Testament episodes of a gamelike nature in which one of the players is assumed to be an omniscient deity. This was followed by *Superior Beings: If They Exist, How Would We Know?* (1983), *Rational Politics* (1985), *Superpower Games* (1985), *Negotiating Games* (1990), and *Theory of Moves* (1994). His latest book, written with Alan D. Taylor, is *Fair Division* (1996), an analysis of cake-cutting and other fair division problems.

Suppose we have a finite, connected map on a plane and a supply of *n* crayons of different colors. The first player, the minimizer, selects any crayon and colors any region on the map. The second player, the maximizer, then colors any other region, using any of the *n* colors. Players continue in this way, alternately coloring a region with any of the *n* colors but always obeying the rule that no two regions of the same color can share any portion of a common border. Like colors may, of course, touch at points.

The minimizer tries to obviate the need for an *n* + 1 color to com-

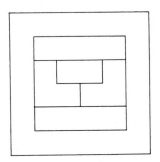

Figure 11.7

plete the map. The maximizer tries to force the use of it. The maximizer wins if either player is unable to play, using one of the n crayons, before the map is fully colored. If the entire map is colored with n colors, the minimizer wins.

The subtle and difficult problem is: What is the smallest value of n such that when the game is played on any map, the minimizer can, if both players play optimally, always win?

To make the problem clearer, consider the simple map in Figure 11.7. It proves that n is at least 5. Of course, if no game is played, the map can be easily colored with four colors, as indeed any map on the plane can be. (This is the famous four-color theorem, now known to be true.) But if the map is used for playing Brams's game, and only four colors are available, the maximizer can always force the minimizer to use a fifth color. If five colors are available, the minimizer can always win.

Brams conjectures that the minimum value of n is 6. A map has been found on which five colors allow the maximizer always to win. Can you construct such a map and give the maximizer's winning strategy? Remember, the minimizer goes first, and neither player is under any compulsion to introduce a new color at any turn if he can legally play a color that has already been used.

Answers

ANSWER 11.1—(August 1968)

Assume the cells are numbered from 1 through 10, taking them left to right and top to bottom. The first player can win only by taking either cell 2 or 6 on his first move. I leave it to you to work out the first player's strategy for all responses.

ANSWER 11.2—(February 1970)

A simple way to win David Silverman's geography game is to name Tennessee. The second player can only prefix Connecticut or Vermont. Since no state begins with E or ends with C or V, the third player is eliminated. It is now your turn to start again. You can win with Maine or Kentucky. Maine eliminates the second player immediately because no state begins with E or ends with M. Kentucky is a quick winner, among several other possibilities. It forces him to name New York. You win by prefixing Michigan, Washington, or Wisconsin.

Three other first moves also win for the first player on his second move: Delaware, Rhode Island, and Maryland. Other states, such as Vermont, Texas, and Connecticut, lead to wins on the first player's third move.

ANSWER 11.3—(March 1965)

The Mudville team could have scored as few as no runs at all even though Casey, the lead-off man, came to bat every inning. In the first inning Casey and the next two batters walk and the next three strike out. In the second inning the first three men walk again, which brings Casey back to bat. But each runner is caught off base by the pitcher, so Casey is back at the plate at the start of the third inning. This pattern is now repeated until the game ends with no joy in Mudville, even though the mighty Casey never once strikes out.

There are, of course, many other ways the game could be played. Robert Kaplan of Cambridge, Massachusetts, wrote the following letter:

Dear Mr. Gardner:

That was indeed an amusing problem concerning Casey and the Mudville nine—amusing, that is, to all save lovers of Mudville. For on the unfortunate day described in your problem, Mudville scored not a run. This is what happened:

In the first inning, Casey and two of his confreres reached base, but batters four through six struck, flied, or otherwise made out. No runs.

In the second inning, batters seven and eight struck out, let us say, but the Mudville pitcher, to the surprise of all, reached base on a bobbled infield roller. Casey came up to bat, frowning mightily. With the count two and two, the perfidious rival pitcher, ignoring the best interests of poetry, baseball mythology and Mudville, whirled toward first and picked off his opposite number, who, dreaming of Cooperstown and the Hall of Fame, had strolled too far from the bag. The crowd sighed, Casey glowered, and the inning was over: no runs.

Now as you know, if an inning ends with a pick-off play at any base, the batter who was in the box at the time becomes the first batter next inning. So it was with Casey; once again Mudville loaded the bases; but once again three outs were made with no runs scoring, so that the inning ended with batter six making the last out.

Life may be linear but fate is cyclic: innings four, six, and eight followed precisely the same pattern as inning two (though you may be sure that after his second miscue in the fourth, the Mudville pitcher was lifted and his relief was responsible not only for the flood of runs the opponents scored, but for similar cloud-gazing on the base-paths). And of course, Casey led off the fifth, seventh, and ninth innings as he had the third—and again, Mudville would load the bases, but could not deliver (if I remember correctly, the gentlemen responsible for this orgy of weak hitting were Cooney, Burrows, Blake and Flynn). Grand total for Mudville: a goose egg.

> Sincerely though sorrowfully yours,
> Robert Kaplan

P.S. I see in rereading the problem that there were no substitutions or changes in the Mudville batting order during the game. How, then, you might justly ask, did the crowd or the manager tolerate such flagrant disdain of first base on the part of their pitcher? The answer is that he was married to the owner's daughter, and no one could say him nay.

ANSWER 11.4—(February 1957)

There are 88 winning first hands. They fall into two categories: (1) four 10's and any other card (48 hands); (2) three 10's and any of the following pairs from the suit not represented by a 10: A–9, K–9, Q–9, J–9, K–8, Q–8, J–8, Q–7, J–7, J–6 (40 hands). The second category was called to my attention by two readers: Charles C. Foster of Princeton, New Jersey, and Christine A. Peipers of New York. I have never seen these hands included in any previously published answer to the problem.

ANSWER 11.5—(June 1964)

When ticktacktoe players are allowed to play either a naught or a cross on each move, the first player can always win by first taking the

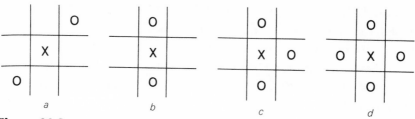

a   b   c   d

Figure 11.8

center cell. Suppose he plays a cross. The second player has a choice of marking either a corner or a side cell.

Assume that he marks a corner cell. To avoid losing on the next move he must mark it with a 0. The first player replies by putting a 0 in the opposite corner, as in diagram a in Figure 11.8. The second player cannot prevent his opponent from winning on his next move.

What if the second player takes a side cell on his second move? Again he must use a 0 to avoid losing on the next move. The first player replies as shown in the next diagram [b]. The second player's next move is forced [c]. The first player responds as shown in the final diagram [d], using either symbol. Regardless of where the second player now plays, the first player wins on his next move.

Wild ticktacktoe (as this game was called by S. W. Golomb) immediately suggests a variant: reverse wild ticktacktoe. The rules are as before except that the first player to get three like-symbols in a row *loses*. Robert Abbott was the first to supply a proof that the game is a draw when played rationally. The first player cannot assure himself a win, but can always tie by using a symmetry strategy similar to his strategy for obtaining a draw in ordinary (or "tame") reverse ticktacktoe. He first plays any symbol in the center. Thereafter he plays symmetrically opposite the second player, always choosing a symbol different from the one previously played.

As Golomb has pointed out, this strategy gives at least a draw in reverse ticktacktoe (wild or not) on all boards of odd-order. On even-order boards the second player can obtain at least a draw by a similar strategy. With the order-3 cubical board, Golomb adds, on which a draw is impossible, the strategy assures a win for the first player in reverse ticktacktoe, wild or tame.

ANSWER 11.6—(April 1969)

To determine the target word, label the six probe words as follows:

EVEN		ODD	
E_1	DAY	O_1	SAY
E_2	MAY	O_2	DUE
E_3	BUY	O_3	TEN

E_1 and E_2 show that the target word's first letter is not D or M, otherwise the parity (odd or even) would not be the same for both words. E_1 and O_1 show that the target word's first letter is either D or S, otherwise the parity could not be different for the two words. The first letter cannot be D and therefore must be S.

Since S is the first letter, E_2 and E_3 are wrong in their first letters. Both end in Y, therefore the second letter of the target word cannot be A or U, otherwise E_2 and E_3 could not have the same parity. Knowing that U is not the second letter and D not the first, O_2 shows that E is the third letter. Knowing that the target word begins with S and ends with E, O_3 shows that E is the second letter. The target word is SEE.

ANSWER 11.7—(March 1965)

In analyzing the topological properties of a network with an unusual pattern it is sometimes helpful to transform the network to a topologically equivalent one that exhibits the network's regularities better. The pattern of the penny-dime game [at top of Figure 11.9] is readily seen to be equivalent to the board at the bottom in the illustration. If the penny moves directly toward the dime, it cannot trap it because the dime has what in chess and checkers is called the "opposition." The meaning of this term is brought out by coloring every other spot. As long as both pieces avoid the triangle at the upper right the dime's move will always carry it to a spot of the same color as the spot occupied by the penny; therefore the penny, on its next move, can never catch the dime. To gain the opposition the penny must move once along the long outside arc that joins the two colored spots numbered 1 and 3. Because this alters the relative parity of the two pieces it is then a simple matter for the penny to corner the dime.

Translating back to the original board, this means that the penny's best strategy is to move either first to 1, then all around the outside

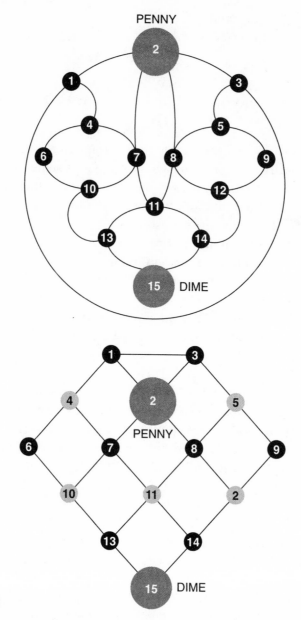

Figure 11.9

circle to 3, or first to 3 and then around to 1. In either case the penny will then have no difficulty trapping the dime, on spot 6, 9, or 15, before the seventh move.

ANSWER 11.8—(February 1960)

Regardless of the number of slips involved in the game of Googol, the probability of picking the slip with the largest number (assuming that the best strategy is used) never drops below .367879. This is the reciprocal of e, and the limit of the probability of winning as the number of slips approaches infinity.

If there are 10 slips (a convenient number to use in playing the game), the probability of picking the top number is .398. The strategy is to turn three slips, note the largest number among them, then pick the next slip that exceeds this number. In the long run you stand to win about two out of every five games.

What follows is a compressed account of a complete analysis of the game by Leo Moser and J. R. Pounder of the University of Alberta. Let n be the number of slips and p the number rejected before picking a number larger than any on the p slips. Number the slips serially from 1 to n. Let $k + 1$ be the number of the slip bearing the largest number. The top number will not be chosen unless k is equal to or greater than p (otherwise it will be rejected among the first p slips), and then only if the highest number from 1 to k is also the highest number from 1 to p (otherwise *this* number will be chosen before the top number is reached). The probability of finding the top number in case it is on the $k + 1$ slip is p/k, and the probability that the top number actually is on the $k + 1$ slip is $1/n$. Since the largest number can be on only one slip, we can write the following formula for the probability of finding it:

$$\frac{p}{n}\left(\frac{1}{p} + \frac{1}{p+1} + \frac{1}{p+2} + \cdots + \frac{1}{n-1}\right)$$

Given a value for n (the number of slips) we can determine p (the number to reject) by picking a value for p that gives the greatest value to the above expression. As n approaches infinity, p/n approaches $1/e$, so a good estimate of p is simply the nearest integer to n/e. The general strategy, therefore, when the game is played with n slips, is to let n/e numbers go by, then pick the next number larger than the largest number on the n/e slips passed up.

This assumes, of course, that a player has no knowledge of the range of the numbers on the slips and therefore no basis for knowing whether a single number is high or low within the range. If one *has* such knowledge, the analysis does not apply. For example, if the game is played with the numbers on 10 one-dollar bills, and your first draw is a bill with a number that begins with 9, your best strategy is to keep the bill. For similar reasons, the strategy in playing Googol is not strictly applicable to the unmarried girl problem, as many readers pointed out, because the girl presumably has a fair knowledge of the range in value of her suitors and has certain standards in mind. If the first man who proposes comes very close to her ideal, wrote Joseph P. Robinson, "she would have rocks in her head if she did not accept at once."

Fox and Marnie apparently hit independently on a problem that had occurred to others a few years before. A number of readers said they had heard the problem before 1958—one recalled working on it in 1955—but I was unable to find any published reference to it. The problem of maximizing the *value* of the selected object (rather than the chance of getting the object of highest value) seems first to have been proposed by the famous mathematician Arthur Cayley in 1875. (See Leo Moser, "On a Problem of Cayley," in *Scripta Mathematica*, September-December, 1956, pages 289–92.)

ANSWER 11.9—(April 1981)

John Horton Conway's strategy for his game of Whim is as follows. Treat the whim move as if it were another pile consisting of one counter if there is a pile of four or more counters, and as if it were another pile consisting of two counters if there is no pile of four or more. Until someone makes the whim move the invisible whim pile remains. Making the whim move removes the whim pile. A whim position is a winning position for the player who would win in ordinary nim if the whim pile were actually there. The winning strategy, therefore, is simply to imagine the whim is present until a player removes it and to play the strategy of ordinary nim. If a move changes a position from one in which at least one pile has four or more counters to one in which there is no such pile, the whim pile acquires its second counter after the move, not before.

ANSWER 11.10—(November 1967)

Ticktacktoe on the Tri-Hex pattern [see Figure 11.10] is a win for the first player, but only if he plays first on one of the black spots. Regardless of his opponent's choice of a spot, the first player can always play so that his opponent's next move is forced, then make a

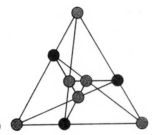

Figure 11.10

third play that threatens a win on two rows, thereby ensuring a win on his last move.

If the opening move is on a corner of the board, the second player can force a draw by seizing another corner. If the opening move is on a vertex of the central equilateral triangle, the second player can force a draw by taking another corner of that triangle. For a more complete analysis see "New Boards for Old Games," by Thomas H. O'Beirne, in *New Scientist*, January 11, 1962.

ANSWER 11.11—(March 1965)

What is the probability of forming one ring by a random joining of pairs of upper ends of six blades of grass, followed by a random joining of pairs of lower ends? Regardless of how the upper ends are joined, we can always arrange the blades as shown in Figure 11.11. We now have only to determine the probability that a random pairing of lower ends will make a ring.

If end *A* is joined to *B*, the final outcome cannot be one large ring. If, however, it is joined to *C*, *D*, *E*, or *F*, the ring remains possible. There is therefore a probability of 4/5 that the first join will not be disastrous. Assume that *A* is joined to *C*. *B* may now join *D*, *E*, or *F*. Only *D* is fatal. The probability is 2/3 that it will join *E* or *F*, and in either case the remaining pair of ends must complete the large ring. The same would hold if *A* had been joined to *D*, *E*, or *F* instead of to *C*.

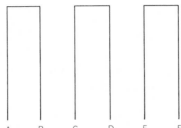

Figure 11.11   A   B   C   D   E   F

Therefore the probability of completing the ring is $4/5 \times 2/3 = 8/15 = .53+$. That the probability is better than half is somewhat unexpected. This means that in the pencil-and-paper version explained earlier the second player has a slight advantage. Since most people would expect the contrary, it makes a sneaky game to propose for deciding who picks up the tab. Of course you magnanimously allow your companion to play first.

The problem generalizes easily. For two blades of grass the probability is 1 (certain), for four blades it is 2/3, for six it is $2/3 \times 4/5$, and for eight blades, $2/3 \times 4/5 \times 6/7$. For each additional pair of blades simply add another fraction, easily determined because the numerators of this series are the even numbers in sequence and the denominators are the odd numbers in sequence! For a derivation of this simple formula, and the use of Stirling's formula to approximate the probability when very large numbers of fractions must be multiplied, see *Challenging Mathematical Problems with Elementary Solutions, Vol. I*, by the Russian twin brothers A. M. and I. M. Yaglom. (It is problem No. 78 in the English translation by James McCawley, Jr., 1964.)

Waldean Schulz included the following additional twist to this problem in a paper titled "Brain Teasers and Information Theory" that he wrote for a philosophy class at the University of Colorado taught by David Hawkins. Suppose you are the second player. How can you join the lower ends in such a way that you can ask a single yes-no question which, if answered by the first player, will tell you if you won or lost?

The answer is to join the two outside lines, the two next-to-outside lines, and the two middle lines. The question is: Did you connect the upper ends in a bilaterally symmetric way? A yes answer means you lost, a no answer means there is a single loop and you win. It is surprising that one "bit" of information is sufficient to distinguish between winning and losing patterns.

ANSWER 11.12—(October 1960)

Figure 11.12 shows the finish of a drawn game of Hip. This beautiful, hard-to-find solution was first discovered by C. M. McLaury, a mathematics student at the University of Oklahoma to whom I had communicated the problem by way of Richard Andree, one of his professors.

Two readers (William R. Jordan of Scotia, New York, and Donald L.

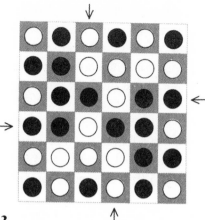

Figure 11.12

Vanderpool of Towanda, Pennsylvania) were able to show, by an exhaustive enumeration of possibilities, that the solution is unique except for slight variations in the four border cells indicated by arrows. Each cell may be either color, provided all four are not the same color, but since each player is limited in the game to 18 pieces, two of these cells must be one color, two the other color. They are arranged here so that no matter how the square is turned, the pattern is the same when inverted.

The order-6 board is the largest on which a draw is possible. This was proved in 1960 by Robert I. Jewett, then a graduate student at the University of Oregon. He was able to show that a draw is impossible on the order-7, no matter how many cells of each color. Since all higher squares contain a 7 × 7 subsquare, draws are clearly impossible on them also.

As a playable game, Hip on an order-6 board is strictly for the squares. David H. Templeton, professor of chemistry at the University of California's Lawrence Radiation Laboratory in Berkeley, pointed out that the second player can always force a draw by playing a simple symmetry strategy. He can either make each move so that it matches his opponent's last move by reflection across a parallel bisector of the board, or by a 90-degree rotation about the board's center. (The latter strategy could lead to the draw depicted.) An alternate strategy is to play in the corresponding opposite cell on a line from the opponent's last move and across the center of the board. Second-player draw strategies were also sent by Allan W. Dickinson of Richmond Heights, Missouri, and Michael Merritt, a student at Texas

A&M College. These strategies apply to all even-order fields, and since no draws are possible on such fields higher than 6, the strategy guarantees a win for the second player on all even-order boards of 8 or higher. Even on the order-6, a reflection strategy across a parallel bisector is sure to win, because the unique draw pattern does not have that type of symmetry.

Symmetry play fails on odd-order fields because of the central cell. Since nothing is known about strategies on odd-order boards, the order-7 is the best field for actual play. It cannot end in a draw, and no one at present knows whether the first or second player wins if both sides play rationally.

In 1963 Walter W. Massie, a civil engineering student at Worcester Polytechnic Institute, devised a Hip-playing program for the IBM 1620 digital computer and wrote a term paper about it. The program allows the computer to play first or second on any square field of orders 4 through 10. The computer takes a random cell if it moves first. On other plays, it follows a reflection strategy except when a reflected move forms a square, then it makes random choices until it finds a safe cell.

On all square fields of order n, the number of different squares that can be formed by four cells is $(n^4 - n^2)/12$. The derivation of this formula, as well as a formula for rectangular boards, is given in Harry Langman, *Play Mathematics* (Hafner, 1962, pages 36–37.)

As far as I know, no studies have been made of comparable "triangle-free" colorings on triangular lattice fields.

ANSWER 11.13—(May 1975)

It is hard to believe, but the best strategy in the integer-choosing game is to limit one's choices of numbers to 1, 2, 3, 4, and 5. The selection is made at random, with the relative frequencies of $1/16$ for numbers 1 and 5, $4/16$ for number 3, and $5/16$ for numbers 2 and 4. One could have in one's lap a spinner designed for picking numbers according to these frequencies.

For a proof of the strategy see "A Psychological Game," by N. S. Mendelsohn (*American Mathematical Monthly* 53, February 1946, pages 86–88) and pages 212–15 of I. N. Herstein and I. Kaplansky's *Matters Mathematical* (Harper & Row, 1974).

Walter Stromquist, in a letter, proposed using a pair of dice as follows: "After a few beers, you cannot be expected to distinguish the fives and sixes, so that if either of these numbers appear on either die, you will have to roll again. Also, since you are really using only 2/3 of each die, it is only natural to multiply the total by 2/3 (dropping all fractions) before writing it down. For example, the largest number you can roll with two dice (without rolling again) is 8, so that the largest number you would ever choose is 2/3 of that, or 5. Out of 16 plays, you should expect to choose with these frequencies: 1 once, 2 five times, 3 four times, 4 five times, 5 once." These are precisely the desired frequencies for playing the best strategy.

ANSWER 11.14—(February 1979)

The first player's strategy for winning John Horton Conway's order-3 game of turnablock requires numbering the nine cells as is shown at the top of Figure 11.13. The player must only make moves that leave black counters on cells whose numbers have a "nim sum" of zero. The nim sum of a collection of numbers is traditionally obtained by writing the numbers in binary form and adding without carrying. If all the digits in the total are zero, then the nim sum is zero. Conway suggests a simpler procedure for finding a nim sum: express each of the numbers to be added as a sum of distinct powers of 2, and then cancel pairs of like powers. If no powers remain, then the nim sum is zero; otherwise the nim sum is simply the sum of the remaining powers. For example, consider 1 + 5 + 12. Writing each number in this

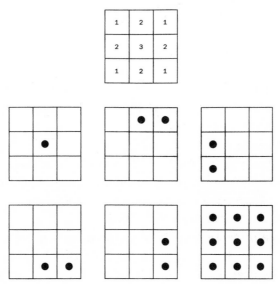

Figure 11.13

1	2	1	4	1	2	1	8
2	3	2	8	2	3	2	12
1	2	1	4	1	2	1	8
4	8	4	6	4	8	4	11
1	2	1	4	1	2	1	8
2	3	2	8	2	3	2	12
1	2	1	4	1	2	1	8
8	12	8	11	8	12	8	13

Figure 11.14

expression as the sum of distinct powers of 2 gives $1 + 4 + 1 + 8 + 4$. The pairs of 1's and 4's cancel, leaving a nim sum of 8. Similarly, consider $1 + 2 + 5 + 6$. This expression becomes $1 + 2 + 4 + 1 + 4 + 2$. All the powers cancel, and the nim sum is zero.

The dots in Figure 11.13 indicate the blocks that can be turned over for a winning first move. In other words, after each of these moves black counters are left on cells whose numbers have a nim sum of zero. Any second move by the other player will necessarily leave black counters on cells with a nonzero nim sum. The first player will always be able to respond with a zero nim-sum move, and by continuing in this way he is sure to win.

Figure 11.14 supplied by Conway gives the cell numbering for all rectangular turnablock boards with sides from one through eight. The numbering of an $a \times b$ board smaller than the order-8 board is given by the $a \times b$ rectangle in the upper left-hand corner of the matrix. The numbers in the cells are obtained by a process called nim multiplication, which you will find explained in Conway's *On Numbers and Games* (Academic Press, 1976, page 52).

ANSWER 11.15—(April 1981)

The map shown in Figure 11.15 found by Lloyd Shapley, a mathematician at the Rand Corporation, proves that when five colors are used in playing Steven J. Brams's map-coloring game, there is a map on which the maximizer can always win.

The map is a projection of the skeleton of a dodecahedron, with an outside region (*A*) that represents the "back" face of the solid. The maximizer's strategy is always to play on the face of the dodecahe-

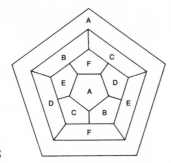

Figure 11.15

dron opposite the face where the opponent last played, using the same color. (In the illustration regions representing pairs of opposite faces are given the same letter.) As you can readily see, this strategy eliminates successive colors from further use, forcing the game to the point where the minimizer cannot play without using a sixth color.

Is there a map on which the maximizer can force the use of a seventh color when the opponent plays optimally? This remains an unsolved problem.

Steven J. Brams's map-coloring game led Robert High, a mathematician with Informatics, Inc., of New York City, to some surprising results. Call the first player Min (who tries to minimize colors) and the second player Max (who tries to maximize colors). I had given a map of six regions on which Max can force five colors. High found that a projection of the cube gives a simpler six-region map for forcing five colors. Max merely places at each turn a new color on the face opposite Min's last play.

The map forcing six colors, is a projection of the skeleton of a dodecahedron. High found that if four corners of a cube are replaced by triangular faces, as shown in Figure 11.16 a projection gives a map

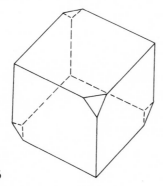

Figure 11.16

of only 10 regions on which Max can force six colors. The strategy is less elegant than the one on the dodecahedral 12-region map and is a bit too involved to give here.

The biggest surprise was High's discovery of a 20-region map on which Max can force seven colors! Imagine each corner of a tetrahedron replaced by a triangular face, then each of the 12 new corners is replaced by a triangular face. Figure 11.17 shows a planar projection of the resulting polyhedron's skeleton. Here the strategy is more complex than the one for High's 10-region map, but he sent a game tree that proves the case. High conjectures that no planar map allows Max to force more than seven colors, but this remains unproved. It is even possible that there is no minimum upper bound.

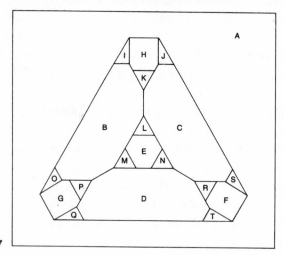

Figure 11.17

chapter 12

Chess

Martin Gardner has always been captivated by chess, and hence it is only natural that he would have included many such problems in his column. The selected problems in this chapter are varied, but all involve chess pieces on a chessboard. Some are like the game puzzles of the last chapter, asking who will win in a given position. Some use a chess piece as a mathematical object with certain defining characteristics of attack and capture. For example, the famous "knight's tour" problem asks if a knight can move about the chessboard visiting every square exactly once. To solve all of these problems you only need to know the rules of chess; expertise in the game is not necessary.

There are two types of chess problems that deserve special mention. The first is the extremal chess problem. "Extremal" is the mathematician's way of saying, Find a configuration that maximizes or minimizes some aspect. For example, How many knights can you simultaneously place on a chessboard? Of course such problems have nothing to do with chess as a game. In fact such problems often are generalized to larger "chessboards."

Second there are the chess logic puzzles of Raymond Smullyan. Smullyan, who also appears in the next chapter, has combined his interests in logic and chess to create "retrograde chess" problems. He did not invent this type of problem but has made it his own. These are questions about chess as a game. A retrograde problem asks unexpected questions about what can be inferred about a chess position. For example, What piece moved last? Most readers will be surprised just how constrained the course of a chess game is to a logician.

Problems

PROBLEM 12.1—Avoiding Checks

Your king is on a corner cell of a chessboard and your opponent's knight is on the corner cell diagonally opposite. No other pieces are on the board. The knight moves first. For how many moves can you avoid being checked? (From David L. Silverman's collection of game problems, *Your Move*.)

PROBLEM 12.2—Superqueens

A "superqueen" is a chess queen that also moves like a knight. Place four superqueens on a 5 × 5 board so that no piece attacks another. If you solve this, try arranging 10 superqueens on a 10 × 10 board so that no piece attacks another. Both solutions are unique if rotations and reflections are ignored. (Thanks to Hilario Fernandez Long.)

PROBLEM 12.3—Don't Mate in One

Karl Fabel, a German chess problemist, is responsible for the outrageous problem depicted in Figure 12.1. It appeared in Mel Stover's delightful column of offbeat chess puzzles in *Canadian Chess Chat* magazine.

You are asked to find a move for White that will *not* result in an immediate checkmate of Black's king.

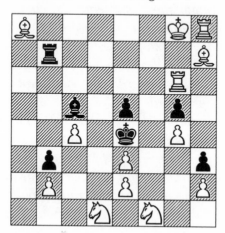

Figure 12.1

ALGORITHMIC PUZZLES & GAMES

PROBLEM 12.4—Twenty-Five Knights

Every square of a 5 × 5 chessboard is occupied by a knight. Is it possible for all 25 knights to move simultaneously in such a way that at the finish all cells are occupied as before? Each move must be a standard knight's move: two squares in one direction and one square at right angles.

PROBLEM 12.5—Lord Dunsany's Chess

Admirers of the Irish writer Lord Dunsany do not need to be told that he was fond of chess. (Surely his story "The Three Sailors' Gambit" is the funniest chess fantasy ever written.) Not generally known is the fact that he liked to invent bizarre chess problems which, like his fiction, combine humor and fantasy.

The problem depicted in Figure 12.2 was contributed by Dunsany to *The Week-End Problems Book*, compiled by Hubert Phillips. Its solution calls more for logical thought than skill at chess, although one does have to know the rules of the game. White is to play and mate in four moves. The position is one that could occur in actual play.

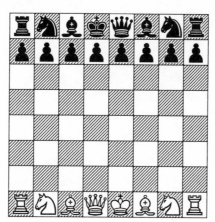

Figure 12.2

PROBLEM 12.6—Minimum and Maximum Attacks

Many beautiful chess problems do not involve positions of competitive play; they use the pieces and board only for posing a challenging mathematical task. Here are two classic task problems that surely belong together:

1. The minimum-attack problem: Place the eight pieces of one color (king, queen, two bishops, two knights, two rooks) on the board so

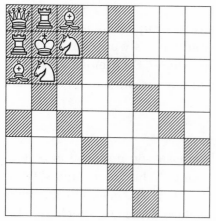

Figure 12.3

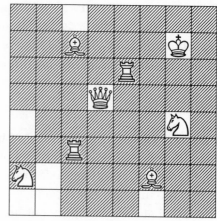

Figure 12.4

that the *smallest* possible number of squares are under attack. A piece does not attack the square on which it rests, but of course it may attack squares occupied by other pieces. In Figure 12.3, 22 squares (gray) are under attack, but this number can be reduced considerably. It is not necessary that the two bishops be placed on opposite colors.

2. The maximum-attack problem: Place the same eight pieces on the board so that the *largest* possible number of squares are under attack. Again, a piece does not attack its own square, but it may attack other occupied squares. The two bishops need not be on opposite colors. In Figure 12.4, 55 squares (gray) are under attack. This is far from the maximum.

There is a proof for the maximum number when the bishops are on squares of the same color. No one has yet proved the maximum when the bishops are on different colors. The minimum is believed to be the same regardless of whether the bishops are on the same or opposite colors, but both cases are unsupported by proof. So many chess experts have worked on these problems that it is not likely any of the conjectured answers will be modified. Should you beat the records, it will be big news in chess-problem circles.

PROBLEM 12.7—Maximizing Chess Moves

When the eight chess pieces of one color (pawns excluded) are placed alone on the board in the standard starting position, 51 different moves can be made. Rooks and bishops can each make 7 different moves, knights and the king can each make 3, the queen can make 14.

By changing the positions of the pieces it is easy to increase the number of possible moves. What is the maximum? In other words, how can the eight pieces of one color be placed on an empty board in such a way that the largest possible number of different moves can be made?

The two bishops should be placed on opposite color squares to conform with standard chess practice, and the move of castling is not considered. Actually neither qualification is necessary because in both cases a violation would only restrict the freedom of pieces to move.

PROBLEM 12.8—Eccentric Chess

On a recent visit to an imaginary chess club I found a game in progress between Mr. Black and Mr. White, the club's two most eccentric players. To my surprise the board appeared as shown in Figure 12.5. My first thought was that each player had started without his king's knight and that Black had moved first, but Mr. Black informed me that he had just completed his fourth move in a standard game that had been played as follows:

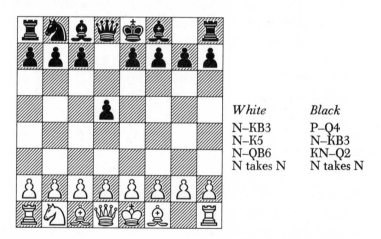

White	Black
N–KB3	P–Q4
N–K5	N–KB3
N–QB6	KN–Q2
N takes N	N takes N

Figure 12.5

An hour later, after losing a game to another player, I came back to see what Black and White were up to. In their second game the board looked exactly the same as it had before except that now *all four* knights were missing! Mr. Black, playing black, looked up and said, "I've just completed my *fifth* move."

1. Can you construct a legitimate game that leads to such a peculiar opening position?

"By the way," said Mr. White, "I've invented a problem that might amuse your readers. Suppose we dump a complete set of chessmen into a hat—all 16 black pieces and all 16 white—shake the hat, then remove the pieces randomly by pairs. If both are black, we put them on the table to form a black pile. If both happen to be white, we put them on the table to form a white pile. If the two pieces fail to match in color, we toss them into their chess box. After all 32 pieces have been removed from the hat, what's the probability that the number of pieces in the black pile will be exactly the same as the number in the white pile?"

"H'm," I said. "Offhand I'd guess the probability would be rather low." Black and White continued their game with subdued chuckles.

2. What is the exact probability that the two piles will be equal?

PROBLEM 12.9—Starting a Chess Game

A full set of 32 chessmen is placed on a chessboard, one piece to a cell. A "move" consists in transferring a piece from the cell it is on to any empty cell. (This has nothing to do with chess moves.) Gilbert W. Kessler, a mathematics teacher in a Brooklyn high school, thought of the following unusual problem: How can you place the 32 pieces so that a maximum number of transfer moves is required to arrange the pieces in the correct starting position for a game of chess?

It is not specified which side of the board is the black side, but the playing sides must, as in regulation chess, be sides with a white square in the bottom right corner, and of course the queen must go on a square matching her own color. One is tempted at first to think that the maximum is 33 moves, but the problem is trickier than that.

PROBLEM 12.10—Indian Chess Mystery

Raymond M. Smullyan's long-awaited collection of chess problems was published in 1979 by Knopf with the title *The Chess Mysteries of Sherlock Holmes*. Just as there has never been a book of logic puzzles quite like Smullyan's *What Is the Name of This Book?*, so there has never been a book of chess problems as brilliant, original, funny, and profound as this one.

Although a knowledge of chess rules is necessary, as Smullyan says in his introduction, the problems in the book actually lie on the borderline between chess and logic. Most chess problems deal with the future, such as how can White move and mate in three. Smullyan's problems belong to a field known as retrograde analysis (retro analysis for short), in which it is necessary to reconstruct the past. This can be done only by careful deductive reasoning, by applying what Smullyan calls "chess logic."

Sherlock Holmes would have had a passion for such problems, and his enthusiasm would surely have aroused the interest of Dr. Watson, particularly after Watson had learned from Holmes some of the rudiments of chess logic. Each problem in Smullyan's book is at the center of a Sherlockian pastiche narrated by Watson in his familiar style. Some of the problems are so singular that it is difficult to believe they have answers. In one, for example, Holmes proves that White has a mate in two but that it is not possible to show the actual mate. In another problem Holmes shows that in the days when chess rules allowed a promoted pawn to be replaced by a piece of the opposite color, a position could arise in which it is impossible to decide if castling is legal even when all the preceding moves are known.

In the second half of the book Holmes and Watson set sail for an island in the East Indies where they hope to find a buried treasure by combining cryptography with retrograde chess analysis. Their first adventure takes place on the ship. Two men from India have been playing a game with pieces that are colored red and green instead of the usual black and white or black and red.

The players have temporarily abandoned the game to stroll around the deck when Holmes and Watson arrive on the scene. The position of the game is shown in Figure 12.6. Several chess enthusiasts are

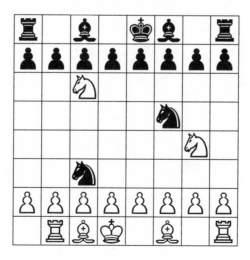

Figure 12.6

studying the position and trying to decide which color corresponds to White, that is, which side had made the first move.

"Gentlemen," says Holmes, "it turns out to be quite unnecessary to guess about the matter. It is *deducible* which color corresponds to White."

In retro-chess problems it is not required that one side or the other play good chess, only that they make legal moves. Your task is to decide which color moved first and prove it by ironclad logic.

Problem 12.11—Where Was the King?

Raymond Smullyan invented this elegant chess problem when he was a student at the University of Chicago in 1957. He showed it to his friend William Browder, now a distinguished mathematician at the university, who passed it on to his father, Earl Browder, former head of the Communist Party in the United States and an ardent chess player. The father sent it to the *Manchester Guardian*, where it was inadvertently published without mentioning Smullyan. A later issue gave proper credit for the problem, and other retrograde problems by Smullyan ran in subsequent issues.

A retrograde chess problem is one that can be solved only by deducing the moves that precede the position shown. In this case we see in Figure 12.7 a position in a legal game just after the white king has been knocked off the board. Where was the king standing? and what was White's last move?

Figure 12.7

PROBLEM 12.12—Monochromatic Chess

Here is another brilliant and unorthodox chess problem by Raymond Smullyan. Figure 12.8 shows the position of an end game with only five men on the board; the black and white kings, two white pawns, and one pawn of unknown color [shown in gray]. During the course of the game no piece has moved from a square of one color to a square of another color. Is the unknown pawn black or white?

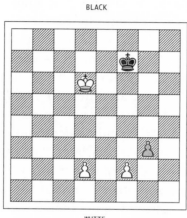

BLACK

WHITE

Figure 12.8

PROBLEM 12.13—Mutilated Score Sheet

Figure 12.9 reproduces the score sheet of a chess game played in a German chess club in 1897. As you can see, a cigar or a cigarette has burned a hole in the sheet, so that Black's moves, which ended in

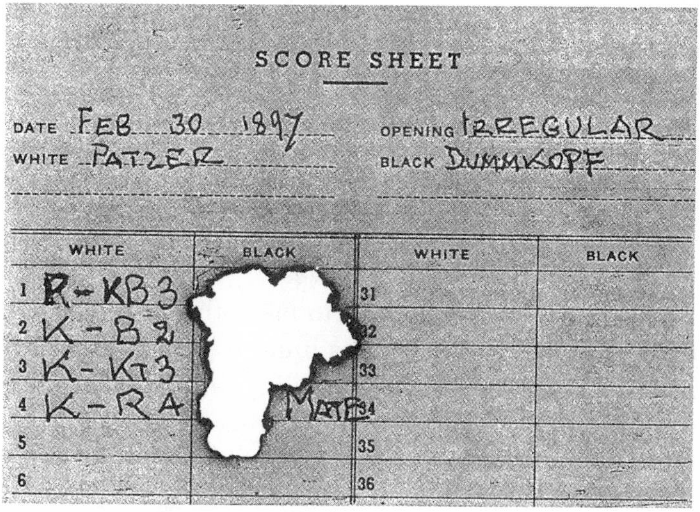

Figure 12.9

Black's win on his fourth move, have been obliterated. Can you reconstruct the game?

I wish to thank Randolph W. Banner of Newport Beach, California, for passing along this amusing problem. He writes that he thinks he saw it in an English periodical published about 1920.

PROBLEM 12.14—Single-Check Chess

The British Chess Magazine, vol. 36, no. 426, June 1916, reported that an American amateur named Frank Hopkins had invented a variation of chess, which he called "Single Check" or "One Check Wins." The game is played exactly like standard chess except that victory goes to the first player who checks (not checkmates) his opponent's king. The magazine, quoting from an article in *The Brooklyn Daily Eagle*, reported that "a suspicion that the white pieces had a sure win turned into a certainty" when U.S. grand master Frank J. Marshall one day "laconically remarked that he could 'bust' the new game." Hopkins was unconvinced until Marshall quickly won by moving *only his two white knights*. Marshall's strategy is not given, except for his opening move, nor does the account of the incident give the number of white moves before he administered the fatal check.

Since 1916 the idea of "single check" chess has occurred independently to many players. I first heard of it from Solomon W. Golomb, who knew it as "presto chess," a name given it by David L. Silverman after learning of the game from a 1965 reinventor. Back in the late 1940s the game had been reinvented by a group of mathematics graduate students at Princeton University. One of the students, William H. Mills, had discovered then what was undoubtedly Marshall's strategy: a way in which White, moving only his knights, can win on or before his fifth move. In 1969 Mills and George Soules together found different five-move wins involving pieces other than knights. Can you duplicate Marshall's feat by solving the problem shown in Figure 12.10. How can White, moving only his knights, check the black king in no more than five white moves?

Attempts have been made to make the game more even by imposing additional rules. Hopkins himself proposed beginning with each player's pawns on the third row instead of the second. Sidney Sackson, who told me of the 1916 reference, suggests that the winner be the first to check a specified number of times, say from 5 to 10,

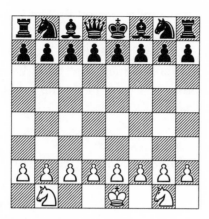

Figure 12.10

depending on how long one wants the game to last. Whether either proposal effectively destroys White's advantage I cannot say.

PROBLEM 12.15—Third-Man Theme

Two kings are the sole occupants of a chessboard [see Figure 12.11]. The task is to add a third man, creating a position that meets these provisos:

1. Neither king is in check.
2. The position can be reached in a legal game.
3. It can be proved, by retrograde analysis of previous legal moves, that neither side has a legal play.

Note carefully the wording. It asks not for a double stalemate but only for a position in which neither side can move. The solution is unique.

Figure 12.11

Chess

This sophisticated little problem appeared in *The Problemist* (September 1974, page 471), where it was credited to G. Husserl of Israel. Newman Guttman called it to my attention. The original problem asked for the minimum number of men that must be added to the board to meet the conditions, but I have made the problem easier by stating that the minimum is one.

PROBLEM 12.16—Uncrossed Knight's Tours

In the *Journal of Recreational Mathematics* for July 1968, L. D. Yarbrough introduced a new variant on the classic problem of the knight's tour. In addition to the rule that a knight touring a chessboard cannot visit the same cell twice (except for a final reentrant move that in certain tours allows the knight to return to the starting square), the knight is also not permitted to cross its own path. (The path is taken to be a series of straight lines joining the centers of the beginning and ending squares of each leap.) The question naturally arises: What are the longest uncrossed knight's tours on square boards of various sizes?

Figure 12.12 reproduces samples of the longest uncrossed tours Yarbrough found for square boards from order-3 through order-8. The order-7 tour is particularly interesting. Seldom does a reentrant tour have maximum length; this is one of the rare exceptions as well as a tour of pleasing fourfold symmetry.

The idea of uncrossed knight's tours caught the interest of Donald E. Knuth. He wrote a "backtrack" computer program that, among other things, found every possible maximum-length uncrossed knight's tour on square boards through order-8. Rotations and reflections are, as usual, not considered different. The computer found two tours for the order-3 board, five for order-4, four for order-5, one for order-6, 14 for order-7, and four for order-8 (the standard chessboard).

It is the unique tour on the 6 × 6 that is most surprising and that provides our problem. Only on the order-6 board did Yarbrough fail to find a maximum-length uncrossed tour. His tour has 16 moves, but there is one uncrossed tour on this board that has 17 moves. You are invited to match your wits against the computer and see whether you can discover the 17-move tour.

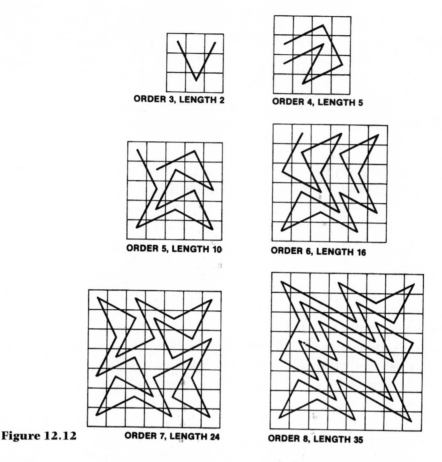

ORDER 3, LENGTH 2

ORDER 4, LENGTH 5

ORDER 5, LENGTH 10

ORDER 6, LENGTH 16

Figure 12.12

ORDER 7, LENGTH 24

ORDER 8, LENGTH 35

PROBLEM 12.17—Lost-King's Tours

Several years ago Scott Kim proposed what he calls the "Lost-King's Tour." This is a king's tour of a small chessboard subject to the following conditions:

1. The king must visit each cell once and only once.
2. The king must change direction after each move; that is, it cannot move twice consecutively in the same direction.
3. The number of spots where the king's path crosses itself must be minimized.

Part A of Figure 12.13 shows the only possible tour on a 3 × 3 board from cell *A* to cell *B*. It has one crossing and is unique, except of course for reflection by the main diagonal. A closed tour is impossible on this board. Closed tours with no crossings are easily found on the

Chess

351

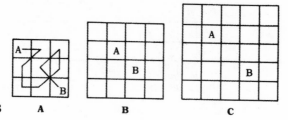

Figure 12.13   A   B   C

4 × 4 board. On the 5 × 5 board a closed tour probably requires two crossings. The problem is less interesting on larger chessboards because crossing-free closed tours and crossing-free open tours from any cell to any other cell are believed to be always possible.

Here are two beautiful tour problems devised by Kim:

On the order-4 board shown in part B of the illustration, find a lost-king's tour from A to B with as few as three crossings. The solution is unique.

On the order-5 board shown in part C of the illustration find a lost-king's tour from A to B with as few as two crossings. This problem is unusually difficult. Kim does not know whether the solution he found is unique or whether the problem can be solved with only one crossing.

PROBLEM 12.18—Queen's Tours

Here is a choice selection of five queen's-tour problems. You do not have to be a chess player to work them out; you need only know that the queen moves an unlimited distance horizontally, vertically, or diagonally. The problems are roughly in order of increasing difficulty.

1. Place the queen on square A [see Figure 12.14]. In four continuous moves traverse all nine of the gray-shaded squares.
2. Place the queen on cell D (the white queen's starting square) and make the longest trip possible in five moves. ("Longest" means the actual length of the path, not the maximum number of cells traversed.) The queen must not visit the same cell twice, and she is not allowed to cross her own path. Assume the path to be through the center points of all cells.
3. Place the queen on cell B. In 15 moves pass through every square once and only once, ending the tour on cell C.

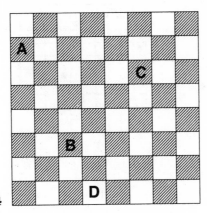

Figure 12.14

4. Place the queen on a corner square. In 14 moves traverse every cell of the board, returning to the starting square on the 14th move. Individual cells may be visited more than once. This "reentrant queen's tour" was first published in 1867 by Sam Loyd, who always considered it one of his finer achievements. The tour, whether reentrant or open at the ends, cannot be made in fewer than 14 moves.

5. Find a similar reentrant tour in 12 moves on a 7 × 7 board. That is, the queen must start and end on the same cell and pass through every cell at least once. As before, cells may be entered more than once.

Answers

ANSWER 12.1—(July 1971)

You can evade check forever. Head toward the board's center, always moving your king to a color opposite to that of the knight. Since a knight changes the color of its cell at every move, whenever the king is on a color different from the knight's, no knight move can check the king. Your only danger lies in being trapped in a corner where you can be forced to move diagonally and be checked by the knight's next move.

ANSWER 12.2—(July 1971)

Figure 12.15 shows the two solutions.

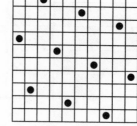

Figure 12.15

Answer 12.3—(June 1961)

In the chess problem, White can avoid checkmating Black only by moving his rook four squares to the west. This checks the black king, but Black is now free to capture the checking bishop with his rook.

When this problem appeared in *Scientific American*, dozens of readers complained that the position shown is not a possible one because there are two white bishops on the same color squares. They forgot that a pawn on the last row can be exchanged for any piece, not just the queen. Either of the two missing white pawns could have been promoted to a second bishop.

There have been many games by masters in which pawns were promoted to knights. Promotions to bishops are admittedly rare, yet one can imagine situations in which it would be desirable—for instance, to avoid stalemating the opponent. Or White may see that he can use either a new queen or a new bishop in a subtle checkmate. If he calls for a queen, it will be taken by a black rook, in turn captured by a white knight. But if White calls for a bishop, Black (not seeing the mate) may be reluctant to trade a rook for a bishop and so lets the bishop remain.

Answer 12.4—(February 1967)

The 25 knights cannot simultaneously jump to different squares. This is easily proved by a parity check. A knight's move carries the piece to a square of a different color from that of the square where it started. A 5 × 5 chessboard has 13 squares of one color, 12 of another. Thirteen knights obviously cannot leap to 12 squares without two of them landing on the same square. The proof applies to all boards with an odd number of squares.

If rooks are substituted for knights, but limited to a move of one square, the same impossibility proof obviously applies. It would also apply, of course, to any mixture of knights and such rooks.

ANSWER 12.5—(May 1959)

The key to Lord Dunsany's chess problem is the fact that the black queen is not on a black square as she must be at the start of a game. This means that the black king and queen have moved, and this could have happened only if some black pawns have moved. Pawns cannot move backward, so we are forced to conclude that the black pawns reached their present positions from the other side of the board! With this in mind, it is easy to discover that the white knight on the right has an easy mate in four moves.

White's first move is to jump his knight at the lower right corner of the board to the square just above his king. If Black moves the upper left knight to the rook's file, White mates in two more moves. Black can, however, delay the mate one move by first moving his knight to the bishop's file instead of the rook's. White jumps his knight forward and right to the bishop's file, threatening mate on the next move. Black moves his knight forward to block the mate. White takes the knight with his queen, then mates with his knight on the fourth move.

ANSWER 12.6—(February 1962)

Figure 12.16 shows how eight chess pieces of one color can be placed on the board so that only 16 squares are under attack. The queen and bishop in the corner can be switched to provide a 16-square minimum with bishops of the same color. This is believed to be the minimum regardless of whether the bishops are the same color or different colors. The position also solves two other minimum problems for the eight pieces: a minimum number of moves (10, and a minimum number of pieces (three) that can move.

Figure 12.17 shows one way to place the eight pieces so that all 64 squares are under attack, obviously the maximum. With bishops of

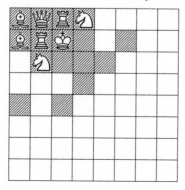

Figure 12.16

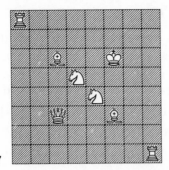

Figure 12.17

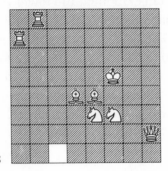

Figure 12.18

opposite color, 63 is believed to be the maximum. There are scores of distinct solutions, one of which is shown in Figure 12.18. The exact number of different solutions is not known.

The maximum-attack problem with bishops of opposite color was first proposed by J. Kling in 1849 with the added proviso that the king occupy the single unattacked square. You may enjoy searching for such a solution, as well as for a pattern (unusually difficult) in which the unattacked square is at the corner of the board. Two readers, C. C. Verbeek of The Hague and Roger Maddux of Arcadia, California, sent identical solutions of the second problem, with the unattacked corner square occupied by a rook. It has been shown that the unattacked square may be at any spot on the board.

ANSWER 12.7—(November 1963)

If eight chess pieces of one color are placed on the board as shown in Figure 12.19, a total of exactly 100 different moves can be made. According to T. R. Dawson, the English chess problemist, this question was first asked in 1848 by a German chess expert, M. Bezzel. His

Figure 12.19

solution, the one shown here, was published the following year. In 1899 E. Landau, in *Der Schachfreund*, September 1899, proved that 100 moves is the maximum and that Bezzel's solution is unique except for the trivial fact that the rook, on the seventh square of the fourth row from the top, could just as well be placed on the first square of that same row.

Among the many readers who solved this chess problem, 14 supplied detailed proof that 100 moves is indeed the maximum.

For a way of placing the eight pieces so that a *minimum* number of moves (10) is possible, see Figure 12.16.

ANSWER 12.8—(February 1970)

1. One possible line of play:

WHITE	BLACK
N–KB3	N–KB3
N–QB3	N–QB3
N–Q4	N–Q4
KN takes N	QP takes N
N takes N	P takes N

Both chess problems were reprinted in the Summer 1969 issue of a mathematics magazine called *Manifold*, published at the University of Warwick in Coventry, England. It cited a 1947 issue of *Chess Review* as its source. The above variation is credited to Larry Blustein, an American player.

Mannis Charosh called my attention to an interesting variant of the problem of the two missing knights. Instead of removing the two king-side knights, remove the two queen-side knights, and instead of advancing the black queen's pawn two squares, advance it only one square. This too has a four-move solution, but it has the merit of being unique. (In the version I gave, Black's first two moves are interchangeable.) The problem appeared in the *Fairy Chess Review*, February 1955, where it was credited to G. Schweig, who had first published it in 1938. I leave the finding of the solution to the reader.

2. The probability is 1. Because the discarded set of pairs must contain an equal number of black and white pieces, the all-black and the all-white pile must be equal.

ANSWER 12.9—(November 1970)

The 32 chess pieces can be placed so that 36 "moves" are needed to transfer the pieces to a correct starting position with Black at the top and White at the bottom [see Figure 12.20].

It was stated in the problem that it was not necessary for Black to be at the top. However, if the final position is Black at the bottom, then 37 moves are required to produce a starting pattern with the queens on the right color.

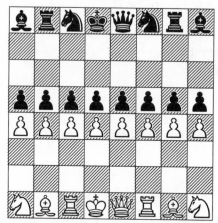

Figure 12.20

ANSWER 12.10—(December 1979)

Here is how Raymond Smullyan proves that Green made the first move (where the light pieces are "red" and the dark pieces are "green"):

Red is now in check, hence Green moved last. It remains to determine who moved first, which can be done by figuring whether an odd or an even number of moves have been made.

The rook on *b*1 has made an odd number of moves; the other three rooks have each made an even number of moves (possibly zero). The Red knights have collectively made an odd number of moves, since they are on squares of the same color, and the Green knights have collectively made an even number of moves. [A knight changes square color on each move.] One king has made an even number of moves (possibly zero), and the other king an odd number. The bishops and pawns have never moved, and both queens were captured before they ever moved. So the grand totality is odd. Thus Green moved first. Hence Green is White and Red is Black.

ANSWER 12.11—(May 1973)

Place the pieces as shown in Figure 12.21 and make the following moves:

WHITE	BLACK
– – –	B–Q4 (check)
P–B4	P takes P
	en passant
	(double check)
K takes P (check)	

Removing the white king will now leave the position given with the statement of the problem.

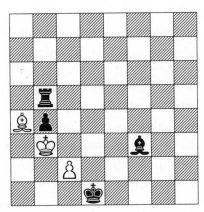

Figure 12.21

In addition to books of philosophical essays and logic problems, Raymond Smullyan published two collections of his marvelous chess problems: *The Chess Mysteries of Sherlock Holmes* (Knopf, 1979) and *The Chess Mysteries of the Arabian Knights* (Knopf, 1981).

Answer 12.12—(April 1974)

The key to Raymond Smullyan's monochromatic chess problem lies in the position of the two white pawns. We were told that no piece has moved to a square of a different color; therefore the only way the white king could have escaped from his home square is by castling. The castling must have been on the king's side; otherwise the white rook would have moved from a black square to a white one. If the pawn of unknown color is white, it must have been a rook's pawn that moved to its present square by capturing. But if this was what happened, the white king could not have reached its present position. The rook's pawn, before it made its capture, would have confined the king to KN1 and in its present position would confine the king to KN1 and KR2. Therefore the pawn in question is black.

The black and white sides of the chessboard were properly identified in Raymond Smullyan's monochromatic chess problem, but several readers asked themselves whether the problem could still be solved if the sides were reversed. Two readers independently sent the following "proof" that the uncolored pawn still must be black. Assume that the pawn is white. To reach the position shown in the problem, the pawn must move at least three times: a first move of two squares, then two captures. The other two white pawns must make at least four moves each to reach their positions. This includes six captures. Thus at least eight captures of black pieces, all on black squares, must be made by the three white pawns. At the start of the game Black has eight pieces on black squares. But one of them, the king's knight, cannot move from its original cell. Black therefore has only seven white pieces on black cells that are available for capture. The initial assumption must be false. The uncolored pawn is black.

Unfortunately the proof is false also. William J. Butler, Jr., sent a legal game, conforming to the monochromatic proviso of the problem, in which the sides are reversed and the position shown in the problem is reached with the uncolored pawn being white. The flaw in the above "proof" is that it overlooks the fact that black pawns on initial white cells can be captured *en passant* by white pawns on black cells.

ANSWER 12.13—(May 1975)

The obliterated chess game is:

1. P–KB3	1. P–K4 or K3
2. K–B2	2. Q–B3
3. K–KT.3	3. QxP (check)
4. K–R4	4. B–K2 (mate)

Several readers thought they had found a second solution. Black's first move is P–Q4. His second move is either P–KN3, P–KR4, or N–KB3. (Black's first two moves may be interchanged.) On his third move Black checks with Q–Q3, then presumably mates with Q–KB5. Unfortunately, the mate can be thwarted by White's fourth move, P–KN4.

ANSWER 12.14—(April 1969)

How can White, using only his knights, win a game of "single check" chess in five or fewer moves?

The opening move must be N (knight) to QB3. Since this threatens several different ways of checking in two moves, Black is forced to advance a pawn that will allow his king to move. If he advances the queen's pawn, N–N5 forces the black king to Q2, then N–KB3 leads to a check on White's fourth move. If Black moves his king-bishop pawn, N–N5 leads to a check on the third move. Black must therefore advance his king's pawn. If he advances it two squares, N–Q5 prevents the black king from moving and White wins on his third move. Black's only good response, therefore, is P–K3.

White's second move is N–K4. Black is forced to advance his king to K2. White's third move, N–KB3, can be met in many ways but none prevents a check on or before White's fifth move. If Black tries such moves as P–Q3, P–KB3, Q–K1, P–Q4, P–QB4, or N–KB3, White responds N–Q4 and wins on his next move. If Black tries P–K4, P–QB4, or N–QB3, White's N–KR4 wins on his next move.

In 1969 William H. Mills discovered that White could also check in five by opening N–QR3. Black must advance his king's pawn one or two squares. N–N5 forces Black's king to K2. White's third move, P–K4, is followed by White's Q–B3 or Q–R5, depending on Black's third move, and leads to a check on White's fifth move.

Two other opening moves leading to five-move checks have since been found by Mills and Georg Soules. They are P–K3 and P–K4.

Against most of Black's replies Q–N4 leads to a three-move check. If Black's second move is N–KR3 or P–KR4, Q–B3 leads to a check in four. If Black's second move is P–K3, White moves Q–R5. Black must respond P–KN3. Then Q–K5 does the trick. (Black's Q–K2 is met by Q takes QBP; B–K2 is met by Q–N7; N–K2 is met by Q–B6.) If Black's second move is N–KB3, White's N–QR3 forces Black to advance his king's pawn one or two squares, then N–N5 forces Black to advance his king, and Q–B3 leads to a check on the next move.

David Silverman has suggested still another way to make single-check chess a playable game. The winner is the first to check with a piece that cannot be taken. So far as I know it is not known which player can always win if both sides play their best.

ANSWER 12.15—(November 1945)

The only solution to the third-man chess problem is to place a white bishop on the board as shown in Figure 12.22. Black obviously cannot move, and White cannot move because it is not his move. To prove that it must be Black's move (therefore Black is stalemated) requires elementary retrograde analysis that I leave to you.

Many readers could not comprehend why the third-man chess problem does not have many other solutions, such as a white pawn in place of the white bishop. The reason is that none of these positions rules out the possibility that it is White's move, and if White can move, the position fails to meet the demand that neither side can move. For instance, assume that the black king is on the QB1 cell. It is checked by an advancing white pawn. The black king moves to the

Figure 12.22

corner. Now it is White's move. The given solution is the only one for which it can be shown by retrograde analysis that it must be Black's move, proving that the position is a stalemate for Black.

ANSWER 12.16—(April 1969)

Figure 12.23 shows the unique maximum-length uncrossed knight's tour on the 6 × 6 board. For similar tours on higher-order square boards and on rectangular boards, see the *Journal of Recreational Mathematics*, vol. 2, July 1969, page 157.

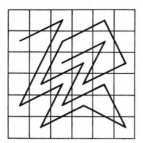

Figure 12.23

ANSWER 12.17—(April 1977)

Solutions to the two lost-king's tours are shown in Figure 12.24. The first tour is unique. The second is almost unique. One other pattern is obtained by modifying the path in the lower left corner as indicated by the dotted lines.

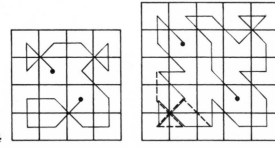

Figure 12.24

ANSWER 12.18—(June 1964)

Answers to the five queen's-tour problems are shown in Figure 12.25. In the fourth and fifth problems there are solutions other than those shown, but none in fewer moves. If you solved the second problem by going first to the lower right corner, up to the upper right cor-

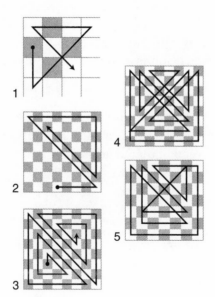

Figure 12.25

ner, along a main diagonal to the lower left corner, up to the upper left, and then right seven squares, you found a path almost (but not quite) as long as the one shown.

part four

Other Puzzles

chapter 13

Logic

Logic puzzles, the purest of mathematical puzzle forms, involve keen reasoning. And Gardner, whose University of Chicago degree was in philosophy, is well versed in the history of logic from Aristotle to Boole to Gödel.

You might easily solve the first few puzzles in this chapter. Keep reading! You will soon find more difficult problems riddled with the truth and falsity of assertions, the most famous of these about truth-tellers and liars. Raymond Smullyan, author of many advanced logic textbooks, is notably the greatest proponent of such "knights and knaves" puzzles, as he called them. Smullyan is also a magician, and Gardner knew him in that context in Chicago in the 1940s and became reacquainted with Smullyan when he started to send material to be used in the Scientific American *column. Some of that material is in this volume.*

Problems

PROBLEM 13.1—Omaha Weather

Many years ago, on a sultry July night in Omaha, it was raining heavily at midnight. Is it possible that 72 hours later the weather in Omaha was sunny?

PROBLEM 13.2—Taciturn Parrot

"I guarantee," said the pet-shop salesman, "that this parrot will repeat every word it hears." A customer bought the parrot but found it would not speak a single word. Nevertheless, the salesman told the truth. Explain.

PROBLEM 13.3—Swiss Barber

Why would a barber in Geneva rather cut the hair of two Frenchmen than of one German?

PROBLEM 13.4—Clean Cut

A logician with some time to kill in a small town decided to get a haircut. The town had only two barbers, each with his own shop. The logician glanced into one shop and saw that it was extremely untidy. The barber needed a shave, his clothes were unkempt, his hair was badly cut. The other shop was extremely neat. The barber was freshly shaved and spotlessly dressed, his hair neatly trimmed. The logician returned to the first shop for his haircut. Why?

PROBLEM 13.5—Name the Name?

What was the name of the Secretary General of the United Nations 35 years ago?

PROBLEM 13.6—Three Errors

Among the assertions made in this problem there are three errors. What are they?

$$a. \quad 2 + 2 = 4$$
$$b. \quad 4 \div 1/2 = 2$$
$$c. \quad 3\tfrac{1}{5} \times 3\tfrac{1}{8} = 10$$
$$d. \quad 7 - (-4) = 11$$
$$e. \quad -10(6 - 6) = -10$$

PROBLEM 13.7—Feemster's Books

"Feemster owns more than a thousand books," said Albert.

"He does not," said George. "He owns fewer than that."

"Surely he owns at least one book," said Henrietta.

If only one statement is true, how many books does Feemster own?

PROBLEM 13.8—Who Lies?

A boy and a girl are sitting on the front steps of their commune.

"I'm a boy," said the one with black hair.

"I'm a girl," said the one with red hair.

If at least one of them is lying, who is which? (Adapted from a problem by Martin Hollis, in *Tantalizers*.)

OTHER PUZZLES

PROBLEM 13.9—Curious Statement

Make a statement about *n* that is true for, and only true for, all values of *n* less than 1 million. (Thanks to Leo Moser.)

PROBLEM 13.10—Find the Pattern

The puzzle shown in Figure 13.1 is from a special issue of the French magazine *Science et Vie* (September 1978) that was devoted entirely to recreational mathematics. In each row the third pattern is obtained from the first two by applying a rule. What is the rule, and what pattern goes in the blank space in the third row?

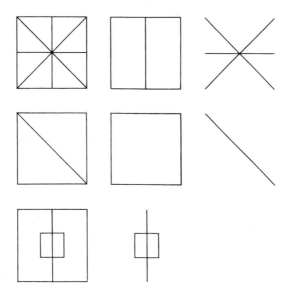

Figure 13.1

PROBLEM 13.11—Will Two Questions Do?

A woman either always answers truthfully, always answers falsely, or alternates true and false answers. How, in two questions, each answered by yes or no, can you determine whether she is a truther, a liar, or an alternater?

PROBLEM 13.12—What Symbol Comes Next?

Can you sketch the sixth figure in the sequence of five symbols shown in Figure 13.2?

Figure 13.2

PROBLEM 13.13—How Many Children?

"I hear some youngsters playing in the backyard," said Jones, a graduate student in mathematics. "Are they all yours?"

"Heavens, no," exclaimed Professor Smith, the eminent number theorist. "My children are playing with friends from three other families in the neighborhood, although our family happens to be largest. The Browns have a smaller number of children, the Greens have a still smaller number, and the Blacks the smallest of all."

"How many children are there altogether?" asked Jones.

"Let me put it this way," said Smith. "There are fewer than 18 children, and the product of the numbers in the four families happens to be my house number which you saw when you arrived."

Jones took a notebook and pencil from his pocket and started scribbling. A moment later he looked up and said, "I need more information. Is there more than one child in the Black family?"

As soon as Smith replied, Jones smiled and correctly stated the number of children in each family.

Knowing the house number and whether the Blacks had more than one child, Jones found the problem trivial. It is a remarkable fact, however, that the number of children in each family can be determined solely on the basis of the information given above!

PROBLEM 13.14—Which Symbol Is Different?

Every person who ever took an I.Q. test has surely been annoyed by questions that involve a row of symbols and a request to identify the symbol that "does not belong" to the set. Sometimes there are so many different ways a symbol can be different from the others that brighter students are penalized by having to waste time deciding which symbol is "most obviously different" to the person who designed the test.

It was irritation over such ambiguity that prompted Tom Ransom, a puzzle expert of Toronto, Ontario, to devise a delightful parody of such a test. You are asked to inspect the five symbols in Figure 13.3 and pick out the one that is "most different."

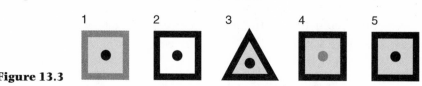

Figure 13.3

PROBLEM 13.15—White, Black, and Brown

Professor Merle White of the mathematics department, Professor Leslie Black of philosophy, and Jean Brown, a young stenographer who worked in the university's office of admissions, were lunching together.

"Isn't it remarkable," observed the lady, "that our last names are Black, Brown, and White and that one of us has black hair, one brown hair, and one white."

"It is indeed," replied the person with black hair, "and have you noticed that not one of us has hair that matches his or her name?"

"By golly, you're right!" exclaimed Professor White.

If the lady's hair isn't brown, what is the color of Professor Black's hair?

PROBLEM 13.16—Ten Statements

Evaluate each of the 10 statements as to its truth or falsity:

1. Exactly one statement on this list is false.
2. Exactly two statements on this list are false.
3. Exactly three statements on this list are false.
4. Exactly four statements on this list are false.
5. Exactly five statements on this list are false.
6. Exactly six statements on this list are false.
7. Exactly seven statements on this list are false.
8. Exactly eight statements on this list are false.
9. Exactly nine statements on this list are false.
10. Exactly ten statements on this list are false.

PROBLEM 13.17—Tricky Track

Three high schools—Washington, Lincoln, and Roosevelt—competed in a track meet. Each school entered one man, and one only, in each event. Susan, a student at Lincoln High, sat in the bleachers to cheer her boyfriend, the school's shot-put champion.

When Susan returned home later in the day, her father asked how her school had done.

"We won the shot-put all right," she said, "but Washington High won the track meet. They had a final score of 22. We finished with 9. So did Roosevelt High."

"How were the events scored?" her father asked.

"I don't remember exactly," Susan replied, "but there was a certain number of points for the winner of each event, a smaller number for second place, and a still smaller number for third place. The numbers were the same for all events." (By "number" Susan of course meant a positive integer.)

"How many events were there altogether?"

"Gosh, I don't know, Dad. All I watched was the shot put."

"Was there a high jump?" asked Susan's brother.

Susan nodded.

"Who won it?"

Susan didn't know.

Incredible as it may seem, this last question can be answered with only the information given. Which school won the high jump?

PROBLEM 13.18—Tennis Match

Miranda beat Rosemary in a set of tennis, winning six games to Rosemary's three. Five games were won by the player who did not serve. Who served first?

PROBLEM 13.19—Three Cards

Gerald L. Kaufman an architect and the author of several puzzle books, devised this logic problem.

Three playing cards, removed from an ordinary bridge deck, lie facedown in a horizontal row. To the right of a king there's a queen or queens. To the left of a queen there's a queen or queens. To the left of a heart there's a spade or spades. To the right of a spade there's a spade or spades.

Name the three cards.

PROBLEM 13.20—Child with the Wart

A: "What are the ages, in years only, of your three children?"

B: "The product of their ages is 36."

A: "Not enough information."

B: "The sum of their ages equals your house number."

A: "Still not enough information."

B: "My oldest child—and he's at least a year older than either of the others—has a wart on his left thumb."

A: "That's enough, thank you. Their ages are. . . ."

OTHER PUZZLES

Complete A's sentence. (Mel Stover of Winnipeg was the first of several readers to send this problem, the origin of which I do not know.)

PROBLEM 13.21—Rotating Disk

Six players—call them A, B, C, D, E, and F—sit around a circular table divided into six equal parts. At the center of the table is a disk mounted on a central pin around which it can rotate [see Figure 13.4]. The disk is marked with arrows and digits.

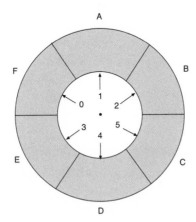

Figure 13.4

The wheel is spun five times. After each spin each player scores the number of points within his segment of the table. (If the wheel stops with its arrows exactly between adjacent players, the spin is not counted.) The players keep a running total of points, and the one with the largest total after the fifth spin is the winner. If there are ties for the highest score, no one wins and the game is played again.

The outcome of the first spin is shown in the illustration. C is ahead with five points. After the second spin D is ahead. After the fifth spin A is the winner. What was each player's final score? The information seems to be insufficient, yet the question can be answered accurately by deductive reasoning. This unusual logic problem is adapted from a puzzle in one of D. St. P. Barnard's popular "Brain-twister" columns in the British *Observer*.

PROBLEM 13.22—Fork in the Road

Here's a twist on an old type of logic puzzle. A logician vacationing in the South Seas finds himself on an island inhabited by the two proverbial tribes of liars and truth-tellers. Members of one tribe always tell the truth, members of the other always lie. He comes to a fork in a road and has to ask a native bystander which branch he should take to reach a village. He has no way of telling whether the native is a truth-teller or a liar. The logician thinks a moment, then asks *one* question only. From the reply he knows which road to take. What question does he ask?

PROBLEM 13.23—Truthers, Liars, and Randomizers

Logic problems involving truth-tellers and liars are legion, but the following unusual variation—first called to my attention by Howard De Long of West Hartford, Connecticut—had not to my knowledge been printed before it appeared in *Scientific American*.

Three men stand before you. One always answers questions truthfully, one always responds with lies, and one randomizes his answers, sometimes lying and sometimes not. You do not know which man does which, but the men themselves do. How can you identify all three men by asking three questions? Each question may be directed toward any man you choose, and each must be a question that is answered by yes or no.

PROBLEM 13.24—Lunch at the TL Club

Every member of the TL Club is either a truther, who always tells the truth when asked a question, or a liar, who always answers with a lie. When I visited the club for the first time, I found its members, all men, seated around a large circular table, having lunch. There was no way to distinguish truthers from liars by their appearance, and so I asked each man in turn which he was. This proved unenlightening. Each man naturally assured me he was a truther. I tried again, this time asking each man whether his neighbor on the left was a truther or a liar. To my surprise each told me the man on his left was a liar.

Later in the day, back home and typing up my notes on the luncheon, I discovered I had forgotten to record the number of men at the table. I telephoned the club's president. He told me the number was 37. After hanging up I realized that I could not be sure of this figure because I did not know whether the president was a truther or a liar. I then telephoned the club's secretary.

"No, no," the secretary said. "Our president, unfortunately, is an unmitigated liar. There were actually 40 men at the table."

Which man, if either, should I believe? Suddenly I saw a simple way to resolve the matter. Can you, on the basis of the information given here, determine how many men were seated at the table? The problem is derived from a suggestion by Werner Joho, a physicist in Zurich.

PROBLEM 13.25—Knights and Knaves

Raymond Smullyan is responsible for the following four charming logic puzzles involving knights and knaves and perhaps some other people. In all four problems a knight always tells the truth, a knave always lies.

> A says: "B is a knight."
> B says: "A is not a knight."

Prove that one of them is telling the truth but is not a knight.

> A says: "B is a knight."
> B says: "A is a knave."

Prove either that one of them is telling the truth but is not a knight or that one is lying but is not a knave.

In the above problems we must consider the possibility that a speaker is neither a knight nor a knave. In the next two problems each of the three people involved is either a knight or a knave.

> C says: "B is a knave."
> B says: "A and C are of the same type [both knights or both knaves]."

What is A?

> A says: "B and C are of the same type."
> C is asked: "Are A and B of the same type?"

What does C answer?

Answers

ANSWER 13.1—(July 1971)

No, because after 72 hours it would have been midnight again.

ANSWER 13.2—(August 1968)

The parrot was deaf.

ANSWER 13.3—(July 1971)

He makes twice as much money.

ANSWER 13.4—(August 1968)

Each barber must have cut the other's hair. The logician picked the barber who had given his rival the better haircut.

ANSWER 13.5—(July 1971)

The same as it is now.

"Can you answer this?" Golomb wrote in 1971. "No, U Thant!"

ANSWER 13.6—(August 1968)

Only equations b and e are false, therefore the statement that there are three errors is false. This is the third error.

ANSWER 13.7—(July 1971)

There are three permissible combinations of true and false for the three statements: *TFF*, *FTF*, and *FFT*. The only noncontradictory combination is *FTF*, which means that Feemster owns no books at all.

When George said that Feemster "owns fewer than that," I meant him to mean fewer than the amount specified by Albert. If "that" is taken to refer to "a thousand" instead of "more than a thousand," however (as many readers pointed out), Feemster could own exactly 1,000 books as well as none.

ANSWER 13.8—(July 1971)

The four possible true-false combinations for the two statements are *TT*, *TF*, *FT*, and *FF*. The first is eliminated because we were told that one statement is false. The second and third are eliminated because in each case, if one person lied, the other cannot have spoken truly. Therefore both lied. The boy has red hair, the girl black hair.

ANSWER 13.9—(July 1971)

One answer: "The value of n is less than 1 million."

This theorem is related to a paradox of induction that I came across in Karl Popper's *Conjectures and Refutations* where he attributes it to

J. Agassi. "All events occur before the year 3000." Since this statement has so far been confirmed by every event in the history of the universe, some theories of induction are forced to regard it as strongly confirmed, thus suggesting that it is highly probable the world will end before 3000.

ANSWER 13.10—(February 1979)

The rule for obtaining the third pattern in each row is to superpose the first two patterns and eliminate any lines they have in common. Hence the pattern to be placed at the end of the third row is simply a square.

ANSWER 13.11—(August 1969)

Ask the woman "Are you an alternater?" twice. Two no answers prove she is a truther, two yes answers prove she is a liar, and a yes-no or no-yes response proves she is an alternater.

After the above answer appeared, several readers sent the following solution. I quote from a letter from Joseph C. Crowther, Jr.:

> You can discover the inclination of the lady if you simply ask two questions about an obvious truth, such as "Do you have two ears?" or "Is water wet?" A truther will answer yes both times, and a liar will answer no. An alternater will not only say one of each, but the order in which she says them will determine which alternation she happens to be on, a fact which may prove useful in further conversation.

Ralph Seifert, Jr., sent a one-question solution that he attributed to his friend M. A. Zorn. "If someone asked you the same question twice, would you falsely answer 'no' exactly once?" The truther will say no, the liar yes, and the alternater will be so hopelessly befuddled that she won't be able to answer at all.

ANSWER 13.12—(November 1975)

The sixth figure is given in Figure 13.5, with the vertical lines of symmetry shown in gray. On the right of each gray line are the numerals 1 through 6; on the left are their mirror reflections.

Figure 13.5

ANSWER 13.13—(November 1957)

When Jones began to work on the professor's problem he knew that each of the four families had a different number of children and that the total number was less than 18. He further knew that the product of the four numbers gave the professor's house number. Therefore his obvious first step was to factor the house number into four different numbers which together would total less than 18. If there had been only one way to do this, he would have immediately solved the problem. Since he could not solve it without further information, we conclude that there must have been more than one way of factoring the house number.

Our next step is to write down all possible combinations of four different numbers which total less than 18, and obtain the products of each group. We find that there are many cases where more than one combination gives the same product. How do we decide which product is the house number?

The clue lies in the fact that Jones asked if there was more than one child in the smallest family. This question is meaningful only if the house number is 120, which can be factored as $1 \times 3 \times 5 \times 8$, $1 \times 4 \times 5 \times 6$, or $2 \times 3 \times 4 \times 5$. Had Smith answered "No," the problem would remain unsolved. Since Jones did solve it, we know the answer was "Yes." The families therefore contained 2, 3, 4, and 5 children.

This problem was originated by Lester R. Ford and published in the *American Mathematical Monthly*, March 1948, as Problem E776.

ANSWER 13.14—(November 1975)

The first symbol of Tom Ransom's I.Q. test differs from the other four in having a gray border. The second symbol differs in not being shaded. The third differs in not being square, and the fourth in having a gray spot. The fifth is the only symbol that is not unique with respect to some property, and therefore it is the one symbol "most different" from the others. To put it differently, the first four are each different in the same general way, whereas the fifth is different in a different way, which makes it radically different.

One is reminded of the story about the report of a psychiatrist on the most remarkable case in his experience. The patient was a person who showed no trace of any kind of neurosis. In my opinion Ransom has found a simple, elegant model of a semantic paradox that is frequently encountered but seldom explicitly recognized.

ANSWER 13.15—(February 1960)

The assumption that the "lady" is Jean Brown, the stenographer, quickly leads to a contradiction. Her opening remark brings forth a reply from the person with black hair, therefore Brown's hair cannot be black. It also cannot be brown, for then it would match her name. Therefore it must be white. This leaves brown for the color of Professor Black's hair and black for Professor White. But a statement by the person with black hair prompts an exclamation from White, so they cannot be the same person.

It is necessary to assume, therefore, that Jean Brown is a man. Professor White's hair can't be white (for then it would match his or her name), nor can it be black because he (or she) replies to the black-haired person. Therefore it must be brown. If the lady's hair isn't brown, then Professor White is not a lady. Brown is a man, so Professor Black must be the lady. Her hair can't be black or brown, so she must be a platinum blonde.

ANSWER 13.16—(May 1973)

Only the ninth statement is true.

David L. Silverman contributed the problem to the *Journal of Recreational Mathematics*, January 1969, page 29, presenting it in the form of 1,969 statements. Underwood Dudley answered it in the October issue, page 231, as follows: "At most one of the statements can be true because any two contradict each other. All the statements cannot be false, because this implies that the list contains exactly zero false statements. Thus exactly one statement can be true. Thus exactly $n - 1$ are false, and the $(n - 1)$st (the 1,968th) statement is true."

Alan Brown pointed out that if the word "exactly" is removed from each of the 10 statements in the logic problem, there is a different and unique solution: The first five statements are true; the last five, false.

The problem obviously generalizes to as many statements as you care to add. What happens if you *decrease* the number to just one?

> 1. Exactly one statement on this list is false.

Norman Pos wrote to point out that the problem then reduces to the traditional liar paradox: "This sentence is false." To circumvent the paradox, Pos added a zero statement at the beginning:

> 0. Exactly none of the statements on this list is false.

Pos was surprised to discover that this shifts the one true statement from position $n - 1$ to position n. That adding such a statement at the top of, say, 1,000 numbered statements would shift the unique true sentence from next-to-last to last he found an amusing case of syntactical "action at a distance."

ANSWER 13.17—(October 1960)

There is space only to suggest the procedure by which it can be shown that Washington High won the high jump event in the track meet involving three schools. Three different positive integers provide points for first, second, and third place in each event. The integer for first place must be at least 3. We know there are at least two events in the track meet and that Lincoln High (which won the shot put) had a final score of 9, so the integer for first place cannot be more than 8. Can it be 8? No, because then only two events could take place and there is no way that Washington High could build up a total of 22 points. It cannot be 7 because this permits no more than three events, and three are still not sufficient to enable Washington High to reach a score of 22. Slightly more involved arguments eliminate 6, 4, and 3 as the integer for first place. Only 5 remains as a possibility.

If 5 is the value for first place, there must be at least five events in the meet. (Fewer events are not sufficient to give Washington a total of 22, and more than five would raise Lincoln's total to more than 9.) Lincoln scored 5 for the shot put, so its four other scores must be 1. Washington can now reach 22 in only two ways: 4, 5, 5, 5, 3 or 2, 5, 5, 5, 5. The first is eliminated because it gives Roosevelt a score of 17, and we know that this score is 9. The remaining possibility gives Roosevelt a correct final tally, so we have the unique reconstruction of the scoring shown in the table [Figure 13.6].

EVENTS	1	2	3	4	5	SCORE
WASHINGTON	2	5	5	5	5	22
LINCOLN	5	1	1	1	1	9
ROOSEVELT	1	2	2	2	2	9

Figure 13.6

Washington High won all events except the shot put, consequently it must have won the high jump.

Many readers sent shorter solutions than the one just given. Two readers (Mrs. Erlys Jedlicka of Saratoga, California, and Albert Zoch, a student at Illinois Institute of Technology) noticed that there was a shortcut to the solution based on the assumption that the problem had a unique answer. Mrs. Jedlicka put it this way:

Dear Mr. Gardner:

Did you know this problem can be solved without any calculation whatever? The necessary clue is in the last paragraph. The solution to the integer equations must indicate without ambiguity which school won the high jump. This can only be done if one school has won all the events, not counting the shot-put; otherwise the problem could not be solved with the information given, even after calculating the scoring and number of events. Since the school that won the shot-put was not the over-all winner, it is obvious that the over-all winner won the remaining events. Hence without calculation it can be said that Washington High won the high jump.

ANSWER 13.18—(February 1962)

The solution I gave in *Scientific American* was so long and awkward that many readers provided shorter and better ones. W. B. Hogan and Paul Carnahan each found simple algebraic solutions, Peter M. Addis and Martin T. Pett each made use of a simple diagram, and Thomas B. Gray, Jr., solved the problem by an ingenious use of the binary system. The shortest solution came from Goran Ohlin, an economist at Columbia University, who expressed it as follows:

"Whoever served first, served five games, and the other player served four. Suppose the first server won x of the games she served and y of the other four games. The total number of games *lost* by the player who served them is then $5 - x + y$. This equals 5 [we were told that the non-server won five games], therefore $x = y$, and the first player won a total of $2x$ games. Because only Miranda won an even number of games, she must have been the first server."

ANSWER 13.19—(November 1965)

The first two statements can be satisfied only by two arrangements of kings and queens: KQQ and QKQ. The last two statements are met by only two arrangements of hearts and spades: SSH and SHS. The two sets combine in four possible ways:

KS, QS, QH
KS, QH, QS
QS, KS, QH
QS, KH, QS

The last set is ruled out because it contains two queens of spades. Since each of the other three sets consists of the king of spades, queen of spades and queen of hearts, we can be sure that those are the three cards on the table. We cannot know the position of any one card, but we can say that the first must be a spade and the third a queen.

ANSWER 13.20—(November 1970)

Only the following eight triplets have a product of 36: 1, 1, 36; 1, 2, 18; 1, 3, 12; 1, 4, 9; 1, 6, 6; 2, 2, 9; 2, 3, 6, and 3, 4, 3. Speaker *A* certainly knew his own house number. He would therefore be able to guess the correct triplet when he was told it had a sum equal to his house number—unless the sum was 13, because only two triplets have identical sums, $1 + 6 + 6$ and $2 + 2 + 9$, both of which equal 13. As soon as *A* was told that *B* had an oldest child he eliminated 1, 6, 6, leaving 2, 2, 9 as the ages of *B*'s three children.

ANSWER 13.21—(April 1972)

The first two rows of the chart [see Figure 13.7] show the results of the first two spins. The first spin was given, and we were told that *D* had the highest total after the second spin. This could happen only if

SPINS	A	B	C	D	E	F
1	1	2	5	4	3	0
2	0	1	2	5	4	3
3	5	4	3	0	1	2
4	5	4	3	0	1	2
5	4	3	0	1	2	5
FINAL SCORES	15	14	13	10	11	12

Figure 13.7   Solution to D. St. P. Barnard's problem

OTHER PUZZLES

the wheel distributed the points as shown in the second row. Now comes the tricky part. Every opposite pair of digits on the disk used in the game add to 5. This means that every spin will give a combined sum of five points to each pair of players seated opposite each other—namely *AD*, *BE*, and *CF*. At the end of the game, which has five spins, each of these pairs of players will have a combined sum of 25 points.

We know that *A* won the game. Since his was the highest score, *D* (who sits opposite) must have ended with the lowest score. *D*'s final score must be less than 13, otherwise *A*'s final score would be smaller. *D*'s final score cannot be 12. True, *A* would score 13, but then a player of the pair *BE*, as well as a player of the pair *CF*, would necessarily score 13 or better, preventing *A* from being the highest scorer.

As we have seen, *D* cannot score more than 11. He already has nine points at the end of the second spin, therefore, at least one of the three remaining spins must give him zero. Since the order of the results of each spin cannot affect the final scores, we can assume that *D* scored zero after the third spin. This determines the points for the other players as indicated in the third row of the chart.

On the next two spins, *D*'s points can only be 0–0, 0–1, 1–1, or 0–2. We test each in turn. If 0–0 or 0–2, *A* will tie with someone on his final score. If 1–1, *F* gets 5–5 and wins with a score of 15. Only 0–1 remains for *D*. This makes *A* the winner, with 15 points, and enables us to complete the chart as indicated. We do not know the order of the last three spins, but the final scores are accurate. The problem is No. 2 in D. St. P. Barnard's first puzzle book, *Fifty Observer Brain-Twisters* (Faber and Faber, Ltd., 1962).

ANSWER 13.22—(February 1957)

If we require that the question be answerable by yes or no, there are several solutions, all exploiting the same basic gimmick. For example, the logician points to one of the roads and says to the native, "If I were to ask you if this road leads to the village, would you say 'yes'?" The native is forced to give the right answer, even if he is a liar! If the road does lead to the village, the liar would say no to the direct question, but as the question is put, he lies and says he would respond yes. Thus the logician can be certain that the road does lead to the village, whether the respondent is a truth-teller or a liar. On the other hand, if the road actually does not go to the village, the liar is forced in the same way to reply no to the inquirer's question.

A similar question would be, "If I asked a member of the other tribe whether this road leads to the village, would he say 'yes'?" To avoid the cloudiness that results from a question within a question, perhaps this phrasing (suggested by Warren C. Haggstrom, of Ann Arbor, Michigan) is best: "Of the two statements, 'You are a liar' and 'This road leads to the village,' is one and only one of them true?" Here again, a yes answer indicates it is the road, a no answer that it isn't, regardless of whether the native lies or tells the truth.

Dennis Sciama, Cambridge University cosmologist, and John McCarthy of Hanover, New Hampshire, called my attention to a delightful additional twist on the problem. "Suppose," Mr. McCarthy wrote (in a letter published in *Scientific American*, April 1957), "the logician knows that 'pish' and 'tush' are the native words for 'yes' and 'no' but has forgotten which is which, though otherwise he can speak the native language. He can still determine which road leads to the village.

"He points to one of the roads and asks, 'If I asked you whether the road I am pointing to is the road to the village would you say pish?' If the native replies, 'Pish,' the logician can conclude that the road pointed to is the road to the village even though he will still be in the dark as to whether the native is a liar or a truth-teller and as to whether 'pish' means yes or no. if the native says, 'Tush,' he may draw the opposite conclusion."

H. Janzen of Queens University, Kingston, Ontario, and several other readers informed me that if the native's answer does not have to be yes or no, there is a question which reveals the correct road regardless of how many roads meet at the intersection. The logician simply points to all the roads, including the one he has just traveled, and asks, "Which of these roads leads to the village?" The truth-teller points to the correct one, and the liar presumably points to all the others. The logician could also ask, "Which roads do not lead to the village?" In this case the liar would presumably point only to the correct one. Both cases, however, are somewhat suspect. In the first case the liar might point to only one incorrect road and in the second case he might point to several roads. These responses would be lies in a sense, though one would not be the strongest possible lie and the other would contain a bit of truth.

The question of how precisely to define "lying" enters of course even into the previous yes and no solutions. I know of no better way to make this clear than by quoting in full the following letter which

Scientific American received from Willison Crichton and Donald E. Lamphiear, both of Ann Arbor, Michigan:

It is a sad commentary on the rise of logic that it leads to the decay of the art of lying. Even among liars, the life of reason seems to be gaining ground over the better life. We refer to puzzle number 4 in the February issue, and its solution. If we accept the proposed solution, we must believe that liars can always be made the dupes of their own principles, a situation, indeed, which is bound to arise whenever lying takes the form of slavish adherence to arbitrary rules.

For the anthropologist to say to the native, "If I were to ask you if this road leads to the village, would you say 'yes'?" expecting him to interpret the question as counter-factual conditional in meaning as well as form, presupposes a certain preciosity on the part of the native. If the anthropologist asks the question casually, the native is almost certain to mistake the odd phraseology for some civility of manner taught in Western democracies, and answer as if the question were simply, "Does this road lead to the village?" On the other hand, if he fixes him with a glittering eye in order to emphasize the logical intent of the question, he also reveals its purpose, arousing the native's suspicion that he is being tricked. The native, if he is worthy the name of liar, will pursue a method of counter-trickery, leaving the anthropologist misinformed. On this latter view, the proposed solution is inadequate, but even in terms of strictly formal lying, it is faulty because of its ambiguity.

The investigation of unambiguous solutions leads us to a more detailed analysis of the nature of lying. The traditional definition employed by logicians is that a liar is one who always says what is false. The ambiguity of this definition appears when we try to predict what a liar will answer to a compound truth functional question, such as, "Is it true that if this is the way to town, you are a liar?" Will he evaluate the two components correctly in order to evaluate the function and reverse his evaluation in the telling, or will he follow the impartial policy of lying to himself as well as to others, reversing the evaluation of each component before computing the value of the function, and then reversing the computed value of the function? Here we distinguish the *simple liar* who always utters what is simply false from the *honest liar* who always utters the logical dual of the truth.

The question, "Is it true that if this is the way to town, you are a liar?" is a solution if our liars are honest liars. The honest liar and the truth-teller both answer "yes" if the indicated road is not the way to town, and "no" if it is. The simple liar, however, will answer "no" regardless of where the village is. By substituting equivalence for implication we obtain a solution which works for both simple and honest liars. The question becomes, "Is it true that this is the way to town if and only if you are a liar?" The answer is uniformly "no" if it is the way, and "yes" if it is not.

But no lying primitive savage could be expected to display the scrupulous consistency required by these conceptions, nor would any liar capable of such acumen be so easily outwitted. We must therefore consider the case of the *artistic liar* whose principle is always to deceive. Against such an opponent the anthropologist can only hope to maximize the probability of a favorable outcome. No logical question can be an infallible solution, for if the liar's principle is to deceive, he will counter with a strategy of deception which circumvents logic. Clearly the essential feature of the anthropologist's strategy must be its psychological soundness. Such a strategy is admissible since it is even more effective against the honest and the simple liar than against the more refractory artistic liar.

We therefore propose as the most general solution the following question or its moral equivalent, "Did you know that they are serving free beer in the village?" The truth-teller answers "no" and immediately sets off for the village, the anthropologist following. The simple or honest liar answers "yes" and sets off for the village. The artistic liar, making the polite assumption that the anthropologist is also devoted to trickery, chooses his strategy accordingly. Confronted with two contrary motives, he may pursue the chance of satisfying both of them by answering, "Ugh! I hate beer!" and starting for the village. This will not confuse a good anthropologist. But if the liar sees through the ruse, he will recognize the inadequacy of this response. He may then make the supreme sacrifice for the sake of art and start down the wrong road. He achieves a technical victory, but even so, the anthropologist may claim a moral victory, for the liar is punished by the gnawing suspicion that he has missed some free beer.

ANSWER 13.23—(March 1965)

Label the three men, A, B, C, and let T stand for truth-teller, L for liar, and R for randomizer. There are six possible permutations of T, L, and R:

A	B	C
T	L	R
T	R	L
L	R	T
L	T	R
R	T	L
R	L	T

Ask A "Is B more likely to tell the truth than C?" If he answers yes, lines 1 and 4 are eliminated and you know that C is not the randomizer. If he answers no, lines 2 and 3 are eliminated and you know that B is not the randomizer. In either case, turn to the man who is not the randomizer and ask any question for which you both know the answer. For example: "Are you the randomizer?" His answer will establish whether he is the truth-teller or the liar. Knowing this, you can ask him if a certain one of his companions is the randomizer. His answer will establish the identities of the other two men.

Many readers sent different solutions. The most unusual, by Kenneth O'Toole, was passed along to me by Mary S. Bernstein. A man is asked, "If I asked each of you if I had on a hat, and your two companions gave the same answer, would your answer agree with theirs?" The truther says no, the liar yes, and the randomizer cannot reply because he knows his companions cannot agree. Here we encounter ambiguity because, in a sense, any answer by the randomizer would be a "lie." Assuming, however, that the randomizer remains silent, the question need be asked of only two men to identify all three.

ANSWER 13.24—(November 1967)

If each man at a circular table is either a truther or a liar, and each says that the man on his left is a liar, there must be an even number at the table, arranged so that truthers and liars alternate. (No arrangement of an odd number of truthers and liars is possible without at least one man describing the man on his left as a truther.) Consequently the club's president lied when he said the number was 37.

Since the secretary called the president a liar, he must have been a truther. Therefore he spoke truthfully when he gave the number as 40.

ANSWER 13.25—(April 1977)

Raymond Smullyan's four logic problems are answered as follows:

1. *A* is either telling the truth or not. Suppose he is. Then *B* is a knight and telling the truth when he says *A* is not a knight. In this case *A* is telling the truth but is not a knight.

   Suppose *A* is lying. Then *B* is not a knight, *B* however, is telling the truth when he says *A* is not a knight. Hence in this case *B* is telling the truth but is not a knight.

2. *B* is either telling the truth or not. Suppose he is. Then *A* is a knave and must be lying when he says *B* is a knight. In this case *B* is telling the truth but is not a knight.

   Suppose *B* is lying. Then *B* is surely not a knight; therefore *A* must be lying when he says *B* is a knight. Since *B* is lying, *A* is not a knave. In this case *A* is lying but is not a knave.

3. *B* is either a knight or a knave. Suppose he is a knight. *A* and *C* must then be the same type, as *B* says. *C* is lying when he says *B* is a knave; therefore *C* is a knave. If *C* is a knave, *A* must be also.

   Suppose *B* is a knave. Then *A* and *C* are different. *C* is telling the truth when he says *B* is a knave, so that *C* must be a knight. Because *A* and *C* are different, *A* must be a knave. In either case *A* is a knave.

4. Smullyan's solution of this problem is somewhat lengthy, and I shall content myself with a summary. *A* and *B* are either knight-knight, knave-knave, knight-knave, or knave-knight. In each case analysis shows that *C*, whether he is a knight or a knave, must answer yes.

chapter 14

Cryptarithms

A cryptarithm takes a normal grade-school arithmetic problem, like addition or long division, and consistently substitutes symbols for the digits. A classic example is "SEND + MORE = MONEY" which is solved by discovering that it corresponds to the sum 9567 + 1085 = 10652. The problems in this chapter are like this but with delightful twists.

Strictly speaking these puzzles are logic problems and could have been included in the last chapter. The allure of cryptarithms is that they can be solved by deducing what the digits are rather than by performing an exhaustive search. It is not clear why there were so many of these in Gardner's column—other than that they were popular—but they certainly merit a separate chapter.

Problems

PROBLEM 14.1—Consecutive Digits

```
  A B C D
  D C B A
  • • • •
  ───────
1 2 3 0 0
```

ABCD are four consecutive digits in increasing order. *DCBA* are the same four in decreasing order. The four dots represent the same four digits in an unknown order. If the sum is 12300, what number is represented by the four dots? (From W. T. Williams and G. H. Savage, *The Strand Problems Book*.)

PROBLEM 14.2—Lonesome 8

The most popular problem ever published in *The American Mathematical Monthly*, its editors disclosed, is the following. It was contributed by P. L. Chessin of the Westinghouse Electric Corporation to the April 1954, issue.

"Our good friend and eminent numerologist, Professor Euclide Paracelso Bombasto Umbugio, has been busily engaged in testing on his desk calculator the 81×10^9 possible solutions to the problem of reconstructing the following exact long division in which the digits were indiscriminately replaced by X save in the quotient where they were almost entirely omitted:

```
                8
        _____
 X X X )X X X X X X X X X
        X X X
        _____
          X X X X
          X X X
          _____
            X X X X
            X X X X
            _____
```

"Deflate the Professor! That is, reduce the possibilities to $(81 \times 10^9)^0$."

Because any number raised to the power of zero is one, your task is to discover the unique reconstruction of the problem. The 8 is in correct position above the line, making it the third digit of a five-digit answer. The problem is easier than it looks, yielding readily to a few elementary insights.

PROBLEM 14.3—Pair of Cryptarithms

In most cryptarithms a different letter is substituted for each digit in a simple arithmetical problem. The two remarkable cryptarithms shown in Figure 14.1 are unorthodox in their departure from this practice, but each is easily solved by logical reasoning and each has a unique answer.

In the multiplication problem at the left in the illustration, newly devised by Fitch Cheney of the University of Hartford, each E stands for an even digit, each O for an odd digit. The fact that every even

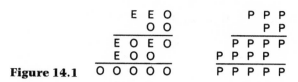

```
      E E O           P P P
          O O           P P
      ─────         ───────
      E O E O         P P P P
      E O O           P P P P
      ───────       ─────────
Figure 14.1  O O O O    P P P P P
```

digit is represented by *E* does not mean, of course, that all the even digits are the same. For example, one *E* may stand for 2, another for 4, and so on. Zero is considered an even digit. You are asked to reconstruct the numerical problem.

In the multiplication problem at the right, each *P* stands for a prime digit (2, 3, 5, or 7). This charming problem was first proposed by Joseph Ellis Trevor, a chemist at Cornell University. It has since become classic of its kind.

PROBLEM 14.4—Three Cryptarithms

Of the three remarkable cryptarithms in Figure 14.2 the first [left] is easy, the second [middle] is moderately hard, and the third [right] is so difficult that I do not expect any reader to solve it without the use of a computer.

Problem 1: Each dot represents one of the 10 digits from 0 to 9 inclusive. Some digits may appear more than once, others not at all. As you can see, a two-digit number multiplied by a two-digit number yields a four-digit product, to which is added a three-digit number starting with 1. Replace each dot with the proper digit. The solution is unique.

Problem 2: As in the first cryptarithm, a multiplication is followed by an addition. In this case, however, each dot is a digit from 1 to 9 inclusive (no 0) and each digit appears once. The answer is unique.

Problem 3: Each dot in this multiplication problem stands for a

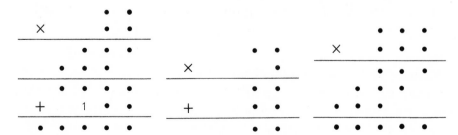

Figure 14.2

digit from 0 to 9 inclusive. Each digit appears exactly *twice*. Again, the answer is unique.

PROBLEM 14.5—Nora L. Aron

A college girl has the unusual palindromic name Nora Lil Aron. Her boyfriend, a mathematics major, bored one morning by a dull lecture, amuses himself by trying to compose a good number cryptogram. He writes his girl's name in the form of a simple multiplication problem:

$$
\begin{array}{r}
\text{NORA} \\
\underline{\text{L}} \\
\text{ARON}
\end{array}
$$

Is it possible to substitute one of the 10 digits for each letter and have a correct product? He is amazed to discover that it is, and also that there is a unique solution. You should have little trouble working it out. It is assumed that neither four-digit number begins with zero.

PROBLEM 14.6—Talkative Eve

This cryptarithm (or alphametic, as some puzzlists prefer to say) is an old one of unknown origin, surely one of the best and, I hope, unfamiliar to most of you:

$$
\frac{\text{EVE}}{\text{DID}} = .\text{TALKTALKTALK.}\ldots
$$

The same letters stand for the same digits, zero included. The fraction EVE/DID has been reduced to its lowest terms. Its decimal form has a repeating period of four digits. The solution is unique. To solve it, recall that the standard way to obtain the simplest fraction equivalent to a decimal of *n* repeating digits is to put the repeating period over *n* 9's and reduce the fraction to its lowest terms.

PROBLEM 14.7—Two Cryptarithms

Here are two elegant cryptarithms by Alan Wayne that have not been published before. The first is in French, the second in German. Each letter represents just one decimal digit, and we adopt the usual convention that zero must not begin a number. Both have unique solutions.

```
      V I N G T            E I N
        C I N Q            E I N
    +   C I N Q            E I N
      ─────────        +   E I N
      T R E N T E      ─────────
                          V I E R
```

PROBLEM 14.8—Find the Square Root

The Square Root of Wonderful was the name of a play on Broadway. If each letter in WONDERFUL stands for a different digit (zero excluded) and if OODDF, using the same code, represents the square root, then what *is* the square root of wonderful?

PROBLEM 14.9—Rookwise Chain

There are many ways in which the nine digits (not counting zero) can be arranged in square formation to represent a sum. In the example shown at left in Figure 14.3, 318 plus 654 equals 972. There are also many ways to place the digits on a square matrix so that, taken in serial order, they form a rookwise connected chain. An example is at right in the illustration. You can start at 1, then, moving like a chess rook, one square per move, you can advance to 2, 3, 4, and so on to 9.

The problem is to combine both features in the same square. In other words, place the digits on a 3 × 3 matrix so that they form a rookwise connected chain, from 1 to 9, and also in such a way that the bottom row is the sum of the first two rows. The answer is unique.

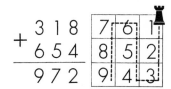

Figure 14.3

Answers

ANSWER 14.1—(July 1971)

If *ABCD* = 1234, it is impossible to obtain a sum as large as 12300. If *ABCD* = 3456, it is impossible to obtain a sum as small as 12300. Therefore *ABCD* = 2345, from which it is easy to determine that the four dots stand for 4523.

Hans Marbet, of Switzerland, pointed out that if *ABCD* are replaced by any four consecutive digits, and the four dots replaced by the same digits in the order *CDAB*, there is a valid solution for a number system with the base $A + B + D$. For example, in base 16:

$$
\begin{array}{r}
4567 \\
7654 \\
6745 \\
\hline
12300
\end{array}
$$

ANSWER 14.2—(May 1959)

In long division, when two digits are brought down instead of one, there must be a zero in the quotient. This occurs twice, so we know at once that the quotient is *x080x*. When the divisor is multiplied by the quotient's last digit, the product is a four-digit number. The quotient's last digit must therefore be 9, because eight times the divisor is a three-digit number.

The divisor must be less than 125 because eight times 125 is 1000, a four-digit number. We now can deduce that the quotient's first digit must be more than 7, for seven times a divisor less than 125 would give a product that would leave more than two digits after it was subtracted from the first four digits in the dividend. This first digit cannot be 9 (which gives a four-digit number when the divisor is multiplied by it), so it must be 8, making the full quotient 80809.

The divisor must be more than 123 because 80809 times 123 is a seven-digit number and our dividend has eight digits. The only number between 123 and 125 is 124. We can now reconstruct the entire problem as follows:

$$
\begin{array}{r}
80809 \\
\hline
124\,)\,10020316 \\
992 \\
\hline
1003 \\
992 \\
\hline
1116 \\
1116 \\
\end{array}
$$

ANSWER 14.3—(October 1962)

Fitch Cheney's cryptarithm has the unique answer

$$
\begin{array}{r}
285 \\
39 \\
\hline
2565 \\
855 \\
\hline
11115
\end{array}
$$

The unique answer to Joseph Ellis Trevor's cryptarithm is

$$
\begin{array}{r}
775 \\
33 \\
\hline
2325 \\
2325 \\
\hline
25575
\end{array}
$$

Trevor's problem, the more difficult of the two, is perhaps best approached by searching first for all three-digit numbers composed of prime digits that yield four prime digits when multiplied by a prime. There are only four:

$$775 \times 3 = 2325$$
$$555 \times 5 = 2775$$
$$755 \times 5 = 3775$$
$$325 \times 7 = 2275$$

No three-digit number has more than one multiplier, therefore the multiplier in the problem must consist of two identical digits. Thus there are only four possibilities that need to be tested.

ANSWER 14.4—(November 1963)

The three cryptarithms have the unique solutions shown in Figure 14.4. The left one was devised by Stephen Barr, the middle one is the work of the English puzzlist Henry Ernest Dudeney; the right one is

Figure 14.4

Cryptarithms

from Frederik Schuh's *Wonderlijke Problemen*. For a translation of Schuh's analysis see pages 287–91 in F. Gobel's *Master Book of Mathematical Recreations*.

The third cryptarithm, which I thought no one could solve without a computer, was solved with pencil and paper by no fewer than 53 *Scientific American* readers. Few of the solvers went on to show that no other solution was possible. Six readers, however, programed computers to check all possibilities, and they confirmed the uniqueness of the answer.

ANSWER 14.5—(February 1967)

NORA × L = ARON has the unique solution 2178 × 4 = 8712. Had Nora's middle initial been A, the unique solution would have been 1089 × 9 = 9801. The numbers 2178 and 1089 are the only two smaller than 10,000 with multiples that are reversals of themselves (excluding trivial cases of palindromic numbers such as 3443 multiplied by 1). Any number of 9's can be inserted in the middle of each number to obtain larger (but dull) numbers with the same property; for instance, 21999978 × 4 = 87999912.

For a report on such numbers, in all number systems, see "Integers That Are Multiplied When Their Digits Are Reversed," by Alan Sutcliffe, in *Mathematics Magazine,* vol. 39, no. 5, November 1966, pages 282–87.

Larger numbers can also be fabricated by repeating each four-digit number: thus, 217821782178 × 4 = 871287128712, and 108910891089 × 9 = 980198019801. Of course numbers such as 21999978 may also be repeated to produce reversible numbers. Leonard F. Klosinski and Dennis C. Smolarski, in their paper "On the Reversing of Digits," *Mathematics Magazine,* vol. 42, September 1969, pages 208–10, show that 4 and 9 are the only numbers which can serve as multipliers for reversing non-palindromic numbers. This can be put another way. If an integer is a factor of its reversal, the larger of the two numbers divided by the smaller must equal 4 or 9.

The fact that 8712 and 9801 are the only four-digit numbers that are integral multiples of their reversals is cited by G. H. Hardy, in his famous *Mathematician's Apology*, as an example of nonserious mathematics. For those who are fascinated by such oddities, I pass along the following chart, sent by Bernard Gaiennie, which points up the curious relationship between the two numbers:

```
1089    6534
2178    7623
3267    8712
4356    9801
5445
```

The nine numbers are, of course, the first nine multiples of 1089. Note the consecutive order of the digits when you go up and down the columns. If the first five numbers are multiplied respectively by 9, 4, 2⅓, 1½, and 1, the products give the last five numbers in reverse order. Now 1, 4, and 9 are the first three square numbers, but those other two multipliers, 2⅓ and 1½, seem to come out of left field!

ANSWER 14.6—(February 1970)

To obtain the simplest fraction equal to a decimal of n repeated digits, put the repeating period over n 9's and reduce to its lowest terms. In this instance TALK/9999, reduced to its lowest terms, must equal EVE/DID. DID, consequently, is a factor of 9999. Only three such factors fit DID: 101, 303, 909.

If DID = 101, then EVE/101 = TALK 9999, and EVE = TALK/99. Rearranging terms, TALK = 99 × EVE. EVE cannot be 101 (since we have assumed 101 to be DID) and anything larger than 101, when multiplied by 99, has a five-digit product. And so DID = 101 is ruled out.

If DID = 909, then EVE/909 = TALK/9999, and EVE = TALK/11. Rearranging terms, TALK = 11 × EVE. In that case the last digit of TALK would have to be E. Since it is not E, 909 also is ruled out.

Only 303 remains as a possibility for DID. Because EVE must be smaller than 303, E is 1 or 2. Of the 14 possibilities (121, 141, . . . , 292) only 242 produces a decimal fitting TALKTALK . . . , in which all the digits differ from those in EVE and DID.

The unique answer is 242/303 = .798679867986. . . . If EVE/DID is not assumed to be in lowest terms, there is one other solution, 212/606 = .349834983498 . . . , proving, as Joseph Madachy has remarked, that EVE double-talked.

ANSWER 14.7—(November 1975)

The French cryptarithm is solved with 94851 + 6483 + 6483 = 107817, and the German one with 821 + 821 + 821 + 821 = 3284.

Many readers pointed out that the numeral 1 is commonly called *eins* in German, not *ein* as it appeared in the German cryptarithm. William C. Giessen was the first to add that if this change is made, the cryptarithm still has a unique solution: 1329 + 1329 + 1329 + 1329 = 5316.

ANSWER 14.8—(June 1961)

If OODDF is the square root of WONDERFUL, what number does it represent? O cannot be greater than 2 because this would give a square of 10 digits. It cannot be 1 because there is no way that a number, beginning with 11, can have a square in which the second digit is 1. Therefore O must be 2.

WONDERFUL must be between the squares of 22000 and 23000. The square of 22 is 484; the square of 23 is 529. Since the second digit of WONDERFUL is 2, we conclude that WO = 52.

What values for the letters of 22DDF will make the square equal 52NDERFUL? The square of 229 is 52441; the square of 228 is 51984. Therefore OODD is either 2299 or 2288.

We now use a dodge based on the concept of digital root. The sum of the nine digits in WONDERFUL (we were told zero is excluded) is 45, which in turn sums to 9, its digital root. Its square root must have a digital root that, when squared, gives a number with a digital root of nine. The only digital roots meeting this requirement are 3, 6, 9, therefore OODDF must have a digital root of 3, 6, or 9.

F cannot be 1, 5, or 6, because any of those digits would put an F at the end of WONDERFUL. The only possible completions of 2299F and 2288F that meet the digital root requirement are 22998, 22884, and 22887.

The square of 22887 is 523814769 the only one that fits the code word WONDERFUL.

ANSWER 14.9—(June 1961)

The timesaving insight in this problem is the realization that if the nine digits are placed on a 3 × 3 matrix to form a rookwise connected chain from 1 to 9, the odd digits must occupy the central and four corner cells. This is easily seen by coloring the nine cells like a checkerboard, the center cell black. Since there is one more black cell than white, the path must begin and end on black cells, and all even digits will fall on white cells.

There are 24 different ways in which the four even digits can be

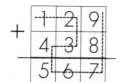

Figure 14.5

arranged on the white cells. Eight of these, in which 2 is opposite 4, can be eliminated immediately because they do not permit a complete path of digits in serial order. The remaining 16 patterns can be quickly checked, keeping in mind that the sum of the two upper digits on the left must be less than 10 and the sum of the two upper digits on the right must be more than 10. The second assertion holds because the two upper digits in the middle are even and odd, yet their sum is an even digit. This could happen only if 1 is carried over from the sum of the right column. The only way to form the path so that the bottom row of the square is the sum of the first and second rows is shown in Figure 14.5.

When this solution appeared in *Scientific American*, Harmon H. Goldstone of New York City, and Scott B. Kilner of Corona, California, wrote to explain a faster method they had used. There are only three basically different rook paths (ignoring rotations and reflections) on the field: the one shown in the solution, a spiral path from corner to center, and an "S" path from corner to diagonally opposite corner. On each path the digits can run in order in either direction, making six different patterns. By considering each in its various rotations and reflections, one quickly arrives at the unique answer.

Note that if the solution is mirror inverted (by a mirror held above it), it forms a square, its digits still in rookwise serial order, such that the middle row subtracted from the top row gives the bottom row.

Charles W. Trigg, in a detailed analysis of solutions to ABC + DEF = GHK (in *Recreational Mathematics Magazine*, no. 7, February 1962, pages 35–36), gives the only three solutions, in addition to the one shown here, on which the digits 1 through 9 are in serial order along a *queenwise* connected path.

Cryptarithms

chapter 15

Wordplay

Why is there a chapter on wordplay in a book on math puzzles? For most people wordplay is either concerned with puzzles like crosswords or with jokes like puns. However, there are other forms of wordplay where the emphasis is less on the meaning of a word and more on a word being a sequence of letters. For example, an anagram of a word or phrase is another word or phrase formed by rearranging the letters (such as ASTRONOMER and MOON STARER). An anagram is elevated to a higher level when the meanings are considered (especially if they are similar, as in our example) but fundamentally a word is a sequence of symbols and an anagram is a mathematical concept.

Soon after the column began, Gardner edited Bombaugh's lost classic Oddities and Curiousities of Words and Literature *(Dover, 1961). Later he edited Dudeney's* 300 Best Word Puzzles *(Scribner, 1968). In putting these books together Gardner learned a lot about the history of wordplay and its underlying mathematical flavor. He received much help from Dmitri Borgmann, an indefatigable proponent of "logology," who went on to found the journal* Word Ways *to which Gardner has been a frequent contributor.*

Problems

PROBLEM 15.1—Translation, Please!

What well-known quotation is expressed by this statement in symbolic logic?

$$2B \lor \sim 2B = ?$$

PROBLEM 15.2—Proverbial Mare

An intelligent horse learns arithmetic, algebra, geometry, and trigonometry but is unable to understand the Cartesian coordinates of analytic geometry. What proverb does this suggest? (From Howard W. Eves, in *Mathematical Circles*, vol. 1.)

PROBLEM 15.3—Match This!

Rearrange the 12 matches in Figure 15.1 to spell what matches are made of.

Figure 15.1

PROBLEM 15.4—Baby Crossword

Solve the crossword puzzle in Figure 15.2 with the help of these clues:

1	5	6	7
2			
3			
4			

HORIZONTAL
1. Norman Mailer has two
2. Is indebted
3. Chicago vehicles
4. Relaxation

VERTICAL
1. Skin blemish
5. Works in the dark
6. Character in *Winds in the Willows*
7. Famous white Dixieland trombonist

Figure 15.2

PROBLEM 15.5—Fore "und" Aft

What familiar English word begins and ends with *und*?

PROBLEM 15.6—Two Many!

Translate: "He spoke from 222222222222 people."

PROBLEM 15.7—Number Person

Change 11030 to a person by adding two straight line segments.

PROBLEM 15.8—Double Up

If *AB*, *BC*, *CD*, and *DE* are common English words, what familiar word is DCABE? (Thanks to David L. Silverman, *Word Ways*, August 1969.)

Wordplay

PROBLEM 15.9—Consecutive Letters

What four consecutive letters of the alphabet can be arranged to spell a familiar four-letter word? (From Murray R. Pearce, *Word Ways*, February 1971.)

PROBLEM 15.10—Littlewood's Footnotes

Every now and then a magazine runs a cover picture that contains a picture of the same magazine, on the cover of which one can see a still smaller picture of the magazine, and so on presumably to infinity. Infinite regresses of this sort are a common source of confusion in logic and semantics. Sometimes the endless hierarchy can be avoided, sometimes not. The English mathematician J. E. Littlewood, commenting on this topic in his *A Mathematician's Miscellany* (London: Methuen, 1953), recalls three footnotes that appeared at the end of one of his papers. The paper had been published in a French journal. The notes, all in French, read:

1. I am greatly indebted to Prof. Riesz for translating the present paper.
2. I am indebted to Prof. Riesz for translating the preceding footnote.
3. I am indebted to Prof. Riesz for translating the preceding footnote.

Assuming that Littlewood was completely ignorant of the French language, on what reasonable grounds did he avoid an infinite regress of identical footnotes by stopping after the third footnote?

PROBLEM 15.11—Who's Behind the Mad Hatter?

The following problem, by John F. Collins of Santa Monica, California, appeared in the August 1968 issue of *Word Ways*.

"The March Hare and the Mad Hatter were sipping their eggnog and watching the crowd when Alice happened to glance in the Hare's direction and ask, 'Why are you giving me such an angry look?'

"'I'm not *giving* it to you, I'm giving it *back*,' replied the Hare.

"'I didn't look crossly at you.'

"'Well, *somebody* did,' the Hare said, turning to glare at the Hatter.

"Just then, someone came up from behind and put his hands over the Hatter's eyes.

"'Guess who!,' said the newcomer in a thin, flat voice.

"The Hatter froze for a moment and declared, rather coldly, 'I have no use for practical jokers.'

"'Ha! Neither have I,' retorted the stranger, still keeping his hands over the Hatter's eyes.

"At that, the Hatter seemed to accept the challenge of the game and started asking a series of questions in a manner that mingled hope with care.

"Question: 'Ahem. Would you, by chance, be in a black suit this evening?'

"Answer: 'I would, but not by chance, by design.'

"Q. 'I presume you're a member of all the posh clubs?'

"A. 'Afraid not. Never even been invited.'

"Q. 'Surely you're better than average?'

"A. 'Yes, indeed!'

"Q. 'Not spotted, I hope?'

"A. 'Knock wood.'

"Q. 'Married?'

"A. 'No, happy.'"

Who is behind the Mad Hatter?

PROBLEM 15.12—Circle Squared

Word puzzlists have long been fascinated by a type of puzzle called the word square. The best way to explain this is to provide an example:

```
M E R G E R S
E T E R N A L
R E G A T T A
G R A V I T Y
E N T I T L E
R A T T L E R
S L A Y E R S
```

Note that each word in the above order-7 square appears both horizontally and vertically. The higher the order, the more difficult it is to devise such squares. Word square experts have succeeded in forming many elegant order-9 squares, but no order-10 squares have been constructed in English without the use of unusual double words such as Pango-Pango.

Charles Babbage, the nineteenth-century pioneer in the design of computers, explains how to form word squares in his autobiography, *Passages from the Life of a Philosopher*, and adds: "The various ranks of the church are easily squared; but it is stated, I know not on what authority, that no one has succeeded in squaring a bishop." Readers of *Eureka*, a mathematics journal published by students at the University of Cambridge, had no difficulty squaring *bishop* when they were told of Babbage's remarks. The square shown below (from the magazine's October 1961 issue) was one of many good solutions received:

```
B  I  S  H  O  P
I  L  L  U  M  E
S  L  I  D  E  S
H  U  D  D  L  E
O  M  E  L  E  T
P  E  S  E  T  A
```

As far as I know, no one has yet succeeded—perhaps even attempted—to square the word *circle*. Only words found in an unabridged English dictionary may be used. The more familiar the words, the more praiseworthy the square.

PROBLEM 15.13—Dudeney's Word Square

Charles Dunning, Jr., of Baltimore, Maryland, set himself the curious task of placing letters in the nine cells of a 3 × 3 matrix so as to form the largest possible number of three-letter words. The words may be read from left to right or right to left, up or down, and in either direction along each of the two main diagonals. Dunning's best result,

```
T  E  A
U  R  N
B  A  Y
```

Figure 15.3

shown in Figure 15.3, gives 10 words: tea, urn, bay, tub, but, era, are, any, try, bra.

How close it is possible to come in English to the theoretical maximum of 16 words? A letter may be used more than once, but words must be different in order to count. They should be dictionary words. I have on hand a specimen from one of H. E. Dudeney's puzzle books that raises the number of words to 12 but perhaps you can do better.

Problem 15.14—Anagram Dictionary

Nicholas Temperley, while a student at Cambridge, proposed that devotees of wordplay produce, as a working tool, an anagram dictionary. Every word in English is first converted to its "alphabetical anagram," in which its letters appear in alphabetical order. *Scientific*, for example, becomes CCEFIIINST. These alphabetical anagrams are then arranged in alphabetical sequence to form the dictionary. Each entry in the volume is an alphabetical anagram, and under it are listed all the English words that can be formed with those letters. Thus BDE-MOOR will be followed by *bedroom*, *boredom*, and all other words formed by those letters. AEIMNNOST will be followed by *Minnesota* and *nominates*. Some entries even have mathematical interest. AEGILNRT for instance, will be followed by such mathematical terms as *integral*, *relating*, and *triangle* as well as other words such as *altering*. EIINNSTXY will be followed by both *ninety-six* and *sixty-nine*. AGHILMORT will have among its anagrams both *algorithm* and *logarithm*. If a crossword puzzle gives the clue *bean soup* and an indication that this is an anagram, someone armed with the anagram dictionary need only alphabetize its letters and look up ABENOPSU to find *subpoena*. If the clue is *the classroom*, it takes only a moment to discover *schoolmaster*.

Most of the entries will begin with letters near the front of the alphabet. Temperley estimated that more than half will start with *A*, which is to say that more than half of all English words contain *A*. (This is not true of common words, but rarer words tend to be longer and are more likely to have an *A*.) After *I* the number of entries will drop sharply. Beyond *O* the list is extremely short.

Can you answer these questions?

1. What will be the dictionary's last entry? (Place names, such as *Uz*, the home of Job, are not included.)

2. What will be the first and second entries?

3. What will be the last entry starting with *A*?

4. What will be the first entry starting with *B*?

5. An entry, ABCDEFLO, begins with the first six letters of the alphabet. What is the word?

6. What will be the longest entry that is itself a word? (Short examples include *adder*, *aglow*, *beefy*, *best*, *dips*, *fort*.)

7. What will be the longest entry that does not repeat any letter?

PROBLEM 15.15—Scrambled Quotation

Letters in the sentence "Roses are red, violets are blue" are scrambled by the following procedure. The words are written one below the other and flush at the left:

    ROSES
    ARE
    RED
    VIOLETS
    ARE
    BLUE

The columns are taken from left to right and their letters from the top down, skipping all blank spaces, to produce this ordering:

    RARVABOREIRLSEDOEUELESETS.

The task is to find the line of poetry that, when scrambled by this procedure, becomes

    TINFLABTTULAHSORIOOASAWEIKOKNARGEKEDYEASTE.

Walter Penney of Greenbelt, Maryland, contributed this novel word problem to the February 1970 issue of *Word Ways*.

PROBLEM 15.16—Lewis Carroll's "Sonnet"

In 1887 Lewis Carroll included in a letter to Maud Standen, a "child-friend," a six-line poem that he called an "anagrammatic" sonnet. (He was using "sonnet" in an older sense, meaning any short piece of verse.) "Each line has four feet," Carroll wrote to Maud, "and each foot is an anagram, *i.e.*, the letters of it can be rearranged so as to make one word." Most of the anagrams, he said, had been devised "for some delicious children" he had met the previous summer at East-bourne. The words vary in length from four through seven letters, and it is assumed that proper names are not allowed.

Here is the "sonnet":

> *As to the war, try elm. I tried.*
> *The wig cast in, I went to ride*
> *'Ring? Yes.' We rang. 'Let's rap.'*
> > *We don't.*
> *'O shew her wit!' As yet she won't.*
> *Saw eel in Rome. Dry one: he's wet.*
> *I am dry. O forge! Th' rogue! Why*
> > *a net?*

In most cases there is little doubt about the correct word. For example, the first foot, "As to," could be "oast" or "stoa," but more likely Carroll meant the commoner word "oats." For several of the feet, however, the intended word is not clear. No solution by Carroll has survived, and to this day there is contention among Carrollians over the precise set of 24 words Carroll had in mind. You are invited to make your own list to compare with the conjectures in the answer section.

PROBLEM 15.17—Sigil of Scoteia

One of James Branch Cabell's finest novels, *The Cream of the Jest*, involves a series of hypnotic dreams that Felix Kennaston induces by staring at half of the Sigil of Scoteia. At the end of the novel, Kennaston discovers that the Sigil is nothing more than the broken top of a cold-cream jar designed by someone who "just made it up out of his head." It "is in no known alphabet. It blends meaningless curlicues and dots and circles with an irresponsible hand." The artist had sketched a crack across it "just to make it look ancient like."

A picture of the complete Sigil, which appears in most editions of the book, is reproduced in Figure 15.4. I remember puzzling over it many years ago and vainly trying to decode it. Designed by Cabell himself, it is not a cipher at all. Can you read it?

Cabell is still out of favor with most literary critics, but he has a loyal band of admirers who support the James Branch Cabell Society and its official organ, *Kalki*. It was in the sixth issue of *Kalki* (1968) that I learned the Sigil's secret.

Figure 15.4

PROBLEM 15.18—Limerick Paradox

An amusing variant of the old liar paradox appeared in the British monthly *Games & Puzzles*. It is presented here as the last of four "limericks."

> There was a young girl in Japan
> Whose limericks never would scan.
>   When someone asked why,
>   She said with a sigh,
> "It's because I always attempt to
> get as many words into the last
> line as I possibly can."

> Another young poet in China
> Had a feeling for rhythm much fina.
>   His limericks tend
>   To come to an end
> Suddenly.

> There was a young lady of Crewe
> Whose limericks stopped at line two.

> There was a young man of Verdun.

PROBLEM 15.19—Nevermore

Three remarkable parodies of Edgar Allan Poe's "The Raven" appeared in issues of *Word Ways*. Each parody is based on a specific form of wordplay familiar to readers of *Word Ways*. All three poems parody the entire Poe poem, but I shall quote only the first stanzas of each.

(1)

Midnight intombed December's
* naked icebound gulf.*
Haggard, tired, I nodded, toiling
* over my books.*
Eldritch daguerreotyped dank
* editions cluttered even my bed;*
Exhaustion reigned.
Suddenly, now, a knocking, echoing
* door I cognized:*
"Eminent Boreas, open up no door!

Go, uninvited lonely frigid haunt!
Avaunt, grim guest—and roar!"

(2)

On one midnight, cold and dreary,
* while I, fainting, weak and weary,*
Pondered many a quaint and ancient
* volume of forgotten lore,*
While I studied, nearly napping,
* suddenly there came a tapping,*
Noise of some one gently rapping,
* rapping at the chamber door.*
"Oh, some visitor," I whispered,
* "tapping at the chamber door,*
Only one, and nothing more."

(3)

On a midnight, cool and foggy,
* as I pondered, light and groggy,*
Ancient books and musty ledger
* not remembered any more,*

As I nodded, all but napping,
there I sensed a muffled tapping,
Very much a hushful rapping,
just behind my attic door.
"'Tis a guest, mayhap," I muttered,
"knocking at my attic door—
I can't judge it's any more."

The first parody is by Howard W. Bergerson, author of the Dover paperback *Palindromes and Anagrams*. He set himself such a difficult task that it was impossible to retain Poe's original meter and rhyme scheme. The other two parodies are by Ross Eckler. The last was harder to write, but in both cases he was able to preserve the meter and rhyme scheme of the original. What curious linguistic structures underlie the parodies?

Answers

ANSWER 15.1—(July 1971)

The well-known quotation is "To be or not to be, that is the question."

The Hamlet rebus, "To be or not to be," was invented by Solomon Golomb, a fact I did not know when I gave it.

Jim Levy wrote to say that strictly speaking the symbol for "or" should be one that represented exclusive disjunction (either but not both) rather than inclusive disjunction (either or both), otherwise the statement implies that a person can be and not be simultaneously.

ANSWER 15.2—(July 1971)

Proverb: Do not put Descartes before the horse.

ANSWER 15.3—(July 1969)

The matchstick solution is shown in Figure 15.5.

Figure 15.5

ANSWER 15.4—(August 1968)

The solution to the crossword puzzle is shown in Figure 15.6.

M	M	M	M
O	O	O	O
L	L	L	L
E	E	E	E

Figure 15.6

ANSWER 15.5—(July 1971)

Underground begins and ends with *und*. Solomon W. Golomb further proposed "underfund" and "underwound."

ANSWER 15.6—(April 1969)

The answer I gave to this was, "He spoke from 22 to 2 to 2:22 to 2,222 people." Readers sent other interpretations. David B. Eisendrath, Jr., wrote that if the speaker had been called a colonel it would have implied military time and the translation "He spoke from 22 to 22 to 22:22 to 22 people."

ANSWER 15.7—(July 1971)

My solution was "HOBO." Walter C. Eberlin and David Dunlap independently added two strokes to 11030 so that when it is viewed in a mirror it spells "peon," a word closely related to "hobo" in meaning.

Richard Ellingson took advantage of the fact that I did not specify that the lines of 11030 could not be rearranged. His solution was:

ANSWER 15.8—(July 1971)

"House" remains the best answer, but less familiar words such as "ye" and "el" allow other solutions. George A. Miller sent a computer printout of 269 alphabetized answers, and all the words (from "abhor" to "wavey") are found in standard dictionaries.

ANSWER 15.9—(July 1971)

The consecutive letters *R, S, T, U* will spell "rust" or "ruts."

P. H. Lyons commented: "I hope some readers tried other languages, such as Hawaiian. If the letters of the alphabet need not be in alphabetical order, I have a fair-sized list of other answers in English."

ANSWER 15.10—(October 1962)

"However little French I know," says J. E. Littlewood (in explaining why he was not obliged to write an infinite regress of footnotes to an article that a friend translated), "I am capable of *copying* a French sentence."

ANSWER 15.11—(April 1974)

Once you guess that the stranger is a card—and what a card!—the rest is easy. "Thin, flat voice" is the first hint. The dialogue eliminates first the Joker, then the suits of hearts, diamonds, and clubs. A lack of spots makes the stranger a face card, and being unmarried eliminates the king and queen. Only the jack of spades is left.

ANSWER 15.12—(November 1963)

I confess that what I thought was a new problem turns out, as Dmitri Borgmann informed me, to be one of the first English word squares ever published! In a letter to the British periodical *Notes and Queries* for July 21, 1859, a reader signing himself "W. W." spoke of the word-squaring game "which has of late been current in society" and proceeded to give the following example: Circle, Icarus, Rarest, Create, Lustre, Esteem. "There are very probably," he wrote, "other ways of squaring the circle."

Yes, when I published the problem in *Scientific American* about 1,000 readers found more than 250 different ways of doing it. I despair of summarizing the variations. The most popular choice for a second word was Inures, with Iberia, Icarus, and Isohel following in that order. The square composed by the most (227) people was: Circle, Inures, Rudest, Crease, Lesser (or Lessor), Esters. Almost as many (210) sent essentially the same square, with Lessee and Esteem as the last two words. "This was done with *ease*," wrote Allan Abrahamse, in punning reference to the fact that a main diagonal of this square consists entirely of *E*'s. Fifty-six readers found Circle, Inures, Rumens, Create, Lenten, Essene.

The most popular square with Iberia as the second word was Circle, Iberia, Recent (or Relent, Repent, and so on), Create, Linter, Eaters (or Eatery). The most popular with Icarus second: Circle, Icarus, Rarest,

Create, Luster, Esters. With Isohel second: Circle, Isohel, Roband (or Roland), Chaise (or Chasse), Lenses, Eldest. Each of these three squares was arrived at by more than a hundred readers.

Of some 40 other words chosen for the second spot, Imaret was the favorite. More than 40 readers used it, mostly as follows: Circle, Imaret, Radish, Crissa, Lesson, Ethane. Many squares with unusual words were found by one reader only; the following are representative:

Circle, Imoros, Romist, Crimea, Losest, Estate (Frederick Chait).

Circle, Isolux, Rosace, Claver, Lucent, Exerts (Ross and Otis Schuart).

Circle, Iterum, Refine, Cringe, Lunger, Emeers (Ralph Hinrichs).

Circle, Isaian, Rained, Cingle, Laelia, Endear (Robert Utter).

Circle, Ironer, Rowena, Cnemis, Lenite, Eraser (Ralph Beaman).

Circle, Inhaul, Rhymed, Camise, Lueses, Eldest (Riley Hampton).

Circle, Irenic, Regime, Cnidus, Limuli, Ecesis (Mrs. Barbara B. Pepelko).

A number of readers tried the more difficult task of squaring the square. All together about 24 different squared squares came in, all with esoteric words such as Square, Quaver, Uakari, Avalon, Rerose, Erinea (Mrs. P. J. Federico). Several readers tried to square the triangle, but without success. Edna Lalande squared the ellipse: Ellipse, Lienees, Lecamas, Inagile, Pemican, Sealane, Essenes.

Four readers (Quentin Derkletterer, Solomon Golomb, John McClellan, and James Topp) independently hit on this delightful squared cube:

```
C U B E
U G L Y
B L U E
E Y E S
```

Derkletterer took off from this square, along a third co-ordinate, and managed to cube the cube: Cube, Ugly, Blue, Eyes; Ugly, Glue, Lull, Yelp; Blue, Lull, Ulus, Else; Eyes, Yelp, Else, Sped [see Figure 15.7]. Patrick O'Neil and Charles Keith cubed the cube this way: Cube, Upon, Bold, Ends; Upon, Pole, Olio, Neon; Bold, Olio, Liar, Dora; Ends, Neon, Dora, Snap. Benjamin F. Melkun and Glenn A. Larson found still another cubed cube, then vanished along a fourth co-ordinate and came back with hypercubes for the words Pet and Eat.

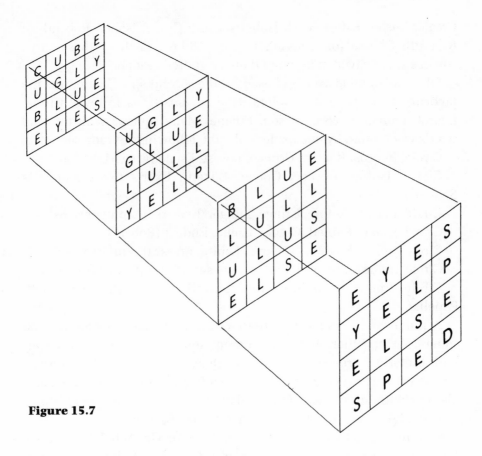

Figure 15.7

They were unable to cube the sphere. R. J. Rea was able to cube eggs; and H. M. Thomas and H. P. Thomas cubed root, dice, beef, and ice, but were unable to cube sugar.

Leigh Mercer, the London expert on wordplay, sent me the best-known squares in which the words, taken in order, form sentences:

> Just, Ugly, Slip, Type.
> Might, Idler, Glide, Hedge, Trees?
> Crest, Reach, Eager (Scene Three).
> Leave, Ellen, Alone, Venom, Enemy.

ANSWER 15.13—(June 1964)

Dudeney, in his posthumously published *A Puzzle-Mine*, was able to achieve 12 good English words by placing letters on the 9-cell square like this:

```
G  E  T

A  I  A

S  U  P
```

The words are: get, teg, sup, pus, pat, tap, gas, sag, pig, gip, sit, aia. If the contraction "'tis" is permitted, the number is 13.

More than 50 readers sent in 12-word squares, most of them superior to Dudeney's 12-worder. Many readers showed how 12 words could be obtained from a cross of A's in the center of the square, as shown [number 1] in Figure 15.8. Twenty-six readers sent in 13-word squares, in most cases with words that could all be found in *Webster's New Collegiate Dictionary*. The typical square [numbered 2 in the illustration] was independently discovered by Vaughn Baker, Mrs. Frank H. Driggs, William Knowles, and Alfred Vasko.

Vaughn Baker, David Grannis, Horace Levinson, H. P. Luhn, Stephen C. Root, Hugh Rose, Frank Tysver, C. Brooke Worth, and George Zinsmeister all produced 14-word squares. Baker's square [numbered 3 in the illustration] has only one word—"wey"—that is not usually found in short dictionaries. Frederick Chait, James Garrels, B. W. Le Tourneau, Marvin Weingast, and Arnold Zeiske devised 15-worders, but none with more than 12 short dictionary words.

Five readers hit the jackpot with 16 words: Dmitri Borgmann, L. E. Card, Mrs. D. Harold Johnson, Peter Kugel, and Wylie Wilson. The five squares [numbered 4 through 8] are reproduced in the order in which the alphabetized names appear above. There is no way to

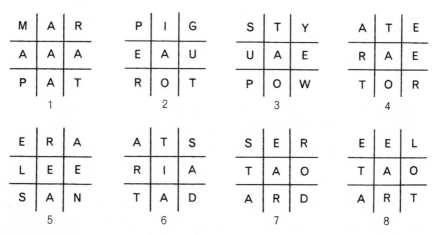

Figure 15.8

decide which square is best, since all exploit obscure words and even the meaning of "word" is hazy.

Several readers experimented with order-4 squares. L. E. Card, of Urbana. Illinois, achieved the maximum (20 dictionary words), with:

```
S   N   A   P
A   E   R   A
R   A   I   L
T   R   A   P
```

"Tras" is the plural of "tra," a Malaysian coin.

ANSWER 15.14—(November 1965)

Dmitri Borgmann, author of *Language on Vacation*, is my authority for the following answers to the questions about the anagram dictionary.

1. Among common words, SU, *us*, is probably the last entry. But this is followed by TTU, *tut*, TTUU, *tutu* (a short ballet skirt), TUX, *tux* (short for tuxedo) and ZZZZ, *zzzz* (to snore), which Borgmann says is in the second edition of *The American Thesaurus of Slang*, by Lester V. Berrey and Melvin Van Den Bark.
2. The first and second entries are A, *a*, and AA, *aa* (a kind of lava). Is the third entry AAAAABBCDRR, *abracadabra*?
3. The last entry is AY, *ay*, unless we accept AYY, *yay* (an obsolete variant of "they").
4. Among common words, the first entry beginning with *B* is probably BBBCDEEOW, *cobwebbed*. It is preceded by the less common BBBBBEEHLLUU, *hubble-bubble* (a bubbling sound, also a hookah).
5. ABCDEFOL, *boldface*.
6. The longest entry that is itself a common word (that is, a word with letters in alphabetical order) is *billowy*. In *Language on Vacation* Borgmann supplies a longer word, *aegilops* (a genus of grasses).
7. The longest common English word that does not repeat a letter is *uncopyrightables*. But Borgmann supplies some longer coined words, such as *vodkathumbscrewingly*, with 20 letters, which means to apply thumbscrews while under the influence of vodka. The longest such word ever created, he says, is the 23-letter mon-

ster *pubvexingfjordschmaltzy*, which means "as if in the manner of the extreme sentimentalism generated in some individuals by the sight of a majestic fjord, which sentimentalism is annoying to the clientele of an English inn."

In 1964, shortly after Temperley proposed the compiling of an anagram dictionary, Follett Publishing Company published the *Follett Vest Pocket Anagram Dictionary*, compiled by Charles A. Haertzen. It contains 20,000 words of seven or fewer letters, with an informative introduction and useful bibliography. *Unscrambler*, an anagram dictionary of 13,867 words through seven letters, was published in 1973 by The Computer Puzzle Library, Fort Worth, Texas. An anagram dictionary of more than 3,200 Old Testament names was privately published in 1955 by Lucy H. Love, Darien, Connecticut. It is titled *Bible Names "De-Koder."* Those of you interested in anagrams will enjoy Howard W. Bergerson's monograph, *Palindromes and Anagrams*, a 1973 Dover paperback.

ANSWER 15.15—(November 1970)

The scrambled quotation is "There is no frigate like a book/To take us lands away." It is the first two lines of a poem by Emily Dickinson.

ANSWER 15.16—(November 1975)

The first published attempt to solve Lewis Carroll's anagram poem was made by Sidney H. Williams and Falconer Madan in their *Handbook of the Literature of the Rev. C. L. Dodgson* (Oxford University Press, 1931). Some of their errors were corrected in a revised edition, the *Lewis Carroll Handbook* (Oxford University Press, 1962), by Roger L. Green. For further criticism and speculation see Philip S. Benham, "Sonnet Illuminate," in *Jabberwocky*, the journal of the Lewis Carroll Society (Summer 1974).

Drawing on the above sources, and with help from Spencer D. Brown, here is how matters stand on the words Carroll most likely had in mind.

1. As to: *oats* (not *oast, stoa*).
2. The war: *wreath* (not *thawer*).
3. Try elm: *myrtle*.
4. I tried: *tidier*.
5. The wig: *weight*.

6. Cast in: *antics* (not *sciant, actins,* or *nastic*).

7. I went: *twine.*

8. To ride: *editor* or *rioted.*

9. Ring yes: *syringe.*

10. We rang: *gnawer.*

11. Let's rap: *plaster* (not *stapler, persalt, palters, psalter,* or *platers*).

12. We don't: *wonted.*

13. O shew: *whose.*

14. Her wit: *writhe* (not *wither, whiter*).

15. As yet: *yeast.*

16. She won't: *snoweth.*

17. Saw eel: *weasel.*

18. In Rome: *merino* (not *moiren, minore*).

19. Dry one: *yonder.*

20. He's wet: *seweth* (not *thewes, hewest*).

21. I am dry: *myriad.*

22. O forge: *forego.*

23. Th' rogue: *tougher* (not *rougeth*).

24. Why a net: *yawneth.*

Carroll's letter containing the puzzle poem first appeared in *Six Letters by Lewis Carroll*, privately printed in 1924. It is reprinted as Letter XLV in *A Selection from the Letters of Lewis Carroll to His Child-Friends*, edited by Evelyn M. Hatch (Folcroft, 1973).

Now for some hot news concerning number 16 above. Can you rearrange the letters of "she won't" to make another common two-word phrase?

Answer 15.17—(May 1975)

As David M. Keller disclosed in his article "The Sigil of Scoteia" (*Kalki* 2, 1968), one simply turns the Sigil upside down. "Additional difficulties are found in the division of words at the ends of lines," Keller writes, "and in the substitution of odd characters for some of the letters." The Sigil reads: "James Branch Cabell made this book so that he who wills may read the story of mans eternally unsatisfied hunger in search of beauty. Ettarre stays inaccessible always and her lovliness is his to look on only in his dreams. All men she must evade at the last and many ar the ways of her elusion."

The paradox of the fourth limerick arises when the limerick is completed in one's mind: "Whose limericks stopped at line one." To complete it is to contradict what the limerick is asserting. The four paradoxical limericks prompted J. A. Lindon, the British comic versifier, to improvise the following new ones:

> A most inept poet of Wendham
> Wrote limericks (none would defend 'em).
>   "I get going," he said,
>   "Have ideas in my head.
> Then find I just simply can't."

> That things were not worse was a mercy!
>   You read bottom line first
>   Since he wrote all reversed—
> He did every job arsy-versy.
> A very odd poet was Percy!

> Found it rather a job to impart 'em.
>   When asked at the time,
>   "Why is this? Don't they rhyme?"
> Said the poet of Chartham, "Can't start 'em."

> So quick a verse writer was Tuplett,
> That his limerick turned out a couplet.

> A three-lines-a-center was Purcett,
> So when he penned a limerick (curse it!)
> The blessed thing came out a tercet!

>   Absentminded, the late poet Moore,
>   Jaywalking, at work on line four,
>   Was killed by a truck.

>   So Clive scribbled only line five.

When I ended the column with limericks of decreasing length, I referred to the one-line limerick as the "last of four." Draper L. Kauffman, John Little, John McKay, Thomas D. Nehrer, and James C. Vibber were the first of many who told me I should have called it the last-but-one of five. The fifth, of course, has *no* lines, which is why other readers failed to notice it.

Tom Wright of Ganges, British Columbia, wrote: "I was interested in the limerick paradox, particularly in the decreasing two-line and one-line limericks. I wondered if you had, in fact, added the no-line limerick (about the man from Nepal), and I looked minutely to see if it wasn't there. On examination, my first impulse was to assume that it was indeed not there, since no space was provided, but further cogitation suggested that a no-line poem, requiring no space, might indeed be there. Unable to resolve this paradox by any logical proof, I am abjectly reduced to asking you whether or not a no-line limerick was not printed in the space not provided, or not."

ANSWER 15.19—(February 1979)

The first parody of Edgar Allan Poe's "The Raven" is by Howard W. Bergerson and is called an automynorcagram. The first letters of each word in the parody, taken in the order they appear, spell out the first verse of the parody. For the complete parody see *Word Ways* (8, 1975, pp. 219–22).

The second parody of Poe's poem is by A. Ross Eckler, editor and publisher of *Word Ways*. It is homoliteral, that is, each consecutive pair of words have at least one letter in common. The full parody appears in *Word Ways* (9, 1976, pp. 96–98).

The third parody, also by Eckler, is found in *Word Ways* (9, 1976, November 1976, pp. 231–33). It is heteroliteral, that is, each pair of consecutive words has no letter in common.

A well-known American poet who asked me not to mention his name took issue with my statement that it is not possible to write a parody of "The Raven" in which the initial letters of each word spell out the original stanzas and at the same time retain Poe's rhyme and meter scheme. Here is how he managed it:

One November evening, nodding over volumes ever-plodding,
My bedeviled eyes regretting every volume, evermore
Nincompoopishly I noodled, garnered nonsense-rhymes, or doodled,
Doing interlineations nine-times gerrymandered o'er:
Vain, elusive ruminations, variations on "Lenore,"
       Uttering mutterings encore.

Shh! Egad! Verandah-tapping! Ectoplasmical rap-rapping
Preternaturally lapping one's dilapidated door!
Inchwise noise gone madly yowling by each dulled ear, vampire-howling
Implings licking every doorway, every yielding entrydoor,

Shades ranged everywhere, grim rappers ever tougher to ignore,
                    Noisily gnashing extempore.

"Visitors!" exclaimed, rewoken, Yours Voluminously. "Oaken
Laminationwork unbroken metalclad exterior,
Even vault-thick engine-metal reinforcement may outfettle
Regulation Elementals, not Immeasurables nor
Cosmic Omnicompetences. My poor ostel's open. Pour
                    In, sweet heebiejeebs, like yore."

chapter 16

Physical Objects

The puzzles in this chapter involve objects. They can all be done mentally but you may enjoy manipulating the objects directly. Many of these problems come from Gardner's own files of magic tricks. In 1949 he wrote "Mathematical Tricks with Common Objects," for Scripta Mathematica. *The next chapter deals with questions about objects as well but the focus will be on physics.*

Problems

PROBLEM 16.1—Manhole Covers
Why are manhole covers circular instead of square?

PROBLEM 16.2—Match Square
Move one match in Figure 16.1 to make a square. (The old joke solution of sliding the top match up a trifle to form a square hole at the center is not permitted; here the solution is a different kind of joke.)

Figure 16.1

PROBLEM 16.3—Remove the Matches
In Figure 16.2 remove 6 matches, leaving 10.

Figure 16.2

PROBLEM 16.4—Heads and Tails

Draw a vertical line on a sheet of paper. The challenge is to position three pennies in such a way that the surfaces of two heads are wholly to the right of the line and the surfaces of two tails are wholly to the left of the line.

PROBLEM 16.5—Three-Penny Bet

First, arrange three pennies as shown in Figure 16.3 and challenge someone to put penny *C* between *A* and *B* so that the three coins lie in a straight line—without moving *B* in any way, without touching *A* with his hands, with any part of his body, or with any object, and without moving *A* by blowing on it.

FIGURE 16.3   A   B   C

PROBLEM 16.6—Reverse the Fish

I found this charming brainteaser for children in a Japanese puzzle book by Kobon Fujimura. Arrange eight toothpicks and one button as shown in Figure 16.4. Now see if you can change the position of just

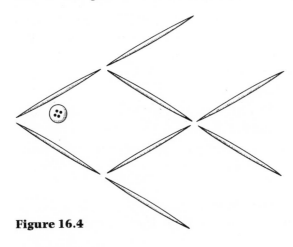

Figure 16.4

Physical Objects

423

three toothpicks and the button so that the fish looks exactly the same as before except it is now swimming in the opposite direction.

PROBLEM 16.7—Form a Swastika

During World War II a gag problem that made the rounds was: How can you make a Nazi cross with five matches? One answer was "Push four of them up his rear end and light them with the fifth." Here is a somewhat similar problem, although one that does not hinge on wordplay.

You are asked to take four cigarettes and eight sugar cubes, place them on a dark-surfaced table top and form the best possible replica of the swastika (a mirror image of the Nazi symbol) shown in Figure 16.5. All 12 objects must be used and none must be damaged in any way.

Figure 16.5

PROBLEM 16.8—Charles Addams's Skier

Single-panel gag cartoons, like Irish bulls, are often based on outrageous logical or physical impossibilities. Lewis Carroll liked to tell about a man who had such big feet that he had to put his pants on over his head. Almost the same kind of impossibility is the basis of a famous *New Yorker* cartoon by Charles Addams of a woman skier going down a slope. Behind her you see her parallel ski tracks approaching a tree, going around the tree with a track on each side and then becoming parallel again.

Suppose you came on a pair of such ski tracks on a snowy slope, going around a tree exactly as in Addams's cartoon. Assume that they are, in truth, tracks made by skis. Can you think of at least six explanations that are physically possible?

PROBLEM 16.9—Functional Fixedness

Past experience sometimes has a negative effect on creative thinking. When this involves a difficulty in seeing how a familiar object can be used in an unorthodox way, psychologists call it a manifestation of "functional fixedness." Here are two problems, familiar to psychologists, that illustrate the concept:

You are seated at a bare table and given six objects: a board, pliers open to maximum extent, two small metal angle irons with screw holes, a peg and a length of wire that has been used to bind the peg firmly to the board [see Figure 16.6]. How can you arrange these objects so that the board becomes a horizontal stand several inches above the tabletop and firm enough to support a vase of flowers?

You are in a bare room. Two strings hang from the ceiling [see Figure 16.7]. Your problem is to tie the ends together. When you grasp one end of the string, however, the other dangles many feet beyond your reach. You are not allowed to use anything you are wearing or have on your person (such as your stockings for the purpose of swinging them to catch a string), but you may use any or all of three objects on the floor: a table-tennis ball, a small horseshoe magnet, and a postage stamp.

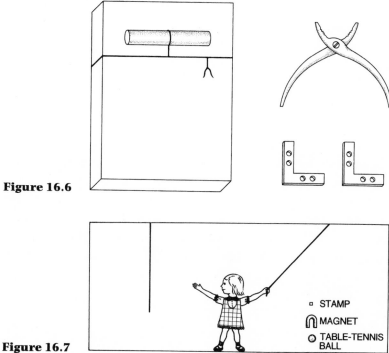

Figure 16.6

Figure 16.7

□ STAMP

Ⴖ MAGNET

○ TABLE-TENNIS
 BALL

Physical Objects

PROBLEM 16.10—Chicken-Wire Trick

This strange parlor trick comes from Tan Hock Chuan, a Chinese professional magician who lives in Singapore. He described it in a letter to Johnnie Murray, an amateur conjuror of Portland, Maine, who passed it on to me.

A blank sheet of paper about eight by five inches (half a sheet of standard-sized paper works nicely) is initialed by an onlooker so that later it can be identified. The magician holds it behind his back (or under a table) for about 30 seconds. When he brings it back into view, it is covered with creases that form a regular hexagonal tessellation [see Figure 16.8]. How is it done? The performer is usually accused of pressing it against a piece of chicken wire, but the creasing actually is done without using anything except the hands.

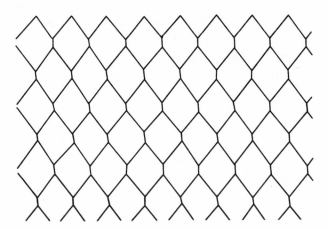

Figure 16.8

PROBLEM 16.11—Gear Paradox

The mechanical device shown in Figure 16.9 was constructed by James Ferguson, an eighteenth-century Scottish astronomer well known in his time as a popular lecturer, author, and inventor, and for the remarkable fact that although he was a member of the Royal Society his formal schooling had consisted of no more than three months in grammar school. (One of his biographies is called *The Story of the Peasant-Boy Philosopher.*) His device is given here as a puzzle that, once solved, will be seen to be a most curious paradox.

Wheel *A* and its axis are firmly fixed so that wheel *A* cannot turn. When the device is rotated clockwise around wheel *A* by means of the handle, wheel *B* will of course rotate in the same direction. The teeth

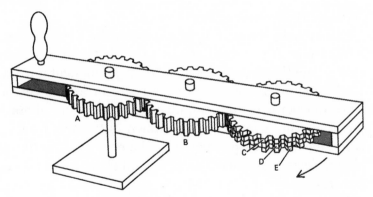

Figure 16.9

of B engage the teeth of three thinner wheels C, D, and E, each of which turns independently. A, B, and E each have 30 teeth. C has 29 teeth, D has 31. All wheels are of the same diameter.

As seen by someone looking down on the device as it is turned clockwise, each of the thin wheels C, D, and E must turn on its axis (with respect to the observer) either clockwise, counterclockwise, or not at all. Without constructing a model, describe the motion of each wheel. If you wish to build a model eventually, it is not necessary that the wheels have the exact number of teeth given. It is only necessary that A and E have the same number of teeth, C at least one less, and D at least one more.

PROBLEM 16.12—Twiddled Bolts

Two identical bolts are placed together so that their helical grooves intermesh [Figure 16.10]. If you move the bolts around each other as you would twiddle your thumbs, holding each bolt firmly by the head so that it does not rotate and twiddling them in the direction shown, will the heads (a) move inward, (b) move outward, or (c) remain the same distance from each other? The problem should be solved without resorting to actual test. (I am indebted to Theodore A. Kalin for calling this problem to my attention.)

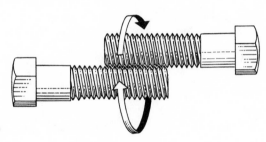

Figure 16.10

Physical Objects

PROBLEM 16.13—Can of Beer

On a picnic Walter van B. Roberts of Princeton, New Jersey, was handed a freshly opened can of beer. "I started to put it down," he writes, "but the ground was not level and I thought it would be well to drink some of the beer first in order to lower the center of gravity. Since the can is cylindrical, obviously the center of gravity is at the center of a full can and will go down as the beer level is decreased. When the can is empty, however, the center of gravity is back at the center. There must therefore be a point at which the center of gravity is lowest."

Knowing the weight of an empty can and its weight when filled, how can one determine what level of beer in an upright can will move the center of gravity to its lowest possible point? When Roberts and his friends worked on this problem, they found themselves involved with calculus: Expressing the height of the center of gravity as a function of the height of the beer, differentiating, equating to zero, and solving for the minimum value of the height of the center of gravity. Later Roberts thought of an easy way to solve the problem without calculus. Indeed, the solution is simple enough to get in one's head.

To devise a precise problem assume that the empty can weighs $1\frac{1}{2}$ ounces. It is a perfect cylinder and any asymmetry introduced by punching holes in the top is disregarded. The can holds 12 ounces of beer, therefore its total weight, when filled, is $13\frac{1}{2}$ ounces. The can is eight inches high. Without using calculus, determine the level of the beer at which the center of gravity is at its lowest point.

PROBLEM 16.14—Touching Cigarettes

Four golf balls can be placed so that each ball touches the other three. Five half-dollars can be arranged so that each coin touches the other four [see Figure 16.11].

Is it possible to place six cigarettes so that each touches the other five? The cigarettes must not be bent or broken.

Figure 16.11

PROBLEM 16.15—Rubber Rope

A worm is at one end of a rubber rope that can be stretched indefinitely [see Figure 16.12]. Initially the rope is one kilometer long. The worm crawls along the rope toward the other end at a constant rate of one centimeter per second. At the end of each second the rope is instantly stretched another kilometer. Thus, after the first second, the worm has traveled one centimeter and the length of the rope has become two kilometers. After the second second, the worm has crawled another centimeter and the rope has become three kilometers long, and so on.

The stretching is uniform, like the stretching of a rubber band. Only the rope stretches. Units of length and time remain constant. We assume an ideal worm, of point size, that never dies, and an ideal rope that can stretch as long as needed.

Does the worm ever reach the end of the rope? If it does, estimate how long the trip will take and how long the rope will be. This delightful problem, which has the flavor of a Zeno paradox, was devised by Denys Wilquin of New Caledonia. It first appeared in December 1972 in Pierre Berloquin's lively puzzle column in the French monthly *Science et Vie*.

Figure 16.12

Answers

ANSWER 16.1—(July 1971)

A square manhole cover, turned on edge, could slip through its hole and fall into the sewer. A case in point: John Bush told of a recent explosion near his home in Brooklyn that blew off a Consolidated Edison square manhole cover. After the smoke cleared the cover was found at the bottom of the manhole. *"Geometria invincibilis est,"* Bush concluded.

Since I failed to ask why manhole covers *and holes* are round instead of square, scores of readers sensibly replied that the covers are round to fit the holes. John W. Stack cited as his authority for this answer M. A. Nhole's *Comprehensive Review of Equilateral Rectangular Beams and Circular Receptacles*, pages 31–4207, published in 1872 by the Sewer and Street Company, Inc. P. H. Lyons had another answer: To reduce the decisions a sewer worker has to make in replacing the cover.

For related material see my "Curves of Constant Width," in *The Colossal Book of Mathematics* (2002).

ANSWER 16.2—(July 1969)

The match can be moved to make a square as shown in Figure 16.13.

Figure 16.13

ANSWER 16.3—(May 1969)

The matches can be removed as shown in Figure 16.14.

Figure 16.14

ANSWER 16.4—(February 1966)

Three pennies can be placed with two heads on one side of a line and two tails on the other as shown in Figure 16.15.

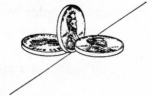

Figure 16.15

ANSWER 16.5—(February 1966)

To put penny C between two touching pennies A and B without touching A or moving B, place a fingertip firmly on B and then slide C

against *B*. Be sure, however, to let go of *C* before it strikes *B*. The impact will propel *A* away from *B*, so that *C* can be placed between the two previously touching coins.

ANSWER 16.6—(April 1974)

The fish swims the other way if you move three toothpicks and the button as shown in Figure 16.16.

Sharon Cammel and Jonathan Schonsheck, in a joint letter, pointed out that there are just two ways to move three toothpicks and make the fish swim the other way: one way sending it a trifle higher in the water, the other sending it lower. Tom Kellerman, age eight, found that by moving two toothpicks he could make the fish swim either up or down, although the fish became shorter and fatter.

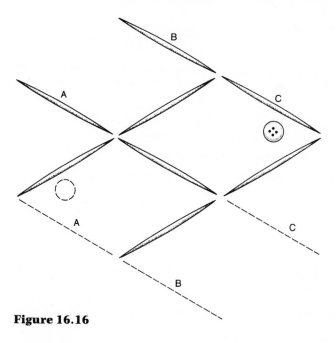

Figure 16.16

ANSWER 16.7—(March 1965)

Four cigarettes and eight sugar cubes can be placed on a dark surface to form an excellent replica of a swastika, as shown in Figure 16.17.

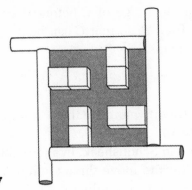

Figure 16.17

ANSWER 16.8—(April 1972)

Here are six possible explanations of the ski tracks:

1. The skier bumped into the tree but protected himself with his hands. Keeping one ski in place, he carefully lifted his other foot and moved to the lower side of the tree. With his back against the tree, he replaced his raised foot and ski on the other side, then continued down the slope.
2. The skier slammed into the tree with such force that his skis came off and continued down the slope without him.
3. Two skiers went down the hill, each wearing only one ski.
4. One skier went down the hill twice, each time with one ski on one foot.
5. A skier went down a treeless slope, moving his legs apart at one spot. Shortly thereafter a tree with a sharpened trunk base was plunged into the snow at that spot.
6. The skier wore stilts that were high enough and sufficiently bowed to allow him to pass completely over the tree.

So many readers sent other preposterous explanations that I can give only a sampling:

A small, supple tree that bent as the skier went over it was proposed by John Ferguson, John Ritter, Brad Schaefer, Oliver G. Selfridge, and James Weaver. Ferguson also suggested (among his 23 possibilities) a pair of skis pulled uphill by long ropes and two toboggan teams of *very* small midgets, four on each ski. Selfridge included this one: The skier, aware of his ineptness, wore a protective lead suit. His impact on the tree sheared out a cylindrical section. The dazed

Figure 16.18

skier passed between top and bottom parts of the tree before the top fell down and balanced perfectly on the base.

Manfred R. Schroeder, director of the Drittes Physikalisches Institut at the University of Göttingen, reported an actual experience he had in 1955 while skiing down a mountain in New Hampshire. "I hit a small but sturdy tree with my right shinbone. The binding came loose and the ski and leg went around different sides of the tree. Below the tree, leg and ski came together again. However, the binding did not engage (no automatic step-in bindings then!) and the tracks ended in a spill about ten yards farther down the slope. Even then, in spite of considerable pain in the leg, I thought it was a worthwhile experience."

Johnny Hart, in his *B.C.* comic strip, has played with the theme. Thor, speeding toward a tree on his stone wheel unicycle, once went around the tree leaving two tracks. What happened on a later occasion is reproduced in Figure 16.18.

ANSWER 16.9—(April 1974)

The board is supported by the pliers and wooden peg [see Figure 16.19]. To tie together the ends of the two hanging cords, tie the magnet to one end and start the cord swinging. Hold the end of the other string and catch the swinging magnet.

Figure 16.19

Physical Objects

For the shelf-making problem E. N. Adams, Bill Kruger, and Susan Southall each showed how the pliers could be opened and wired to the peg to make a tripod, and also how the angle irons could be wired flat to each end of the peg to make a stand. The second solution was also proposed by R. C. Dahlquist, P. C. Eastman, and Ronald C. Read. Don L. Curtis threaded the wire through end holes of the angle irons and with the pliers tightened the wire around the board so that the angle irons were rigidly perpendicular to the board. He sent a photograph to prove that the stand supported a heavy vase of flowers. Robert Rosenwald and Allan Kiron each thought of using the pliers as a hammer for knocking corners of the angle irons into the board to make supports.

Several readers spotted a mistake in the illustration for the problem of knotting the ends of two hanging strings. The strings are too short to be tied. Paul Nelles considered the situation in which a side wall is so close to each string that if the magnet were tied to one end it would collide with the wall when the string is swung. He suggested tying the table-tennis ball to the cord (or fastening it with the stamp), then swinging the string so that the ball bounces off the wall. Michael McMahon suggested using the stamp to attach the ball to the string and then, with a corner of the magnet, addressing the ball and mailing it to the other side of the room.

ANSWER 16.10—(May 1973)

To put a chicken-wire pattern of creases into a small sheet of paper, first roll the sheet into a tube about half an inch in diameter. With the thumb and forefinger of your left hand, pinch one end of the tube flat. Keeping pressure on the pinch with your left hand, your right thumb, and forefinger, pinch the tube flat at a spot as close as possible to the first pinch, making the pinch at right angles to the first one. Press firmly with both hands, at the same time pushing the two pinches tightly against each other to make the creases as sharp as possible. Now the right hand retains its pinch while the left hand makes a third pinch adjacent to and perpendicular to the second one. Continue in this way, alternating hands as you move along the tube, until the entire tube has been pinched. (Children often do this with soda straws to make "chains.") Unroll the paper. You will find it hexagonally tessellated in a manner that is most puzzling to the uninitiated.

John H. Coker wrote to say that when he was a child in Yugoslavia in the early 1930s, his schoolteacher rolled and pinched notes to other teachers in this manner. Because it is extremely difficult to unroll such a tube and then reroll it exactly as before, the tube provided security from the eyes of children asked to transmit the notes.

ANSWER 16.11—(March 1965)

When James Ferguson's curious mechanical device is turned clockwise, wheel *C* rotates clockwise in relation to the observer, *D* rotates counterclockwise, and *E* does not rotate at all!

ANSWER 16.12—(August 1958)

The heads of the twiddled bolts move neither inward nor outward. The situation is comparable to that of a person walking up an escalator at the same rate that it is moving down.

ANSWER 16.13—(April 1972)

Walter van B. Roberts answered his beer-can problem this way: "Imagine that the beer is frozen so that the can of beer can be placed horizontally on a knife-edge pivot and balanced with the can's top to the left. If it balances with the pivot under the beer-filled part, adding more beer would make the can tip to the left, whereas removing beer would make it tip to the right. If it balances with the pivot under the empty part, the reverse would be true. But if it balances with the pivot exactly under the beer's surface, any change in the amount of beer will make the can tip to the left [see Figure 16.20]. Since in this case the center of gravity moves toward the can's top when any change is made in the amount of beer, the center of gravity must be at its lowest point when it coincides with the beer's surface.

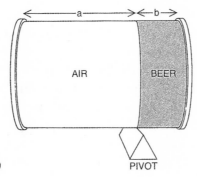

Figure 16.20

"With the can balanced in this condition, imagine that the ends are removed and their mass distributed over the side of the can. This cannot upset the balance because it does not shift the center of gravity of the system, but it allows us to consider the can as an open-ended pipe whose mass per unit length on the empty (left) side is proportional to the weight of an empty can, whereas the mass on the beer-filled right side is proportional to the weight of a full can. The moment of force on the left is therefore proportional to the weight of an empty can multiplied by the square of the length of the empty left side, and the moment on the right side is similarly proportional to the weight of a full can multiplied by the square of the length of the beer-filled right side. Since the can is balanced, these moments must be equal.

"Pencil and paper are now hardly required to deduce that the square of the length of the empty part divided by the square of the length of the full part equals the weight of a full can divided by the weight of an empty can, or, finally, that the ratio of the length of the empty part to the full part is the square root of the ratio of the weight of a full can to an empty one."

Expressed algebraically, let a and b stand for the lengths of the empty and filled parts of the can when the center of gravity is at its lowest point and E and F for the can's weight when empty and full. Then $a^2E = b^2F$, or $a/b = \sqrt{F/E}$.

In the example given, the can weighs nine times as much when it is full as it does when it is empty. Therefore the center of gravity reaches its lowest point when the empty part is three times the length of the full part, in other words, when the beer fills the can's lower fourth. Since the can is eight inches high, the level of the beer is 8/4 = 2 inches.

After the above solution appeared in *Scientific American*, Mark H. Johnson wrote to say that the answer is not strictly accurate. Because the tops and bottoms of the frozen can are at unequal distances from the pivot, they exert unequal moments of force. Distributing their masses over the can's side, to make a uniform and open pipe, would tilt the can slightly to the air side. To solve the problem precisely one needs more data about the can's dimensions and the masses of its top, bottom, and side. Other readers reported that the solution also neglects what naval architects and engineers call the "free surface effect." When liquids are free to move inside containers, a slight raising of the vessel's center of gravity results.

There are several different ways of placing the six cigarettes. Figure 16.21 shows the traditional solution as it is given in several old puzzle books.

Figure 16.21

To my vast surprise, about 15 readers discovered that *seven* cigarettes could also be placed so that each touched all of the others! This of course makes the older puzzle obsolete. Figure 16.22, sent to me by George Rybicki and John Reynolds, graduate students in physics at Harvard, shows how it is done. "The diagram has been drawn," they write, "for the critical case where the ratio of length to diameter of the cigarettes is $7/2\sqrt{3}$. Here the points of contact occur right at the ends of the cigarettes. The solution obviously will work for any length-to-diameter ratio greater than $7/2\sqrt{3}$. Some observations on actual 'regular' size cigarettes give a ratio of about 8 to 1, which is, in fact, greater than $7/2\sqrt{3}$, so this is an acceptable solution." Note that if the center cigarette, pointing directly toward you in the diagram, is withdrawn, the remaining six provide a neat symmetrical solution of the original problem.

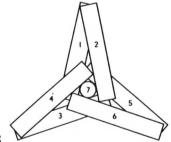

Figure 16.22

Regardless of the parameters (the initial length of the rubber rope, the worm's speed, and how much the rope stretches after each unit of time), the worm will reach the end of the rope in a finite time. This is also true if the stretching is a continuous process, at a steady rate, but the problem is easier to analyze if the stretching is done in discrete steps.

One is tempted to think one can see that the worm will make it. Since the rope expands uniformly, like a rubber band, the expansion is like looking at the rope through increasingly strong magnifying lenses. Because the worm is always making progress, must it not eventually reach the end? Not necessarily. One can progress steadily toward a goal forever without ever reaching it. The worm's progress is measured by a series of decreasing fractions of the rope's length. The series could be infinite and yet converge at a point far short of the end of the rope. Indeed, such is the case if the rope stretches by doubling its length after each second.

The worm, however, does make it. There are 100,000 centimeters in a kilometer, so that at the end of the first second, the worm has traveled 1/100,000th of the rope's length. During the next second, the worm travels (from its previous spot after the stretching) a distance of 1/200,000th of the rope's length, after the rope has stretched to two kilometers. During the third second it covers 1/300,000th of the rope (now three kilometers), and so on. The worm's progress, expressed as fractional parts of the entire rope, is

$$\frac{1}{100,000} \left(1 + \frac{1}{2} + \frac{1}{3} + \frac{1}{4} + \ldots + \frac{1}{n}\right)$$

The series inside parentheses is the familiar harmonic one that diverges and therefore can have a sum as large as we please. The partial sum of the harmonic series is never an integer. As soon as it exceeds 100,000, however, the above expression will exceed 1, which means that the worm has reached the end of the rope. The number of terms, n, in this partial harmonic series will be the number of seconds that have elapsed. Since the worm moves one centimeter per second, n is also the final length of the rope in kilometers.

This enormous number, correct to within one minute, is

$$e^{100,000-y} \pm 1$$

where y is Euler's constant. It gives a length of rope that vastly exceeds the diameter of the known universe, and a time that vastly exceeds present estimates of the age of the universe. (For a derivation of the formula see "Partial Sums of the Harmonic Series," by R. P. Boas, Jr., and J. W. Wrench, Jr., *American Mathematical Monthly* 78, October 1971, pp. 864–70.)

Some notion of the enormous length of the rope when the worm finally reaches the end can be gained from an observation by reader H. E. Rorschach. If the rope starts with a cross-sectional area of one square kilometer, it ends up as a single line of atoms, the space between each adjacent pair of atoms being many times the size of the known universe. The time lapse is comparably greater than the age of the universe.

chapter 17

Physics

This chapter is devoted to problems that rely on elementary physics principles. Gardner originally intended to major in physics in college and went on to write several well-known books about physics, including The Ambidextrous Universe *and* Relativity for the Million *(both of which have been revised with title changes). Of course, Gardner's interest in "physics" puzzles was strengthened by their interplay with magic. As early as 1941 he had written a pamphlet* Magic for the Science Class. *In addition to the problems here, Gardner wrote the "Physics Trick of the Month" column in* Physics Teacher *from 1990 to 2002. (These were later reprinted in* Magic *magazine and collected into books.)*

Problems

PROBLEM 17.1—Two Hundred Pigeons

An old story concerns a truck driver who stopped his panel truck just short of a small, shaky-looking bridge, got out, and began beating his palms against the sides of the large compartment that formed the back of the truck. A farmer standing at the side of the road asked him why he was doing this.

"I'm carrying 200 pigeons in this truck," explained the driver. "That's quite a lot of weight. My pounding will frighten the birds and they'll start flying around inside. That will lighten the load considerably. I don't like the looks of this bridge. I want to keep those pigeons in the air until I get across."

Assuming that the truck's compartment is airtight, can anything be said for the driver's line of reasoning?

PROBLEM 17.2—Rising Hourglass

An unusual toy sold in Paris shops: a glass cylinder, filled with water, at the top of which an hourglass floats [see Figure 17.1]. If the cylinder is inverted, as in the right-hand drawing, a curious thing happens. The hourglass remains at the bottom of the cylinder until a certain quantity of sand has flowed into its lower compartment, then it rises slowly to the top. It seems impossible that a transfer of sand from top to bottom of the hourglass would have any effect on its overall buoyancy. Can you guess the simple *modus operandi*?

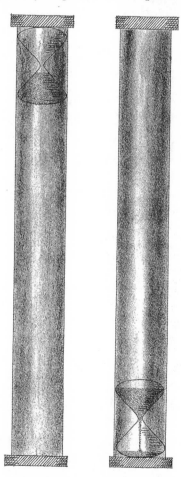

Figure 17.1

PROBLEM 17.3—Iron Torus

A piece of solid iron in the form of a doughnut is heated. Will the diameter of its hole get larger or smaller?

PROBLEM 17.4—Suspended Horseshoe

From a sheet of thin pasteboard cut a horseshoe shape that is a trifle longer than a toothpick [see Figure 17.2]. Lean the pick and horseshoe together on a tablecloth as shown. The problem is to lift both the horseshoe and the toothpick by means of a second toothpick held in one hand. You may not touch the horseshoe or the toothpick on which it leans with anything except the toothpick in your hand. You are not, of course, permitted to break the toothpick and use the pieces like miniature chopsticks. Both objects must be lifted together and held suspended above the table. How is it done?

Figure 17.2

PROBLEM 17.5—Center the Cork

Fill a glass with water and drop a small cork on the surface. It will float to one side, touching the glass. How can you make it float permanently in the center, not touching the glass? The glass must contain nothing but water and the cork.

PROBLEM 17.6—Oil and Vinegar

Some friends were picnicking. "Did you bring the oil and vinegar for the salad?" Mrs. Smith asked her husband.

"I did indeed," replied Mr. Smith. "And to save myself the trouble of carrying two bottles I put the oil and vinegar in the same bottle."

"That was stupid," snorted Mrs. Smith. "I like a lot of oil and very little vinegar, but Henrietta likes a lot of vinegar, and—"

"Not stupid at all, my dear," interrupted Mr. Smith. He proceeded to pour, from the single bottle, exactly the right proportions of oil and vinegar that each person wanted. How did he manage it?

PROBLEM 17.7—Carroll's Carriage

In chapter 7 of Lewis Carroll's *Sylvie and Bruno Concluded*, the German professor explains how people in his country need not go to sea to enjoy the sensations of pitching and rolling. This is accomplished, he says, by putting oval wheels on their carriages. An earl, who is listening, says that he can see how oval wheels could make a carriage pitch backward and forward, but how could they make the carriage roll too?

"They do not *match*, my Lord," the professor replies. "The *end* of one wheel answers to the *side* of the opposite wheel. So first one side of the carriage rises, then the other. And it pitches all the while. Ah, you must be a good sailor, to drive in our boat-carriages!"

Is it possible to arrange four oval wheels on a carriage so that it would actually pitch and roll as described?

Figure 17.3 reproduces a poem written and illustrated by Gelett Burgess (from *The Purple Cow and Other Nonsense*, Dover, 1961). Perhaps it was inspired by Lewis Carroll's carriage.

Remarkable
  Truly is Art!
See—Elliptical
  Wheels on a Cart!
It Looks Very Fair
In the Picture, up There,
But Imagine the
  Ride, when you Start!

Figure 17.3

PROBLEM 17.8—Magnet Testing

You are locked in a room that contains no metal of any sort (not even on your person) except for two identical iron bars. One bar is a magnet, the other is not magnetized. You can tell which is the magnet by suspending each by a thread tied around its center and observing which bar tends to point north. Is there a simpler way?

PROBLEM 17.9—Melting Ice Cube

A cube of ice floats in a beaker of water, the entire system at 0 degrees centigrade. Just enough heat is supplied to melt the cube without altering the system's temperature. Does the water level in the beaker rise, fall, or stay the same?

PROBLEM 17.10—Stealing Bell Ropes

In the tower of a church two bell ropes pass through small holes a foot apart in a high ceiling and hang down to the floor of a room. A skilled acrobat, carrying a knife and bent on stealing as much of the two ropes as possible, finds that the stairway leading above the ceiling is barred by a locked door. There are no ladders or other objects on which he can stand, and so he must accomplish his theft by climbing the ropes hand over hand and cutting them at points as high as possible. The ceiling is so high, however, that a fall from even one-third the height could be fatal. By what procedure can he obtain a maximum amount of rope?

PROBLEM 17.11—Moving Shadow

A man walking at night along a sidewalk at a constant speed passes a street light. As his shadow lengthens, does the top of the shadow move faster or slower or at the same rate as it did when it was shorter?

PROBLEM 17.12—Coiled Hose

A garden hose is coiled around a reel about a foot in diameter placed on a bench as shown in Figure 17.4. One end of the hose hangs down into a bucket and the other end is unwound so that it can be held several feet above the reel. The hose is empty and there are no kinks or obstructions in it. If water is now poured into the upper end

Figure 17.4

by means of a funnel, one would expect that continued pouring would cause water to run out at the lower end. Instead, as water is poured into the funnel it rises in the raised end of the hose until it overflows the funnel; no water ever emerges from the other end. Explain.

Problem 17.13—Egg in Bottle

You cannot push a peeled hardboiled egg through the neck of a glass milk bottle because the air trapped in the bottle keeps it from going through. If, however, you drop a piece of burning paper or a couple of burning matches into the bottle before you stand the egg upright at the opening, the burning will heat and expand the air. When the air cools it contracts, forming a partial vacuum which draws the egg inside. After this has happened a second problem presents itself. Without breaking the bottle or damaging the egg, how can you get it out again?

Problem 17.14—Bathtub Boat

A small boy is sailing a plastic boat in the bathtub. It is loaded with nuts and bolts. If he dumps all this cargo into the water, allowing the boat to float empty, will the water level in the tub rise or fall?

Physics   445

PROBLEM 17.15—Balloon in Car

A family is out for a drive on a cold afternoon, with all vents and windows of the car closed. A child in the backseat is holding the lower end of a string attached to a helium-filled balloon. The balloon floats in the air, just below the car's roof. When the car accelerates forward, does the balloon stay where it is, move backward, or move forward? How does it behave when the car rounds a curve?

PROBLEM 17.16—Hollow Moon

It has been suggested that in the far future it may be possible to hollow out the interior of a large asteroid or moon and use it as a mammoth space station. Assume that such a hollowed asteroid is a perfect, nonrotating sphere with a shell of constant thickness. Would an object inside, near the shell, be pulled by the shell's gravity field toward the shell or toward the center of of the asteroid, or would it float permanently at the same location?

PROBLEM 17.17—Lunar Bird

A bird has a small, lightweight oxygen tank attached to its back so that it can breathe on the moon. Will the bird's flying speed on the moon, where the pull of gravity is less than on the earth, be faster, slower, or the same as its speed on the earth? Assume that the bird carries the same equipment in both instances.

PROBLEM 17.18—Compton Tube

A little-known invention of the physicist Arthur Holly Compton is shown in Figure 17.5. A glass torus several feet across is completely filled with a liquid in which small particles are suspended. The tube is allowed to rest until there is no movement of its liquid, then it is quickly flipped upside down by a 180-degree rotation about the horizontal axis. By viewing the suspended particles through a microscope one can determine whether the liquid is now flowing around the torus.

Assume that the tube is oriented so that its vertical plane extends east and west. As the earth rotates counterclockwise (as seen looking down at the North Pole), the top of the tube moves faster than the bottom because it travels a circular path of larger circumference. Flipping the tube brings this faster-moving liquid to the bottom and the slower-moving liquid to the top, setting up a very weak clockwise circula-

Figure 17.5

tion. The strength of this circulation diminishes as the plane of the tube deviates from east-west, reaching zero when the plane is north-south. One can, therefore, prove that the earth rotates and determine the direction of rotation simply by flipping the tube in different orientations until a maximum circulating speed is produced.

In actual practice, the viscosity of the liquid causes the circulation to decay after 20 seconds or so. Assuming that there is no friction whatever in the tube, and that the flip is made at the equator with an east-west orientation, how long will it take a particle in the liquid to complete one rotation around the tube after the flip is made?

PROBLEM 17.19—Fish Scales

A bowl, three-quarters filled with water, rests on a scale. If you drop a live fish into the water, the scale will show an increase in weight equal to the weight of the fish. Suppose, however, you hold the fish by its tail so that all but the extreme tip of its tail is underwater. Will the scale register a greater weight than it did before you dunked the fish?

PROBLEM 17.20—Bicycle Paradox

A rope is tied to the pedal of a bicycle as shown in Figure 17.6. If someone pulls back on this rope while another person holds the seat lightly to keep the bicycle balanced, will the bicycle move forward, backward, or not at all?

Figure 17.6

PROBLEM 17.21—Inertial Drive

If a rope is tied to the stern of a rowboat, is it possible for a man standing in the boat to propel it forward through quiet water by jerking on the free end of the rope? Could a space capsule drifting in interplanetary space be propelled by a similar method?

PROBLEM 17.22—Worth of Gold

Which is worth more, a pound of $10 gold pieces or half a pound of $20 gold pieces?

PROBLEM 17.23—Switching Paradox

An entertaining curiosity consists of two small 110-volt bulbs (preferably one clear and one frosted) and two on-off switches connected in a simple series circuit plugged into any wall outlet carrying the usual alternating current [see Figure 17.7]. When both switches are on, both bulbs light. If one bulb is unscrewed, the other goes out—as would be expected. When both switches are off, both bulbs are dark. But when switch A is on and B is off, only bulb (a) lights. And when switch B is on and A off, only bulb (b) lights. In short, each switch independently controls its corresponding bulb. Even more inexplicable is the fact that if the two bulbs are interchanged, switch A still controls bulb (a) and switch B still controls bulb (b). Nothing is concealed in the wooden board on which the switches, bulb sockets, and wire are mounted. What is the secret behind the construction of this circuit?

OTHER PUZZLES

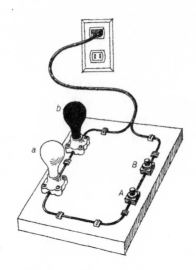

Figure 17.7

Answers

ANSWER 17.1—(August 1966)

The truck driver is wrong. The weight of a closed compartment containing a bird is equal to the weight of the compartment plus the bird's weight except when the bird is in the air and has a vertical component of motion that is accelerating. Downward acceleration reduces the weight of the system, upward acceleration increases it. If the bird is in free fall, the system's weight is lowered by the full weight of the bird. Horizontal flight, maintained by wing-flapping, alternates small up and down accelerations. Two hundred birds, flying about at random inside the panel truck, would cause minute, rapid fluctuations in weight, but the overall weight of the system would remain virtually constant.

ANSWER 17.2—(August 1966)

When the sand is at the top of the hourglass, a high center of gravity tips the hourglass to one side. The resulting friction against the side of the cylinder is sufficient to keep it at the bottom of the cylinder. After enough sand has flowed down to make the hourglass float upright, the loss of friction enables it to rise.

If the hourglass is a trifle heavier than the water it displaces, the toy operates in reverse. That is, the hourglass normally rests at the bottom of the cylinder; when the cylinder is turned over, the hourglass stays at the top, sinking only after the transfer of sand has eliminated the friction. Shops in Paris carried the toy in both versions and in a combined version with two cylinders side by side so that as the hourglass goes up in one it sinks in the other.

The toy, said to have been invented by a Czechoslovakian glassblower, who made them in a shop just outside Paris, is usually more puzzling to physicists than to other people. A common explanation advanced by physicists is that the force of the falling sand grains keeps the hourglass on the bottom, or at least contributes to doing so. It is not hard to show, however, that the net weight of the hourglass remains the same as if the sand were not pouring. See "Weight of an Hourglass," by Walter P. Reid, *American Journal of Physics*, vol. 35, April 1967, pages 351–52.

ANSWER 17.3—(August 1966)

When an iron doughnut expands with heat, it keeps its proportions, therefore the hole also gets larger. The principle is at work when an optician removes a lens from a pair of glasses by heating the frame or a housewife heats the lid of a jar to loosen it.

ANSWER 17.4—(August 1966)

Insert toothpick *A* between the cardboard horseshoe and toothpick *B* and move the horseshoe just enough to let the end of toothpick *B* come to rest on toothpick *A* [see Figure 17.8]. Maneuver the end of *B* under the horseshoe and then lift shoe and toothpick, balancing them as shown in the bottom drawing.

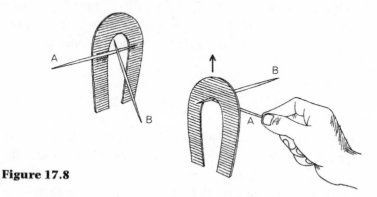

Figure 17.8

ANSWER 17.5—(August 1966)

The cork floats at the center of the surface of the water in the glass only when the glass is filled a bit above its brim. The water's surface tension maintains the slightly convex surface.

Gene Lindberg and M. H. Greenblatt suggested a second method. Rotate a partially filled glass around its vertical axis. This creates a concave surface which centers the cork at the bottom. An even simpler way to create the vortex is by stirring the water with a spoon.

ANSWER 17.6—(August 1966)

Oil floats on vinegar. To pour the oil Mr. Smith had only to tip the bottle. To pour vinegar he corked the bottle, inverted it, then loosened the cork just enough to let the desired amount of vinegar dribble out.

ANSWER 17.7—(August 1966)

On Lewis Carroll's "boat carriage," each pair of oval wheels, on opposite sides of the same axle, turns so that at all times the long axes of the ovals are at right angles to each other. This produces the "roll." If on each side of the carriage the two wheels also had their long axes at right angles, the carriage would neither pitch nor roll. It would simply move up and down, first on two diagonally opposite wheels, then on the other two. However, by gearing the front and back wheels so that on each side of the carriage the two wheels have their long axes at 45-degree angles, the carriage can be made to pitch and roll nicely, with single wheels leaving the ground in a four-beat sequence that is repeated as the carriage moves forward.

Maya and Nicolas Slater wrote from London to say that if one abandons Lewis Carroll's proviso that elliptical wheels opposite each other must have their major axes at right angles, there is a way to make the carriage pitch and roll without any wheel's leaving the ground. The wheels must be geared so that *diagonally* opposite wheels keep their major axes at right angles. Regardless of the angle between the two front wheels, all four wheels remain on the ground at all times. If the front wheels are at a 90-degree angle, the carriage rolls without pitching; if the angle is zero, it pitches without rolling. All intermediate angles combine rolling and pitching. "Our preference is for 45 degrees," the Slaters wrote. "Our only problem is keeping our coachman."

ANSWER 17.8—(August 1966)

Touch the end of one bar to the middle of the other. If there is magnetic attraction, the touching end must be on the magnetized bar. If not, it is on the unmagnetized bar.

ANSWER 17.9—(August 1966)

The water level stays the same. An ice cube floats only because its water has expanded during crystallization; its weight remains the same as the weight of the water that formed it. Since a floating body displaces its weight, the melted ice cube will provide the same amount of water as the volume of water it displaced when frozen.

ANSWER 17.10—(August 1966)

The acrobat first ties the lower ends of the ropes together. He climbs rope *A* to the top and cuts rope *B*, leaving enough rope to tie into a loop. Hanging in this loop with one arm through it, he cuts rope *A* off at the ceiling (taking great care not to let it fall!) and then passes the end of *A* through the loop and pulls the rope until the middle of the tied-together ropes is at the loop. After letting himself down this double rope he pulls it free of the loop, thereby obtaining the entire length of *A* and almost all of *B*.

Many alternate solutions for the rope-stealing problem were received; some made use of knots that could be shaken loose from the ground, others involved cutting a rope partway through so that it would just support the thief's weight and later could be snapped by a sudden pull. Several readers doubted that the thief would get *any* rope because the bells would start ringing.

ANSWER 17.11—(August 1966)

The top of the shadow of a man walking past a streetlight moves faster than the man, but it maintains a constant speed regardless of its length.

ANSWER 17.12—(August 1966)

A quantity of water flows over the first winding of the hose to fall to the bottom and form an air trap. The trapped air prevents any more water from entering the first loop of the hose.

If the funnel end of the empty hose is high enough, water poured into it will be forced over more than one winding to form a series of "heads" in each coil. The maximum height of each head is about equal to the coil's diameter. The diameter, times the number of coils, gives the approximate height the water column at the funnel end must be to force water out at the other end. (This was pointed out by John C. Bryner, Jan Lundberg, and J. M. Osborne.) W. N. Goodwin, Jr., noted that for hoses with an outside diameter of 5/8 inch or less the funnel end can be as low as twice the height of the coils and water will flow all the way through a series of many coils. The reason for this is not yet clear.

ANSWER 17.13—(August 1966)

To remove the egg, tip back your head, hold the bottle upside down with the opening at your lips, and blow vigorously into it. When you take the bottle from your mouth, the compressed air will pop the egg out through the neck of the bottle.

A common misconception about this trick is that the egg was initially drawn into the bottle by a vacuum created by the loss of oxygen. Oxygen is indeed used up, but the loss is compensated by the production of carbon dioxide and water vapor. The vacuum is created solely by the quick cooling and contracting of the air after the flames go out.

ANSWER 17.14—(August 1966)

The nuts and bolts inside the toy ship displace an amount of water equal to their *weight*. When they sink to the bottom of the tub, they displace an amount equal to their *volume*. Since each piece weighs considerably more than the same volume of water, the water level in the tub is lowered after the cargo is dumped.

ANSWER 17.15—(August 1966)

As the closed car accelerates forward, inertial forces send the air in the car backward. This compresses the air behind the balloon, pushing it forward. As the car rounds a curve, the balloon, for similar reasons, moves *into* the curve.

ANSWER 17.16—(August 1966)

Zero gravity prevails at all points inside the hollow asteroid. For an explanation of how this follows from gravity's law of inverse squares see Hermann Bondi's Anchor paperback *The Universe at Large*, page 102.

H. G. Wells, in his *First Men in the Moon*, failed to realize this. At two points in the novel he has his travelers floating near the center of their spherical spaceship because of the gravitational force exerted by the ship itself; this aside from the fact that the gravity field produced by the ship would be too weak to influence the travelers anyway.

ANSWER 17.17—(August 1966)

A bird cannot fly at all on the moon because there is no lunar air to support it. George Milwee, Jr., wrote to ask if I got the idea for this problem from Immanuel Kant! In the Introduction to *The Critique of Pure Reason*, section 3, Kant chides Plato for thinking that he can make better philosophical progress if he abandons the physical world and flies in the empty space of pure reason. Kant bolsters his point with the following delightful metaphor: "The light dove, cleaving the air in her free flight, and feeling its resistance, might imagine that its flight would be still easier in empty space."

ANSWER 17.18—(August 1966)

Consider any particle k in the liquid, before the tube is flipped. The earth carries it counterclockwise (looking down at the North Pole) at a speed of one revolution in 24 hours. But every point in the tube, the same distance from the earth's center, is revolving counterclockwise at the same speed as the particle, therefore the circulatory speed of k, relative to the tube, is zero. After the flip, k continues counterclockwise at the same speed, but the tube's reversal gives k a clockwise motion, relative to the tube, that carries it once around the tube (ignoring friction) in exactly half the time it takes the earth to complete one rotation; namely, 12 hours.

This curious halving of 24 hours can be better understood by considering a specific particle, say one at the top of the tube. Let x be the earth's diameter and y the tube's diameter. Both the particle and the tube's top travel east at a speed of $\pi(x + 2y)$ in 24 hours. After the flip, the particle continues at the same speed, but now it is at the bottom where the tube moves east at the slower rate of πx in 24 hours. Consequently, the particle, relative to the tube, moves east at a speed of $\pi(x + 2y) - \pi x = 2\pi y$ in 24 hours. Since πy is the tube's circumference, we see that the particle is going around the tube at a speed of two circuits in 24 hours, or one in 12. Similar calculations apply to all other particles in the tube.

Most people, in calculating the ideal speed of the water, miss by a factor of 2 and assume it would circulate once around the tube in 24 hours. Compton himself made this error, both in his first paper (*Science*, vol. 37, May 23, 1913, pages 803–6), which describes work with a tube of one-meter radius, and a second paper, reporting results with more elaborate apparatus (*Physical Review*, vol. 5, February

1915, page 109, and *Popular Astronomy*, vol. 23, April 1915, page 199).

When the tube is not at the equator, the maximum effect is achieved by having the plane of the tube tilted from the vertical before the 180-degree flip is made. The nearer to the pole, the greater the tilt needed, which makes it possible to use the tube for determining its latitude. At the North or South Pole, the greatest effect is obtained with the tube horizontal before it is flipped. In all latitudes, when the plane of the tube is properly tilted and oriented for maximum effect, the flip induces a flow of one revolution in 12 hours, assuming of course a total absence of viscosity.

The full story is somewhat complicated, and there is an interesting correspondence between what happens in the Compton tube and the induction of electrical currents that circulate within a conducting ring when it is flipped 180-degrees in a magnetic field. I am indebted to Dave Fultz, of the Hydrodynamics Laboratory of the University of Chicago's department of geophysical sciences, for supplying the information on which this brief account is based. Fultz prefers an explanation in terms of angular velocities and torques, but since this gets into difficult technical terms and concepts, I have let the explanation stand as it is, in the simpler but cruder terms of linear velocities.

When I was an undergraduate at the University of Chicago, I attended a lecture by Compton in which he spoke of his experiment. Afterward, I approached him and asked if it would be possible to build up a stronger circulation in the tube by alternately flipping it around horizontal and vertical axes. Looking puzzled, Compton took a half dollar out of his pocket and began turning it between thumb and finger, muttering to himself. He finally shook his head and said he believed it would not work, but that he would think about it.

ANSWER 17.19—(August 1966)

The bowl increases in weight by the amount of liquid displaced by the dunked fish.

ANSWER 17.20—(August 1966)

Pulling the lower pedal of the bicycle backward causes the pedal to rotate in a way that normally would drive the bicycle forward, but since the coaster brake is not being applied the bicycle is free to move back with the pull. The large size of the wheels and the small gear ratio between the pedal and the wheel sprockets is such that the bicycle moves backward. The pedal moves back also with respect to the ground, although it moves forward with respect to the bicycle. When it rises high enough, the brake sets and the bicycle stops. Those of you who do not believe all this will simply have to get a bicycle and try it. The apparent paradox is explained in many old books. For a recent analysis, see D. E. Daykin, "The Bicycle Problem," *Mathematics Magazine*, vol. 45, January 1972, page 1.

ANSWER 17.21—(August 1966)

A rowboat *can* be moved forward by jerking on a rope attached to its stern. In still water a speed of several miles per hour can be obtained. As the man's body moves toward the bow, friction between boat and water prevents any significant movement of the boat backward, but the inertial force of the jerk is strong enough to overcome the resistance of the water and transmit a forward impulse to the boat. The same principles apply when a boy sits inside a carton and scoots himself across a waxed floor by rapid forward movements inside the box. No such "inertial space drive" is possible inside a spaceship because the near vacuum surrounding the ship offers no resistance.

ANSWER 17.22—(August 1966)

A pound of $10 gold pieces contains twice as much gold as half a pound of $20 gold pieces, therefore it is worth twice as much.

ANSWER 17.23—(August 1966)

Concealed inside the base of each bulb and the base of each on-off switch is a tiny silicon rectifier that allows the current to flow through it in only one direction. The circuit is shown in Figure 17.9 the arrows indicating the direction of current flow permitted by each rectifier. If the current is moving so that a rectifier in the base of a bulb will carry it, the rectifier steals the current and the bulb remains dark. It is easy to see that each switch turns on and off only the bulb whose rectifier points in the same circuit direction as the rectifier in the switch.

Aside from complaints about suddenly tossing in diodes (rectifiers) from left field, readers also complained (understandably) of the difficulty of inserting a tiny diode in the base of a bulb. R. Allen Pelton found it easier to make a working model as follows: "I connected the diode under the base of the porcelain light socket, then cut out the wood base to hold it. I am unable to switch the bulbs, but then again I can insert *any* bulb. My little wood panel has stumped every person yet I have shown it to."

Figure 17.9

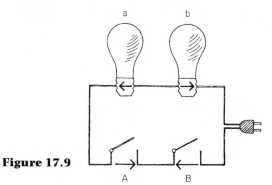

appendix

Twelve More Brain Teasers

Problems

PROBLEM A.1—Three Switches

Three on-off switches are on the wall of a building's first floor. Only one switch operates a single-bulb lamp on the third floor. The other two switches are bogus, unconnected to anything.

You are permitted to set the switches in any desired on-off order. You then go to the third floor to inspect the lamp. Without leaving the lamp's room, how can you determine which switch is genuine?

PROBLEM A.2—Double-Sheet Sum

A 20-page newspaper consists of five double sheets. One of these sheets is selected at random. Its four page numbers are added. Can you name the sum?

PROBLEM A.3—Inscribed Square

The outside square in Figure A.1 has a side of 2. What's the area of the smaller square?

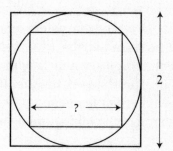

Figure A.1

PROBLEM A.4—Toothpick Giraffe

Five toothpicks form the giraffe shown in Figure A.2. Your task is to change the position of just *one* pick and leave the giraffe in exactly the same form as before. The reformed animal may alter its orientation or be mirror reversed, but must have its pattern unchanged.

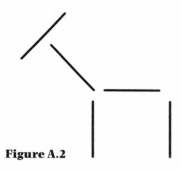

Figure A.2

PROBLEM A.5—Surprising Coincidence

In the row of counting numbers shown below there is an amazing coincidence. Can you spot it?

1234567891011121314151617181920 . . .

PROBLEM A.6—Unattacked Queens

Can you place three white queens and five black queens on a 5 × 5 chessboard so that no queen of one color attacks a queen of the other color? The pattern is unique except, of course, for rotations and reflections.

For the general problem of placing *n* queens on an order-*k* board so as to maximize the number of unattacked cells, see Chapter 34 of *Gardner's Workout* (A.K. Peters, 2001). See also John J. Watkins's book on chess tasks, *Across the Board* (Princeton University Press, 2004).

PROBLEM A.7—Serial Isogons

A serial isogon of 90 degrees is a polygon where adjacent angles meet at right angles, and its sides have lengths 1, 2, 3, 4, . . . as you trace around its perimeter. The simplest such isogon has sides 1 through 8. Can you construct it? The shape is unique. It is the only serial isogon of 90 degrees known to tile the plane.

For the general problem of enumerating higher-order serial isogons, and proofs that their number of sides must be multiples of 8, see Chapter 29 of *Gardner's Workout.*

PROBLEM A.8—Opaque Square

If the two diagonals of a square are drawn as shown in Figure A.3, left, the square is rendered opaque in the following sense. A ray of light entering the square from any direction clearly cannot penetrate the square and emerge through another side or corner. What is the minimum total length of lines drawn within the square that will block all rays of light?

Figure A.3, right, shows how what is called a "Steiner tree" will also make a square opaque. These lines are a bit shorter than the two diagonals, but the square can be made opaque with lines of still shorter total length.

Can you find the pattern? It has not been proved minimal, but almost surely is. For the more general problem of the opaque cube, as well as other opaque regular polygons, see Chapter 1 of *Gardner's Workout.*

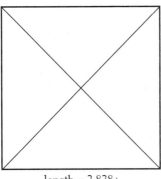 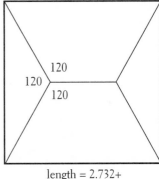

Figure A.3   length = 2.828+   length = 2.732+

PROBLEM A.9—100 to CAT

Arrange nine toothpicks to form 100 as shown in Figure A.4. Can you change 100 to CAT by altering the positions of just two picks?

Figure A.4

PROBLEM A.10—Saving Faces

Imagine that you have a solid pyramid with a square base and sides that are four equilateral triangles. Also on hand is a solid tetrahedron. Its four faces are equilateral triangles congruent with the pyramid's sides. Put another way, all the edges of the two solids are the same length. Now assume that you glue one of the tetrahedron's faces to one of the pyramid's sides. How many faces will the new polyhedron have?

There were nine triangular faces. Because two faces disappear after gluing, the answer must be $9 - 2 = 7$. Right? Wrong!

PROBLEM A.11—Even Words

All five vowels, A, E, I, O, U, as well as Y, which is often a vowel, happen to be found at odd positions in the alphabet. It seems impossible that a three-letter word could be made only of the letters in even positions. However, there is such a word, and one often used by mathematicians. Can you find it?

PROBLEM A.12—Doubled Knights

Raymond Smullyan—we encountered him in the chapter on chess problems—introduces a delightful chess problem in his autobiographical *Some Interesting Memories: A Paradoxical Life* (2002). He says he learned the problem from a high school math teacher. Assume that a game is played with the new rule that a knight can on a single turn move twice. It is not clear if this more powerful knight move is an advantage to the first or second player.

The first player, White, offers to play the game without his queen. How can he be sure to win?

Answers

ANSWER A.1

Call the switches A, B, and C. Turn on A and B, and turn off C. Wait 10 minutes, then turn off A. Go to the third floor. If the lamp's bulb is warm but off, switch A operates the lamp. If the bulb is cold and the lamp on, switch B operates the lamp. If the bulb is cold and unlit, C is the genuine switch.

ANSWER A.2

The sum of the four page numbers on any double sheet is $2(n + 1)$, where n is the number of the newspaper's last page. In the case cited, $n = 20$, therefore the sum of the page numbers on the selected double sheet is $2(20 + 1) = 42$.

Thanks to CalTech physicist Mamikon Mantsakanian for this one. He points out how the solution is related to the discovery by Gauss, when a young boy, of a quick way to sum a sequence of counting numbers.

ANSWER A.3

Imagine the small square rotated to the position shown in Figure A.5. It obviously has an area of 2, or exactly half the area of the outside square.

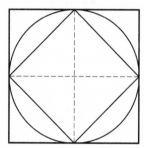

Figure A.5

ANSWER A.4

Figure A.6 shows how moving a single toothpick leaves the giraffe unaltered.

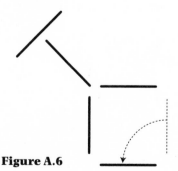

Figure A.6

Chapter

ANSWER A.5

Pi to four decimals appears in the counting sequence as indicated below.

123456789101112131415161718920

ANSWER A.6

Figure A.7 shows the only way to place the eight queens.

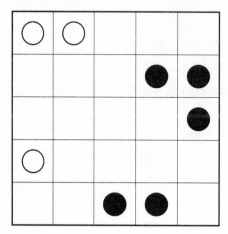

Figure A.7

ANSWER A.7

The unique order-8 serial isogon with right angles is shown in Figure A.8.

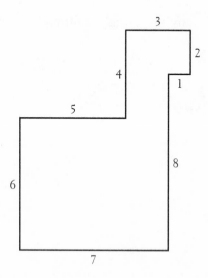

Figure A.8

ANSWER A.8

The best known solution to the opaque square problem is shown in Figure A.9.

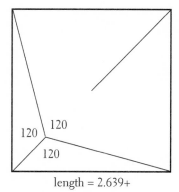

Figure A.9   length = 2.639+

ANSWER A.9

Moving two toothpicks will produce an upside-down CAT as shown in Figure A.10.

Figure A.10

ANSWER A.10

The glued solid only has five faces. At two places a pair of triangles become coplanar to make a single face! If you don't believe it, test it using cardboard models. (Another way to convince yourself of this is to begin with a larger regular tetrahedron with edges twice the size of the original pyramid. By slicing off corners of the larger tetrahedron you can easily produce a pyramid and tetrahedron of the original size. You will see there are two pairs of faces that are coplanar, because each pair came from the same side of the larger tetrahedron.)

ANSWER A.11

The word is "nth."

ANSWER A.12

White jumps queen's knight forward twice. Checkmate!

Selected Titles
on Mathematics
by the Author

If you enjoyed the selection of mathematical puzzles in this volume, you will want to look at these books by Martin Gardner as well.

Martin Gardner's "Mathematical Games" books:

The Scientific American *Book of Mathematical Puzzles and Games,* Simon and Schuster, 1959 (later retitled *The First* Scientific American *Book of Mathematical Puzzles and Games*), republished as *Hexaflexagons and Other Mathematical Diversions: The* Scientific American *Book of Mathematical Puzzles and Games,* University of Chicago Press, 1988.

The Second Scientific American *Book of Mathematical Puzzles and Diversions,* Simon and Schuster, 1961, republished by University of Chicago Press, 1987.

New Mathematical Diversions from Scientific American, Simon and Schuster, 1966; revised edition, *New Mathematical Diversions,* The Mathematical Association of America, 1995.

The Unexpected Hanging and Other Mathematical Diversions, Simon and Schuster, 1969, republished by University of Chicago Press, 1991.

The Numerology of Dr. Matrix, Simon and Schuster, 1967; expanded and republished as *The Incredible Dr. Matrix,* Scribner, 1979. See below.

Martin Gardner's Sixth Book of Mathematical Games from Scientific American, W. H. Freeman and Company, 1971.

Mathematical Carnival, Knopf, 1975, revised with a foreword by John H. Conway, The Mathematical Association of America, 1992.

Mathematical Magic Show, Knopf, 1977 revised with a foreword by Ronald L. Graham, The Mathematical Association of America, 1990.

Mathematical Circus, Knopf, 1979, revised with a foreword by Donald E. Knuth, The Mathematical Association of America, 1992.

The Incredible Dr. Matrix, Scribner, 1979 (see *The Numerology of Dr. Matrix,* above), revised and expanded as *The Magic Numbers of Dr. Matrix,* Prometheus Books, 1985.

Wheels, Life, and Other Mathematical Amusements, W. H. Freeman and Company, 1983.

Knotted Doughnuts and Other Mathematical Entertainments, W. H. Freeman and Company, 1986.

Time Travel and Other Mathematical Bewilderments, W. H. Freeman and Company, 1988.

Penrose Tiles to Trapdoor Ciphers, W. H. Freeman and Company, 1989.

Fractal Music, Hypercards, and More . . . : Mathematical Recreations from Scientific American, W. H. Freeman and Company, 1992.

Last Recreations: Hydras, Eggs, and Other Mathematical Mystifications, Copernicus Books, Springer Verlag, 1997.

Each answer in this book gives the date that each puzzle first appeared in *Scientific American.* However, it may be more convenient to have a reference to the first publication in one of the books of collections, listed above. This table, using abbreviated titles, gives that information.

First	Chapter 3	February 1957
First	Chapter 12	November 1957
Second	Chapter 5	August 1958
Second	Chapter 14	May 1959
New	Chapter 3	February 1960
New	Chapter 12	October 1960
New	Chapter 19	June 1961
Unexpected	Chapter 7	February 1962
Unexpected	Chapter 15	October 1962
Sixth	Chapter 6	November 1963
Sixth	Chapter 12	June 1964
Sixth	Chapter 20	March 1965
Carnival	Chapter 2	February 1966
Carnival	Chapter 14	August 1966

Carnival	Chapter 9	November 1965
Magic Show	Chapter 15	February 1967
Magic Show	Chapter 5	November 1967
Magic Show	Chapter 10	August 1968
Circus	Chapter 15	April 1969
Circus	Chapter 2	July 1969
Circus	Chapter 11	February 1970
Wheels	Chapter 3	November 1970
Wheels	Chapter 8	July 1971
Wheels	Chapter 16	April 1972
Knotted	Chapter 6	May 1973
Knotted	Chapter 15	April 1974
Time	Chapter 9	May 1975
Time	Chapter 16	November 1975
Penrose	Chapter 9	April 1977
Fractal	Chapter 11	February 1979
Last	Chapter 21	December 1979
Last	Chapter 16	April 1981

Note that chapter 17 of this volume is a single original column. All the remaining chapters have been selected from a wide period of time. (The February 1966 column about pennies, and the July 1969 column about matches, were only partially used in this book.)

Index

Page numbers in *italics* refer to illustrations.

Ambiguous Dates, 3, 17

American Journal of Physics, 100, 450

American Mathematical Monthly, 23, 29, 55, 70, 80, 93, 136, 149, 176, 180, 186, 196, 219, 248, 265, 269, 334, 378, 390, 439

Amir-Moéz, Ali R., 134

Ammann, Robert, 189

Amorous Bugs, 117, *117,* 133–34

Amusements in Mathematics (Dudeney), 65, 77–78, 126

Anagram Dictionary, 405–6, 416–17

anagrams, 400

analytic geometry, 113

Anarchism, State and Utopia (Nozick), 236

Andree, Richard, 332

angles:

    acute, 169

    bisecting of, 124

    dissection and, 169–70, *169,* 186–87, *186, 187, 188*

    in Geometry and Divisibility, 109, 124

    right, 169, 460

    in Three Squares, 117, *118,* 134, *134*

Appel, John, 93

Apple Picking, 85, 92

applied mathematics, 233, 270

Archibald, R. C., 136

Archimedeans society, 116, 133

Archimedes, 140, 145–46, 157

area, of a circle, 145

Argyle, P. E., 252

Ars Combinatoria, 25

Asimov, Isaac, 75, 80

Astounding Science Fiction, 158

Austin, A. K., 86, 314

Australian Mathematics Teacher, 187

automynorcagrams, 420

Avoiding Checks, 340, 353

axes of symmetry, 146

Babbage, Charles, 403–4

Babcock, David, 120

Baby Crossword, 401, *401,* 411, *411*

Baker, Alan, 74

Baker, G. C., 138

Baker, Vaughn, 415

Balloon in Car, 446, 453

Banach, Stefan, 121

"Banach-Tarski paradox," 121

Banks, Bruce A., 239, 261

Banner, Randolph W., 348

Barnard, D. St. P., 373, 383

Barr, Stephen, 128, 184, 203–4, 214, 254, 395

Barr's Belt, 203–4, *204,* 214, *214*

Barwell, Brian R., 281

Base for a Square, 62, 73

Base Knowledge, 61–62, 73

Bathtub Boat, 445, 453

B.C., 433, *433*

Beattie, Jay, 75

"Beauty, the Beast, and the Pond" (Schurman and Lodder), 257

Beck, Anatole, 153

Beeler, Michael, 120

Beer Signs on the Highway, 90, 98–99

"bell ringing octopus," 258

Benham, Philip S., 417

Bergan, Rudolf K. M., 160, 161

Bergerson, Howard W., 410, 417, 420

Bergum, G. E., 74

Berloquin, Pierre, 429

Bernhart, Frank R., 118

Bernstein, Mary S., 387

Bezzel, M., 356–57

Bhat, Y. K., 78

Bible Names "De-Koder" (Love), 417

Bicycle Paradox, 447, *448,* 455–56

"Bicycle Problem, The" (Daykin), 456

"Bicycle Tubes Inside Out" (Taylor), 223

Bienenfeld, David, 32

bilateral symmetry, 162, 173, 186, 206

Bills and Two Hats, 43–44, 53–54, *54*

binary fractions, 58–59

binary Gray codes, 55

binary notation, 48, 381

bisecting angles, 124

Bisecting Yin and Yang, 111–12, *111,* 126–27, *126*

Blades of Grass, 319, *319,* 331–32, *331*

Blank Column, The, 6, 20

Bleicher, Michael, 153

Block, James A., 184

"Blue-empty Chromatic Graphs" (Lorden), 29

Blue-Empty Graph, 10–11, 28–29, *28*

Blue-Eyed Sisters, 41, 50–51

"On the Optimum Method of Cutting a Rectangular Box into Unit Cubes" (Putzer and Lowen), 150
"On the Reversing of Digits" (Klosinski and Smolarski), 396
Opaque Square, 461, *461*, 464, *464*
Operations Research (OR), 270
"Optimal Search Procedure, An" (Zimmerman), 265
order-3 polyiamonds, 167
Osborne, J. M., 452
Othello, 321
O'Toole, Kenneth, 387
overhand knots, 147, 205, *205*, 215–16, *215*
Overlap Squares, 113–14, *114*, 130, *130*
Owe, Lars Bertil, 29

Pair of Cryptarithms, 390–91, *391*, 394–95
Palindromes and Anagrams (Bergerson), 410, 417
Pallbearers Review, 182, 253
Palmer, Elton M., 128
Papered Cube, 145, 154–55, *154, 155*
paradoxes:
    "Banach-Tarski paradox," 121
    Bicycle Paradox, 447, *448*, 455–56
    Gear Paradox, 426–27, *427*, 435
    Switching Paradox, 448, *449*, 456–57, *457*
    Zeno's paradox, 94, 429
parallelepipeds, 181, 195
Paraquin, Charles H., 125
parity:
    in Handshakes and Networks, 27
    in Pool-Ball Triangles, 37
parity checks, 20, 270
Parker, George D., 154
"Partial Sums of the Harmonic Series" (Boas and Wrench), 439
Pascal's triangle, 37
Passages from the Life of a Philosopher (Babbage), 404
patchwork quilts, 168, *168*, 183, *183, 184*
Pearce, Murray R., 170–71, 402
Pelc, Andrzej, 265
Pelton, R. Allen, 456–57
Penney, Walter, 236, 406
Penny Bet, 43, 53

Penrose Tiles to Trapdoor Ciphers (Gardner), 263
pentagons:
    in Dodecahedron-Quintomino Puzzle, 13–14, *14*, 32–33, *33*
    in Rigid Square, 139
    in A View of the Pentagon, 40, 49
pentagrams, 187, *188*
pentominoes:
    definition of, 166
    in Pentomino Farms, 174, *174*, 191–92, *192*
    in Two Pentomino Posers, 173–74, *174*, 191, *191*
Pentomino Farms, 174, *175*, 191–92, *192*
Percival, H. S., 290
Perelman, Yakov, 76
perfect polypowers, 67
perfect squares, 62, 73
Perkins, Granville, 157
Perry, R. L., 56
"Persistence of a Number, The" (Sloane), 83
Persistences of Numbers, 71–72, 83–84, *83*
Peters, Edward N., 299–300
Petersen, Jon, 188
Pett, Martin T., 381
Philips Research Reports, 153
Phillips, Hubert, 269, 341
Philpott, Wade, 190, 307, 309
Phipps, Cecil G., 293
physical objects, 422–39
    answers, 429–39
    problems, 422–29
Physical Review, 454–55
physics, 440–57
    answers, 449–57
    author's writings on, 440
    problems, 440–48
pi, in Pied Numbers, 61, 69–70, 81–82, *81*
Pied Numbers, 61, 69–70, 81–82, *81*
Pierce, John R., 265
pigeonhole principle:
    in Counter-Jump "Aha!," 296
    definition of, 33
    in Different Distances, 120
    in Five Couples, 29
    in Rotating Roundtable, 33–34
Pillow Problems (Carroll), 45–46
plane geometry, 107–39

Shieh, T. C., 37
Shimauchi, Takakazu, 191
Shubnikov, A. V., 158
Shufflin' Along (Miller), 69
SIAM News, 266
Sicherman, George, 16, 36–38
Sigil of Scoteia, 407, *408,* 418–19
"Sigil of Scoteia, The" (Keller), 418
Silverman, David L., 17, 47, 56–57, 58, 313,
    324, 340, 348, 362, 379, 401
Simmons, Gustavus J., 93, 206
"Simple Perfect Squared of Lowest Order,
    A" (Duijvestijn), 178
Single-Check Chess, 348–49, *349,* 361–62
Singleton, C. R. J., 179, 194
Singleton, John S., 18
"Single Vacancy Rolling Cube Problems"
    (Harris), 303, 304
Six Letters by Lewis Carroll, 418
*Sixth Book of Mathematical Games from
    Scientific American* (Gardner), 261,
    300
"Skewered!" (Asimov), 80
Skewes, S., 66, 80
Skidell, Akiva, 52
Slater, Maya and Nicolas, 451
Sleator, Daniel, 137
Sliced Cube and Sliced Doughnut, 143–44,
    151, *151, 152*
Slicing a Trapezoid, 167, *167,* 182, *182*
Sliding Pennies, 273, *273,* 290, *290*
Sloane, N. J. A., 71–72, 83
Slocum, Jerry, 307
Sluizer, Allan L., 77
Smith, J. M. Powis, 153
Smith, Philip C., Jr., 124, 141
Smolarski, Dennis C., 396
Smullyan, Raymond, 235, 339, 344–45, 347,
    358, 360, 367, 375, 388
"snowflake curve," 261
soldiers, ten, 10, 26–27
solid geometry, 140–65
    answers, 147–65
    elements of, 140
    problems, 140–47
solids, *see* polyhedra
"Solution of Ulam's Problem on Searching
    with a Lie" (Pelc), 265
"Sonnet Illuminate" (Benham), 417

Sorting and Searching (Knuth), 248
Soules, George, 348, 361
Southall, Susan, 434
South Pole, 148–49, 455
Spencer, Herbert, 137
spheres, 140, 203
    formula for volume of, 145, 157
    in Hole in the Sphere, 146, 157–58, *158*
    in How Many Hemispheres?, 40, 49
    in Three Circles, 136–37
    in Two Spheres in One Pipe, 270, 287
Spira, Siegfried, 154
spirals, 117
Sprague, Roland, 22
squares:
    in Amorous Bugs, 117, *117,* 133–34
    axes of symmetry of, 146
    in Dissecting a Square, 166–67, *166,*
        181–82, *182*
    dissecting into acute triangles, 170, 187
    in Eight-Block Puzzle, 282–83, *283,*
        301–2, *302*
    in Folding a Möbius Strip, 204–5, *204,*
        214, *215*
    in Hidden Cross, 110, *110,* 125, *125*
    in How Many Spots?, 116, *117,* 133, *133*
    in Inscribed Square, 459, *459,* 462, *462*
    in Killing Squares and Rectangles,
        11–13, *12,* 30–32, *31, 32*
    in Leave Four Squares, 109, *109,* 124,
        124
    in Overlap Squares, 113–14, *114,* 130,
        130
    perfect, 62, 73
    polyominoes composed of, 207
    in Rigid Square, 121–22, *121,* 138–39,
        138, 139
    in Square-Triangle Polygons, 173, *173,*
        190, *190*
    in Three Squares, 117, *118,* 134, *134*
    in Trisecting a Square, 108, *108,* 123
    see also pentominoes
Square-Triangle Polygons, 173, *173,* 190,
    190
Square Year, 87, 94
squaring the square, 177–78
S. S. Adams, 287
Stack, John W., 429
Standen, Maud, 406

Stanger, Albert G., 258, 259

Starbuck, George, 160

stars, five-pointed, 146

Starting a Chess Game, 344, 358, *358*

Stealing Bell Ropes, 444, 452

Steiner, Jakob, 113

Steiner Ellipses, 113, 129–30

Steiner Tree, 461, *461*

Steinhaus, Hugo, 17, 247–48

"'Steinmetz Problem' and School Arithmetic, The" (Sutton), 157

Stephens, James, 203

Stern, Marvin, 44

Stewart, Ian, 266

Stillwell, John, 209

Stirling's formula, 332

Stix, Ernest W., Jr., 52

Stockton, Frank, 41

Stolarsky, Kenneth B., 83

Story of the Peasant-Boy Philosopher, The (Ferguson), 426

Stott, Eric, 183

Stover, Mel, 170, 340, 372

straight flushes, 43, *43*, 53

Strand Problems Book, The (Williams and Savage), 389

Stromquist, Walter, 334–35

Stuffing Envelopes, 39, 49

"Subdividing a Box Into Completely Incongruent Boxes" (Cutler), 181

Sugaku Seminar, 191

Sum Is the Product, The, 86, 93–94

Summers, Robert, 254

"Sum of a Digitadition Series, The" (Stolarsky), 83

Superior Mathematical Puzzles (Dinesman), 288

Superqueens, 340, 353, *354*

Surprising Coincidence, 460, 463

Suspended Househoe, 442, *442*, 450, *450*

Sutcliffe, Alan, 396

Sutton, Richard M., 157

swastikas, 146, 424, *424*, 431, *432*

Sweet, John Edson, 119, 136

Swiss Barber, 368, 376

Switching Cars, 277–78, *278*, 296–97

Switching Paradox, 448, *449*, 456–57, *457*

Sylvie and Bruno Concluded (Carroll), 443

Symbols, Signals and Noise (Pierce), 265

symmetry, 262, 308, 334

axes of, 146

bilateral, 162, 173, 186, 206

in Colored-Bowling Pins, 25–26, *26*

in Digit-Placing Problem, 21

in First Black Ace, 56–57

of friezes, 14–15

glide, 14–15

in Leave Four Squares, 109

planes of, 146–47

of rectangles, 187

in Three Coins, 55

two-fold, 214

see also mirror reflections

Symmetry in Science and Art (Shubnikov and Koptsik), 158

Szemerédi, E., 138

Szirtes, Thomas, 244

TAD (triangle of absolute differences), 37–38

Talkative Eve, 392, 397

Talking Money, 61, 73

tangents:

in Three Circles, 118–19, *119*, 136–37

in Triangle of Tangents, 110, *110*, 124

Tan Hock Chuan, 426

Tantalizers (Hollis), 368

Tappay, Robert, 260

Tarski, Alfred, 121

Taylor, Alan D., 322

Taylor, Herbert, 37–38, 210, 223

Teaching of Elementary Science and Mathematics, The (Calandra), 241

Technical Review, 23

Temperley, Nicholas, 405, 417

Templeton, David H., 333

Ten-Digit Numbers, 3, 8, 17, 23

Tennis Match, 372, 381

Ten-Penny Patterns, 205–6, *206*, 216–17, *216*

Ten Soldiers, 10, 26–27

Ten Statements, 371, 379–80

Termite and 27 Cubes, 5–6, *5*, 19–20

ternary notation, 62, 73

tetrads, 171–73, *171*, *172*, 189, *189*

tetrahedrons, 212, 304, 338

in Coplanar Points, 141, 148

definition of, 144

irregular, 141

About the Author

Martin Gardner achieved worldwide fame and respect for the three decades of his highly popular mathematical games column for *Scientific American*, yet he is not just a mathematician. He is the author or editor of more than one hundred books and booklets, including books on mathematics, science, pseudoscience, philosophy, literary criticism, and fiction (including *Visitors from Oz*, based on L. Frank Baum's *The Wonderful Wizard of Oz*). Gardner is also the annotator of a number of classic literary works from *The Night Before Christmas* and Gilbert K. Chesterton's *The Innocence of Father Brown* to Coleridge's *The Ancient Mariner*. Gardner's annotated version of *Alice in Wonderland*, *The Annotated Alice*, has sold more than 50,000 copies worldwide. At ninety-one, Gardner has influenced and inspired generations of scientists, scholars, and nonscientists. Stephen Jay Gould once said that Martin Gardner "has become a priceless national resource," a writer "who can combine wit, penetrating analysis, sharp prose, and sweet reason into an expansive view that expunges nonsense without stifling innovation, and that presents the excitement and humanity of science in a positive way."

About the Editor

Dana Richards is a professor of computer science at George Mason University, researching the mathematical aspects of computer science from algorithms to the theory of computation, with an emphasis on parallel computation. As a member of the National Puzzlers' League, he has had a lifelong interest in mathematical games and has made an extensive study of the history and evolution of puzzles. He is an active Sherlockian, contributing both scholarship and puzzles to publications on the life of Sherlock Holmes and Arthur Conan Doyle. For some twenty-five years Richards has tracked the prolific endeavors that Martin Gardner has pursued, a dedication that has so far produced a book-length bibliography and a close friendship. Like many people, Richards can trace his interest in mathematics to the influence of Gardner's *Scientific American* column. This volume is meant to introduce the excitement and wonder generated by the column to another generation of puzzlers.